Radiation Detection Systems

Devices, Circuits, and Systems
Series Editor - Krzysztof Iniewski

Nano-Semiconductors
Devices and Technology
Krzysztof Iniewski

Atomic Nanoscale Technology in the Nuclear Industry
Taeho Woo

Telecommunication Networks
Eugenio Iannone

Optical, Acoustic, Magnetic, and Mechanical Sensor Technologies
Krzysztof Iniewski

Biological and Medical Sensor Technologies
Krzysztof Iniewski

Graphene, Carbon Nanotubes, and Nanostuctures
Techniques and Applications
James E. Morris and Krzysztof Iniewski

Low Power Emerging Wireless Technologies
Reza Mahmoudi and Krzysztof Iniewski

High-Speed Photonics Interconnects
Lukas Chrostowski and Krzysztof Iniewski

Smart Sensors for Industrial Applications
Krzysztof Iniewski

MEMS: Fundamental Technology and Applications
Vikas Choudhary and Krzysztof Iniewski

Nanoelectronic Device Applications Handbook
James E. Morris and Krzysztof Iniewski

Novel Advances in Microsystems Technologies and Their Applications
Laurent A. Francis and Krzysztof Iniewski

Building Sensor Networks: From Design to Applications
Ioanis Nikolaidis and Krzysztof Iniewski

Embedded and Networking Systems
Design, Software, and Implementation
Gul N. Khan and Krzysztof Iniewski

Multisensor Data Fusion
From Algorithm and Architecture Design to Applications
Hassen Fourati

Electrostatic Discharge Protection
Advances and Applications
Juin J. Liou

Optical Imaging Devices
New Technologies and Applications
Ajit Khosla and Dongsoo Kim

Radiation Detectors for Medical Imaging
Jan S. Iwanczyk

Gallium Nitride (GaN)
Physics, Devices, and Technology
Farid Medjdoub

Mixed-Signal Circuits
Thomas Noulis

MRI
Physics, Image Reconstruction, and Analysis
Angshul Majumdar and Rabab Ward

Reconfigurable Logic
Architecture, Tools, and Applications
Pierre-Emmanuel Gaillardon

Ionizing Radiation Effects in Electronics
From Memories to Imagers
Marta Bagatin and Simone Gerardin

CMOS Time-Mode Circuits and Systems
Fundamentals and Applications
Fei Yuan

Tunable RF Components and Circuits
Applications in Mobile Handsets
Jeffrey L. Hilbert

Cell and Material Interface
Advances in Tissue Engineering, Biosensor, Implant, and Imaging Technologies
Nihal Engin Vrana

Nanomaterials
A Guide to Fabrication and Applications
Sivashankar Krishnamoorthy

Physical Design for 3D Integrated Circuits
Aida Todri-Sanial and Chuan Seng Tan

Wireless Medical Systems and Algorithms
Design and Applications
Pietro Salvo and Miguel Hernandez-Silveira

High Performance CMOS Range Imaging
Device Technology and Systems Considerations
Andreas Süss

Analog Electronics for Radiation Detection
Renato Turchetta

Power Management Integrated Circuits and Technologies
Mona M. Hella and Patrick Mercier

Circuits and Systems for Security and Privacy
Farhana Sheikh and Leonel Sousa

Multisensor Attitude Estimation
Fundamental Concepts and Applications
Hassen Fourati and Djamel Eddine Chouaib Belkhiat

Structural Health Monitoring of Composite Structures Using Fiber Optic Methods
Ginu Rajan and Gangadhara Prusty

Advances in Imaging and Sensing
Shuo Tang and Daryoosh Saeedkia

Semiconductor Devices in Harsh Conditions
Kirsten Weide-Zaage and Malgorzata Chrzanowska-Jeske

Introduction to Smart eHealth and eCare Technologies
Sari Merilampi, Krzysztof Iniewski, and Andrew Sirkka

Diagnostic Devices with Microfluidics
Francesco Piraino and Šeila Selimovic´

Magnetic Sensors and Devices
Technologies and Applications
Kirill Poletkin and Laurent A. Francis

Semiconductor Radiation Detectors
Technology, and Applications
Salim Reza

Noise Coupling in System-on-Chip
Thomas Noulis

High Frequency Communication and Sensing
Traveling-Wave Techniques
Ahmet Tekin and Ahmed Emira

3D Integration in VLSI Circuits
Design, Architecture, and Implementation Technologies
Katsuyuki Sakuma

IoT and Low-Power Wireless: Circuits, Architectures, and Techniques
Christopher Siu and Krzysztof Iniewski

Radio Frequency Integrated Circuit Design
Sebastian Magierowski

**Low Power Semiconductor Devices and Processes for Emerging
Applications in Communications, Computing, and Sensing**
Sumeet Walia and Krzysztof Iniewski

Sensors for Diagnostics and Monitoring
Kevin Yallup and Laura Basiricò

Biomaterials and Immune Response
Complications, Mechanisms and Immunomodulation
Nihal Engin Vrana

High-Speed and Lower Power Technologies
Electronics and Photonics
Jung Han Choi and Krzysztof Iniewski

X-Ray Diffraction Imaging
Technology and Applications
Joel Greenberg and Krzysztof Iniewski

Radiation Detection Systems

Medical Imaging, Industrial Testing and Security Applications

2nd Edition

Edited by

JAN S. IWANCZYK
KRZYSZTOF INIEWSKI

CRC Press
Taylor & Francis Group
Boca Raton London New York

CRC Press is an imprint of the
Taylor & Francis Group, an informa business

CRC Press
Boca Raton and London
Second edition published 2022

by CRC Press
6000 Broken Sound Parkway NW, Suite 300, Boca Raton, FL 33487-2742

and by CRC Press
2 Park Square, Milton Park, Abingdon, Oxon, OX14 4RN

First edition published by CRC Press 2015

CRC Press is an imprint of Taylor & Francis Group, LLC

Library of Congress Cataloging-in-Publication Data

Names: Iwanczyk, Jan S., editor. | Iniewski, Krzysztof, 1960- editor.
Title: Radiation detection systems. Medical imaging, industrial testing and security
    applications / edited by Jan Iwanczyk, Krzysztof Iniewski.
Description: Second edition. | Boca Raton, FL : CRC Press, 2022. | Series: Devices,
    circuits, and systems | Includes bibliographical references and index.
Identifiers: LCCN 2021035690 (print) | LCCN 2021035691 (ebook) |
    ISBN 9781032110875 (hbk) | ISBN 9781032110912 (pbk) | ISBN 9781003218364 (ebk)
Subjects: LCSH: Radiation—Measurement—Instruments. | Radiography, Medical—
    instrumentation. | Radiography—Equipment and supplies. | Radiography, Industrial. |
    Electronic security systems—Equipment and supplies.
Classification: LCC QC795.5 .R38 2022 (print) | LCC QC795.5 (ebook) |
    DDC 539.7/7—dc23
LC record available at https://lccn.loc.gov/2021035690
LC ebook record available at https://lccn.loc.gov/2021035691

ISBN: 978-1-032-11087-5 (hbk)
ISBN: 978-1-032-11091-2 (pbk)
ISBN: 978-1-003-21836-4 (ebk)

DOI: 10.1201/9781003218364

Typeset in Palatino LT Std
by KnowledgeWorks Global Ltd.

Contents

Preface

The advances in semiconductor detectors, scintillators, photodetectors such as silicon photomultiplier (SiPM), and readout electronics in the past decades have led to significant progress in terms of performance and greater choice of the detection tools in many applications. This book presents the state-of-the-art in the design of detectors and integrated circuit design, in the context of medical imaging using ionizing radiation. It addresses exciting new opportunities in x-ray detection, computed tomography (CT), bone dosimetry, and nuclear medicine (positron emission tomography, PET; single photon emission computed tomography, SPECT). In addition to medical imaging, the book explores other applications of radiation detection systems in security applications such as luggage scanning, dirty bomb detection, and border control.

The material in the book has been divided into two volumes. Volume I puts more emphasis on sensor materials, detector, and front end electronics technology and designs as well as system optimization for different applications. Also includes characterization measurements of the developed detection systems. Volume II is devoted to more specific applications of detection systems in medical imaging, industrial testing, and security applications. However, there is an unavoidable certain overlap in topics between both volumes.

A significant portion of a book describes new advances in development of detection systems based on CdZnTe (CZT) and CdTe detectors. The use of these detectors in fast growing medical and security applications is possible due to recent progress in material/detector technologies combined with the availability of application specific integrated circuits (ASIC) that provide a very compact low noise amplification and processing of the signal from individual detector pixels in imaging arrays. These new detectors have already been commercialized for use in surgical probes, gamma cameras, SPECT systems, and bone mineral density scanners. On the way there is a great effort to introduce this technology to the largest medical diagnostic imaging modality namely CT. Spectral x-ray photon counting possible with these new detectors will allow to reduce the radiation dose to the patient and to improve the imaging contrast but also can offer many other advantages. One of the most exciting possibilities is the future use of CT scanners not only as an anatomical modality but also as a functional modality that provides functional information. In security applications CZT and CdTe detection systems are used to detect explosives, hidden radioactive sources in luggage or radiological dispersal devices (dirty bombs) that are transported. There is a great effort to develop new imaging detectors for luggage scanning systems in airports for a direct x-ray transmission and for x-ray diffraction. CT systems for luggage scanning have similar requirements regarding photon counting detectors to that used in medical imaging. However, composition of luggage content is usually more complex than the patient body and for this reason it is advantageous to use in the readout more energy bins than that in medical applications. On the other hand, there are much less concerns about delivered radiation dose. In diffraction systems extremely high-count rate is not the major requirement because detectors are not placed in the direct x-ray beam. In the diffraction applications rather an excellent energy resolution and a large field of view of the detector arrays are the prerequisite.

Currently, SiPMs show very promising results in many fields. SiPMs used for reading the light from scintillators are starting to make a big impact on the design concepts for new nuclear medicine equipment for the gamma cameras used in SPECT and PET applications. These new designs allow for construction of more compact imagers with better performance that are not sensitive to magnetic fields, as are designs utilizing conventional photomultiplier tubes, allowing for the construction of SPECT and PET systems combined with magnetic resonance imaging (MRI) scanners in multimodality systems. Particularly fascinating is the renewed interest in Time-of-Flight (TOF) PET systems now becoming possible with development of combined very fast scintillators with novel SiPM structures making more feasible quest toward 10 ps the coincidence resolving time (CRT) to improve the spatial resolution and to enable the detection of abnormalities at the earliest possible stage.

Individual chapters of the book deal with variety of radiation detection systems beyond mentioned above systems based on CZT, CdTe, and SiPM technologies giving readers a broader view of radiation detection systems.

Jan S. Iwanczyk, PhD,
Los Angeles, CA, USA

Krzysztof (Kris) Iniewski, PhD,
Vancouver, BC, Canada
February 19, 2021

Editors

Dr. Jan S. Iwanczyk is a consultant to universities and private companies since July 2017. He has served as a president and CEO of DxRay, Inc., Northridge, California, from 2005 to 2017. In 2017, DxRay, Inc., has been sold to OSI/Rapican Systems one of the three largest companies in the world that provides equipment for security in the airports. He previously was affiliated with several start-up private and publicly traded companies and centered on bringing novel scientific and medical technologies to the market. During the period from 1979 to 1989, Dr. Iwanczyk was associate professor at the University of Southern California, School of Medicine. He holds Master's degree in Electronics and PhD degree in Physics. His multi-faceted experience combines operations, organizational development with strong scientific research and technical project management qualifications. Dr. Iwanczyk's technical expertise is in the field of x-ray and gamma ray imaging detectors and systems. In recent years he has been developing photon-counting, energy dispersive x-ray imaging detectors based on CdTe, CZT, and Si for medical and security applications. He is the author of over 200 scientific papers, 1 book, several book chapters, and 20 patents. He also lectures at major symposia worldwide as an invited speaker and has received numerous honors and awards including 2002 Merit Award, IEEE Nuclear and Plasma Sciences Society and the 2016 Scientist Award IEEE – RTSD for lifetime achievements.

Dr. Krzysztof (Kris) Iniewski is managing R&D activities at Redlen Technologies, Inc., a detector company based in British Columbia, Canada. During his 15 years at Redlen, he has managed development of highly integrated CZT detector products in medical imaging and security applications. Prior to Redlen, Kris held various management and academic positions at PMC-Sierra, University of Alberta, SFU, UBC, and University of Toronto.

Dr. Iniewski has published over 150+ research papers in international journals and conferences. He holds 25+ international patents granted in the United States, Canada, France, Germany, and Japan. He wrote and edited 75+ books for Wiley, Cambridge University Press, McGraw Hill, CRC Press, and Springer. He is a frequent invited speaker and has consulted for multiple organizations internationally.

Contributors

Carlos D. R. Azevedo
Institute of Nanostructures,
  Nanomodelling and
  Nanofabrication (i3N)
Universidad de Aveiro
Aveiro, Portugal

M.Arimoto
Waseda University
Tokyo, Japan
and
Kanazawa University
Ishikawa, Japan

Marijn Boone
TESCAN XRE
Ghent, Belgium

Chin-Tu Chen
The University of Chicago
Chicago, Illinois

Veerle Cnudde
PPrGRess UCCT
Geoplogy Department
Ghent, Belguim
and
Department of Earth Sciences
Utrecht University
Utrecht, Netherlands

Wei Cunfeng
Chinese Academy of Science
Beijing, China
and
Jinan Laboratory of Applied Science
Jinan, China

Denis Dauvergne
Universite Grenoble Alpes, CNRS/
  IN2P3
Grenoble, France

Diego Gonzalez-Dıaz
Instituto Galego de Fisica de Altas
  Enerxias (IGFAE)
Santiago de Compostela, Spain

Jose R. A. Godinho
Helmholtz-Zentrum Dresden-
  Rossendorf, Helmholtz Institute
  Freiberg for Resource Technology
Freiberg, Germany

Angela Saa Hernandez
Instituto Galego de Fisica de Altas
  Enerxias (IGFAE)
Santiago de Compostela, Spain

Fernando Hueso-González
Instituto de Física Corpuscular
  (IFIC-CSIC/UVEG)
Valencia, Spain

Lubomír Gryc
National Radiation Protection
  Institute (SURO)
Prague, Czech Republic

Bradley M. Guy
Department of Geology/CIMERA
University of Johannesburg
Johannesburg, Republic of South
  Africa

Jan Helebrant
National Radiation Protection
    Institute (SURO)
Prague, Czech Republic

H.Ikeda
Institute of Space and Astronautical
    Science
Kanagawa, Japan

Jussi Liipo
Metso-Outotec
Espoo, Finland

Gabriela Llosá
Instituto de Física Corpuscular
    (IFIC-CSIC/UVEG)
Valencia, Spain

Denis Van Loo
TESCAN XRE
Ghent, Belgium

Wei Long
Chinese Academy of Science
Beijing, China
and
Jinan Laboratory of Applied Science
Jinan, China

Chien-Min Kao
The University of Chicago
Chicago, Illinois

J.Kataoka
Waseda University
Tokyo, Japan

Hiroaki Kiji
Waseda University
Tokyo, Japan

Margarita Merkulova
PProGRess-UGCT
Geology Department
Ghent University
Ghent, Belgium

Li Mohan
Chinese Academy of Science
Beijing, China
and
Jinan Laboratory of Applied Science
Jinan, China

Axel D. Renno
Helmholtz-Zentrum Dresden
    Rossendorf
Helmholtz Institute for Resource
    Technology
Freiberg, Germany
and
PProGRess-UGCT
Geology Department
Ghent University
Ghent, Belgium

Antti Roine
Metso-Outotec
Espoo, Finland

Ana Ros
Instituto de Física Corpuscular
    (IFIC-CSIC/UVEG)
Valencia, Spain

Jorge Roser
Instituto de Física Corpuscular
    (IFIC-CSIC/UVEG)
Valencia, Spain

Thomas De Schryver
TESCAN XRE
Ghent, Belgium

Anna Selivanova
National Radiation Protection
    Institute (SURO)
Prague, Czech Republic

S.Shiota
Hitachi Metals Ltd
Osaka, Japan

Jonathan Sittner
Helmholtz-Zentrum Dresden
    Rossendorf
Helmholtz Institute for Resource
    Technology
Freiberg, Germany
and
PProGRess-UGCT
Geology Department
Ghent University
Ghent, Belgium

Katsuyuki Taguchi
Johns Hopkins University
Baltimore, Maryland

S.Terazawa
Hitachi Metals Ltd
Osaka, Japan

T.Toyoda
Chinese Academy of Sciences
Beijing, China

Aleksandra Wrońska
Marian Smoluchowski Institute of
    Physics
Jagiellonian University
Krakow, Poland

Jiri Zaijcek
Institute of Experimental and
    Applied Physics
Czech Technical University
Prague, Czech Republic

Wang Zhe
Chinese Academy of Science
Beijing, China
and
Jinan Laboratory of Applied Science
Jinan, China

Zhang Zhidu
Chinese Academy of Sciences
Beijing, China

1

Radiation Detection in SPECT and PET

Chin-Tu Chen and Chien-Min Kao

CONTENTS

DOI: 10.1201/9781003218364-1

1.1 Introduction

Nuclear medicine imaging consists of planar scintigraphy (2D gamma-ray projection imaging) and emission computed tomography (ECT) (or 3D tomographic imaging). ECT includes single-photon emission computed tomography (SPECT) and positron emission tomography (PET). This chapter focuses on radiation detection in SPECT and PET, including a brief review of the major early developments, followed by discussions on current state-of-the-art technologies in routine uses, recent advances, and future trends. Note that other aspects of SPECT and PET in image reconstruction, quantitative imaging, radiotracers, or the clinical and research applications will not be covered in this chapter. Information regarding these topics can be found in relevant textbooks (Cherry et al 2012; Wernick and Aarsvold 2004).

1.1.1 Early Developments of SPECT

Tomographic gamma-ray imaging initially evolved in two parallel paths: transaxial section tomography and longitudinal (focal-plane) tomography, both based on NaI(Tl) scintillation crystals and PMTs (photomultiplier tubes) arranged in various configurations with different scanning strategies. Even though longitudinal tomography succeeded first commercially as a clinical imaging product, transaxial tomography eventually took over as the standard clinical SPECT systems.

Longitudinal tomography, which is in essence limited-angle tomography, utilized focused collimators to select a particular plane of interest to image while also capturing blurred images from other out-of-focus planes (Crandall and Cassen 1966). The successful commercial "Pho-Con" scanner was based on a longitudinal tomographic design to produce six slices by using a single gamma camera and 12 slices with dual-head cameras (Anger 1969). Another longitudinal tomographic design utilized a rotating slant-hole collimator to provide limited-angle sampling (Muehllehner 1970). These techniques have been overshadowed by the transaxial tomography in the last three decades. However, with recent advances in applying the concept of "tomosynthesis" and the related computing algorithms for limited-angle tomography, longitudinal tomography may find its way to return to clinics in the future.

Kuhl et al pioneered the concept of both longitudinal and transaxial tomography and developed several generations of scanners from early

1960s to mid-1970s using discrete detectors (Kuhl and Edwards 1963; Kuhl and Edwards 1964; Kuhl et al 1976). In the same time frame, several other multiple-detector based transaxial tomographic scanners were also reported (Patton et al 1969; Bowley et al 1973) with somewhat different configurations and/or scanning trajectories. Investigation of using Anger cameras in transaxial tomography began in early 1960s (Harper et al 1965), with a few incorporating a rotating chair for the patient in conjunction with a stationary camera (Muehllehner 1971; Budinger and Gullberg 1974).

These early SPECT developments advanced this relatively new field significantly. However, the rotating-camera approach (Jaszczak et al 1977; Keyes et al 1977) was the seminal development that led to the broader uses of SPECT in routine clinical practices (Murphy et al 1978). Multiple-camera (especially dual-head or triple-head) whole-body SPECT systems finally became the norm in nuclear medicine clinics (Jaszczak et al 1979). Throughout the 1980s, new or improved approaches for developing rotating-camera SPECT systems and the related image reconstruction and processing methods were reported (Larsson 1980; Tanaka et al 1984). Development of stationary SPECT systems using more complete-sampling configurations without rotating the patients also made progresses (Rogers et al 1988; Genna and Smith 1988), especially for specific applications such as cardiac or brain imaging.

1.1.2 Early Developments of PET

Positron-emitting radioisotopes were first suggested for use in locating brain tumors in the 1950s (Wrenn et al 1951), and a positron imaging device consisting of a pair of NaI(Tl) detectors was soon developed for the brain (Brownell and Sweet 1953). The concept of positron tomograph was also proposed (Anger and Rosental 1959). The first such camera was composed of 32 NaI(Tl) crystals arranged in a circular geometry (Rankowitz et al 1962; Robertson and Niell 1962).

In 1970s, the development of PET flourished. PC-I and PC-II used two opposing banks of multiple detector arrays (Brownell and Burnham 1973; Brownell et al 1977), while PETT-III employed 48 crystals in a hexagonal array (Phelps et al 1975; Ter-Pogossian et al 1975). The use of multiwire proportional chamber (MWPC) was also explored (Lim et al 1975). A high-resolution single-slice system was constructed with the use of a fixed ring array of 280 closely packed NaI(Tl) crystals (Derenzo et al 1975; Budinger et al 1977; Derenzo et al 1977). A system using two large field-of-view (FOV) Anger gamma cameras was developed (Muehllehner et al 1977). PETT-IV and PETT-V were multislice versions of PETT-III for body and brain imaging, respectively (Ter-Pogossian 1977; Mullani et al 1978); while ECAT-II was developed by use of 66 detectors (Phelps et al 1978). All these systems utilized NaI(Tl) crystals except for the MWPC-based device.

In 1977, bismuth germinate (BGO) was suggested for improving detection efficiency (Cho and Farukhi 1977; Derenzo 1977). POSITOME II was the first

PET system developed by using BGO (Thompson et al 1979), which began another era in PET instrumentation including the development of Neuro-PET (Brooks et al 1980), Donner 280-BGO-Crystal Tomograph (Derenzo et al 1981), ECAT-III (Hoffman et al 1983), etc. In this period, two scintillators with relatively fast decay time – cesium fluoride (CsF) and barium fluoride (BaF_2) – were also explored, and CsF was used in building PETT-VI systems (Mullani et al 1980; Ter-Pogossian et al 1982) for brain imaging. These fast scintillators also motivated the development of the first-generation time-of-flight (TOF) systems including SUPER PETT-I (Mullani et al 1980a; Ter-Pogossian et al 1981), TOFPET (Mullani et al 1982), LETI TOFPET (Gariod et al 1982), etc.

For more information, readers can consult several textbooks and review articles (Nutt 2002; Wernick and Aarsvold 2004; Jaszczak 2006; Muehllehner and Karp 2006; Cherry et al 2012; Hutton 2014).

1.2 Overview

1.2.1 Overview of SPECT

Single-photon imaging employs radiotracers labeled by gamma-ray (single-photon) emitting radioisotopes, introduced into live animals or human objects, to follow in vivo function and physiology associated with the specific radiotracer in use. It includes 2D planar scintigraphy and 3D SPECT, both are used in routine clinical nuclear medicine imaging and biomedical research investigations. Currently, planar scintigraphy is used only in a very few specific clinical practices such as thyroid or breast imaging. 3D SPECT constitutes the large majority of the uses of single-photon imaging in both clinical and research applications. Therefore, in this chapter, we will focus mainly on SPECT.

SPECT is considered a functional or molecular imaging method revealing physiological charactristics and/or molecular signatures that the specific radiotracer in use is designed to probe, in contrast to CT and magnetic resonance imaging (MRI) that are uaually considered anatomical imaging methods revealing the structural information. The most commonly used single-photon radioisotope is technitium-99m (Tc-99m), with a half-life of approximately 6 hours by emitting 140 keV gamma-rays. Other single-photon radioisotopes used commonly include thallium-201 (Tl-201), iodine-123 (I-123), indium-111 (In-111), gallium-67 (Ga-67), etc. Radiotracers labeled with these isotopes have been used routinely in imaging of brain, heart, bone, lung, liver, prostate, and other organs.

SPECT detectors capture gamma-rays emitted from the single-photon radiotracers in use and convert the radiation to photonic signals, then in turn, electrical signals which are further processed to form images of the radioactivity distribution. Since gamma-rays emit isotropically and they impinge upon scintillator detectors from all directions, a detected event in the detector without other design considerations would not necessarily correspond to a unique gamma-ray traveling path or direction that can lead to the definition or estimate of the origin of the radiation decay event. In order to better establish the relationship between the detected signals and their original locations of radiation decay, physical collimators with holes or channels only allowing passage of gamma-rays traveling in specific directions are usually designed and used. Physical collimation, even though defines and provides the needed spatial resolution, reduces the system sensitivity significantly because only those photons not stopped by the collimator materials and passing through the holes and open channels would have the chance to be detected; the majority of the gamma-rays are absorbed and stopped by the collimator materials, usually high Z materials with relatively high attenuation coefficients for absorbing gamma-rays.

A typical SPECT imaging system consists of one or more gamma cameras, usually with relatively large FOV. A gantry capable of rotating the gamma cameras around the objects under study, as well as a scanning bed, provides the necessary angular and axial sampling mechanism for the 3D tomographic imaging.

1.2.2 Overview of PET

PET is also a noninvasive functional and molecular imaging technology in which positron-emitting isotopes are used. It is based on the coincidence detection of two gamma-rays that are produced when a positron released by the isotope annihilates with an electron. Since these photons travel in almost opposite directions, the annihilation shall take place on the line connecting their positions of detection – called the line of response (LOR). Thus, unlike SPECT, collimators are not needed to define the line of sight. Conversely, the rate of coincidence detections observed on an LOR, when ignoring attenuation of the gamma-rays by the subject, is proportional to the rate of annihilations, i.e., the total radioactivity, on the LOR. Therefore, PET imaging yields ray-sums of the 3D distributoin of a PET isotope on a set of LORs that are determined by the scanner's geometry. Nowadays, most PET systems are stationary systems that provide a large number of LORs.

As carbon-11 (C-11), nitrogen-13 (N-13), oxygen-15 (O-15), and fluorine-18 (F-18) are all positron emitters, one advantage of PET is the possibility to label many organic molecules for studying a variety of normal or diseased biology. Another advantage is its quantification capability due to its superior sensitivity to other functional imaging modalities and the tractability of its image reconstruction problem.

PET was widely used for studying brain functions before funtional MRI (fMRI) became the method of choice for the task. Currently in the clinic, PET is most used in conjunction with CT, by employing a hybrid PET/CT system, for cancer imaging with flurodeoxyglucose (FDG), which is an F-18 labeled glucose analog. A high uptake of FDG can depict abnormal glucose metabolism associated with the increased aerobic glucolysis in cancer cells, and it is possible to detect cancer with FDG-PET before structural changes appear on CT or MRI images. The utility of PET for evaluting the outcome of cancer treatment is well documented also. With the advent of molecular biology and genetics, the usefulness of PET for studying in vivo the molecular and genetic basis of disease initiation and progression, and for developing therapeutical agents for making molecular level corrections, has been demonstrated. As a result, the potential role of PET in personalized or precision medicine, in which a disease is treated based on a patient's genetic makeup and physiology with in-time, quantitative evaluation of treatment outcome, has been well recognized.

1.3 Current Technologies in Routine Uses

This section provides brief discussions on the main radiation detection technologies currently used for routine SPECT and PET imaging in clinical and research applications. More detailed information can be found in (Wernick and Aarsvold 2004; Cherry et al 2012) and their references.

1.3.1 Scintillation Detectors and Pulse-Height Analysis

Currently, most gamma-ray detectors used in either SPECT or PET are based on inorganic scintillators and PMTs. Scintillator is a material that converts the entire energy or a portion of the energy of the gamma-ray photon it interacts with into visible lights. Typically, the emission of these lights increases quicky to reach a peak and then decays exponentially. Some materials can also have multiple decay components. The light photons are detected and converted to electrical charges by a PMT with a high gain of about 10^6, producing a charge pulse reminiscent of the scintillation light pulse. From the charge pulse, the time when it appears is determined by using a constant fraction disriminator (CFD) and time-to-digital converter (TDC). The charge pulse is also integrated by using a shaper and digitized by an analog-to-digital converter (ADC), to obtain the pulse height, which is related to the amount of gamma-ray energy deposited in the scintillator. In commercial systems, these electronic operations are implemented by application-specific integrated circuit (ASIC). The operation of radiation detectors often relies on a proportional relationship between the pulse height and the gamma-ray energy deposited in the detector; although for some scintillation materials, non-proportional responses are sometimes observed. The pulse time is also determined on the summed pulse.

The pulse-height spectrum is a histogram of the detected events (counts) as a function of the detected energy. Ideally, one would expect that the relative counts are proportional to their relative abundance in the decay sheme at the specific gamma-ray energy levels of the radioisotope in use, forming a set of sharp lines referred as *photopeaks*; and no counts at any other energy levels. However, since the scintillation detector has only limited *energy resolution*, the observed pulse-height spectrum is usually a blurred version of the ideal sharp lines. Considering also gamma-rays only deposited part of their energy inside the scintillator via Compton scattering, there usually is a continuum of detected events of lower energies, which is also subject to the blurring by the limited energy resolution of the detector. Furthermore, including the scattered photons emitted from the object under study that are also detected within the scintillator, the actual measured pulse-height spectrum of a radioisotope inside an object is often a broad-band cotinuum of detected counts with blurred photopeaks corresponding to the gamma-ray energies characteristic to the radioisotope in use.

Those events associated with primary photopeaks contain the most useful information in imaging; while the scattered events are considered less important or even as noise or artifacts. Therefore, a "discrimination window" setting usually centered around the photopeaks is applied to the pulse-height spectrum to accept the desirable primary photons and also to exlude the unwanted scatter counts.

Evidently, the detection scheme described earlier assumes that there is only one incoming gamma-ray photon interacting with the entire detector within the time frame that the output is taken. If there are multiple interactions by multiple incoming photons, some of these photons can be lost or errorneously registered as a single interaction with incorrect pulse height and pulse time generated with imprecisions. Such events are known as *pile-ups*. Also, the electronic processing time for a pulse can be considerably longer than the duration of the pulse. Pulses that occur during the processing time of the preceding one will be ignored, resulting in loss of events called *dead-time loss*.

1.3.2 SPECT

1.3.2.1 Gamma Cameras

Modern gamma cameras are, in essence, still very similar to those first proposed by Anger (Anger 1958; Anger 1964). Figure 1.1 illustrates the major components of such an Anger camera, which includes a relatively large-area continuous NaI(Tl) crystal, an array of PMTs, the collimator, and electronics for pulse-heights analysis, detected event position determination, and other signal processing, which are output and connected to a computer and display.

The thickness of NaI(Tl) used in the typical gamma cameras ranges from approximately 6 mm to 12.5 mm. While the thicker crystals offer higher system sensitivity, the thinner crystals provide better spatial resolution. Since

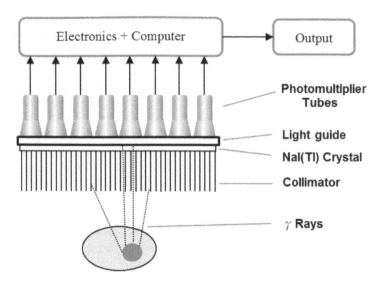

FIGURE 1.1
Major components of an Anger gamma camera.

NaI(Tl) is hygroscopic, the scintillator is usually sealed hermetically in a special housing case. The backside of the NaI(Tl) crystal is attached to the PMT
array, often via a light guide layer. Both rectangular and circular crystals
have been employed in gamma cameras, with the latest ones more in the rectangular configuration. In general, 7 to more than 100 PMTs, mostly round
and square in shape, are used to form the PMTs array.

One or more gamma cameras are typically mounted on a gantry, which
can rotate around the object under imaging to acquire data at various angular positions. Single-head SPECT systems have a simpler and more flexible
configuration and scanning trajectory, but with inferior system sensitivity.
Dual- and triple-head SPECT systems have markedly increased system sensitivity. Non-rotating, stationary ring-based SPECT systems, with the use
of many smaller detector modules, have also been developed and used for
dynamic imaging studies such as those involving cardiac functions.

1.3.2.2 Anger Position Logic

One of the key features in the original Anger camera design was the position
logic circuit, which employed a resistive network to "code" the position of
each PMT and to provide the "weighted" outputs of each detected event so
that both the position and energy deposit of that event could be calculated.
This Anger position logic and its associated circuit designs had long been
utilized in analog gamma cameras. In modern digital cameras, the outputs of
PMTs are digitized and fed into the computer to determine the event position
and deposited energy.

However, this Anger position logic assumes that the scintillation camera responds to radioactive source completely linearly across the entire face of the detector, which is not the case in reality. For example, the light collection efficiency near the center of a PMT is in general better than that toward its edge; also, when a radioactive source is moving near the edge of the scintil-lator, the photon detection tends to respond differently compared to the case when the source is toward the center. These nonlinear responses can produce "pincushion" or "barrel" distortions in the image, which need to be corrected for in the signal processing steps.

1.3.2.3 Collimators

As stated earlier, physical collimators are required to define and provide the spatial resolution in single-photon and SPECT imaging. They also dictate the sensitivity of the SPECT systems since the large majority of the incom-ing photons are stopped and absorbed by the collimators so that only those travel along very selected lines of direction defined by the specific collimator design are permitted to reach the scintillator surface to be detected. Various collimators are designed for different considerations of radioisotopes in use and their corresponding gamma-ray energies, as well as for different study objectives. Materials used for constructing the physical collimators often include lead, tungsten, and other high Z materials with significant stopping power of gamma-rays.

The commonly-used collimators include parallel-hole, diverging, converg-ing, and pinhole collimators. The parallel-hole collimators, most routinely employed in clinics, consist of hexagonal or other shape of holes drilled or cast in lead, or formed by lead foils. The lead walls between holes, called septa, are designed with the thickness necessary to stop the photons trav-eling to the neighboring holes. There is no magnification effect in using a parallel-hole collimator. Diverging collimators have holes diverging from a focal point behind the detector so that a minified image of the object under study is collected, and are used in situation when objects of large sizes, often more extended than the dimension of the gamma camera in use, need to be imaged. On the other hand, converging collimators have holes converging from a focal point in front of the detector so that a magnified image of the object is generated, and are used to enlarge the target regions so that more details can be revealed. Pinhole collimators can consist of either a single pin-hole or multiple pinholes, with the latter usually for providing various angu-lar views in limited-angle tomography and also sometimes for increasing the sensitivity. Each pinhole is an aperture that allows passage of gamma-rays within a certain solid-angle and produces an inverted image of the object. If the source-to-collimator distance is smaller than the distance from the detec-tor face to the collimator aperture opening, then a magnified image can be generated to achieve higher spatial resolution. Conversely, a minified image of an extended volume of object can be produced.

1.3.3 PET

1.3.3.1 Block Detector Modules

As illustrated in Figure 1.2, most PET systems are made of multiple detector rings, each containing a large number of detectors. *Block detectors* are commonly used, which, as illustrated in Figure 1.3 as an example, use a 2×2 PMT array to read a larger, for example 8×8 or 12×12, scintillator array. The scintillator array can be obtained by cutting a monolithic block with a pattern of cut depth that is empirically determined to encode the position of the scintillator within the array in the distribution of light observed by the PMTs. For example, the cut pattern in Figure 1.3 will distribute lights originating in the corner crystals to concentrate in one PMT while distributing lights originating in the central crystals to all PMTs more evenly. Consequently, one can determine the *active* crystal (which produces the signal) by using the relative pulse heights obtained by the PMTs. A popular method is to calculate the x and y coordinates by

$$x = \frac{(B+D)-(A+C)}{A+B+C+D}, \; y = \frac{(A+B)-(C+D)}{A+B+C+D,} \tag{1.1}$$

where A, B, C and D denote the pulse heights of PMTs A, B, C, D. A look-up table, which is determined during system calibration, then maps these values to the identification (ID) of the active scintillator.

1.3.3.2 Coincidence Detection

We have described the detection of a gamma-ray, resulting in a *single event* or *single*. In the context of PET, when two singles are detected within a

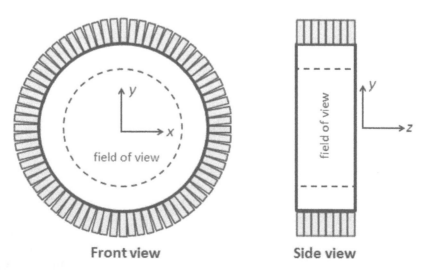

Front view   **Side view**

FIGURE 1.2
A PET system typically consists of multiple rings of small detectors.

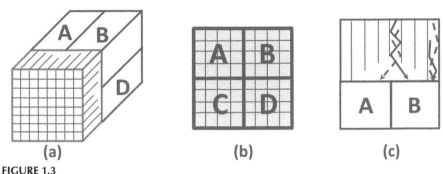

FIGURE 1.3
(a) A PET block detector that uses 2×2 PMTs to read 8×8 crystals, (b) top view shows the PMT arrangement, (c) side view shows crystal position encoding by affecting the light distributions through varying the cut depths.

prescribed *coindence timing window*, they form a *coincidence event* or *coinci-dence*. Generally, only the IDs of the detectors that register the two singles are retained. In TOF systems, the difference in the detection times, called *differential time*, is also stored. Typically, the coincidence logic is implemented by an electronic board that receives singles from a number of detector mod-ules and output coincidences serially, in chronological order. Time tags are inserted into this data stream to provide time information. Other tags can also be introduced to indicate the occurrences of certain events (such as res-piration). The data stream can be stored "as is" to produce *list-mode data*. Or, the events can be histogrammed according to the paired crystal IDs (and also according to the differential-time bin in TOF PET) to obtain *histogrammed-mode data* (also called *sinogram*). For preclinical systems, it is possibe to store singles in list-mode and perform coincidence filtering post-acquisition. This strategy allows the use of more sophisticated software-based coincidence logics but requires larger data bandwidth and storage.

Three types of coincidences can be produced. As illustrated in Figure 1.4, the first is *true coincidence* or *true* (*T*), of which the two gamma-rays are

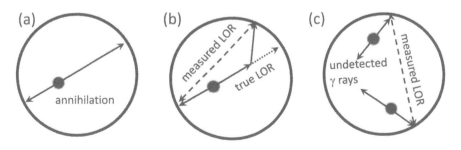

FIGURE 1.4
Three types of coincidence events: (a) true coincidence, (b) scattered coincidence, (c) random coincidence.

associated with the same annihilation and they travel directly from their origination to the detectors. The second is *scattered coincidence* or *scatter* (*S*), of which at least one of the annihilation photons undergoes Compton scattering before reaching the detector. Because scattering alters the traveling direction, the LOR is mispositioned. Evidently, a larger subject leads to more scatters and hence fewer trues (besides scattering, gamma-rays also can be absorbed by the subject via, for example, photoelectric interaction). The third is *random coincidence*, or *accidental coincidence* or *random* (*R*), of which the two gamma-rays are from separate annihilations but are detected in coincidence by chance. For randoms, the measured LORs have no relation to the true LORs. Evidently, using a wider coincidence time window leads to more randoms. Randoms can be measured by artificially delaying one event time by an amount larger than the coincidence timing window so that the resulting coincidences, called *delayed coincidences* or *delays*, can only be randoms. To distinguish, coincidences obtained without delay are called *prompt coincidences* or *prompts*. As already discussed, prompts contain trues, scatters, and randoms.

A PET system shall detect as many trues and as few scatters and randoms as possible. For clinical imaging, the amount of scatters is observed to be substantial. As Compton scattering reduces the gamma-ray energy, the conventional approach to reduce scatters is to reject singles whose energies are below a preset *lower-level discriminator* (LLD) setting. An *upper-level discriminator* (ULD) setting is also used for rejection of pile-ups. Such *energy qualification* is performed before coincidence filtering, as illustrated in Figure 1.5. In practical, approach that increases trues typically also increases scatters and randoms. The tradeoff is assessed by the *noise equivalent count rate* (NECR), defined as $T^2/(T+S+kR)$ where k is a number depending on the scanner design and the random correction method.

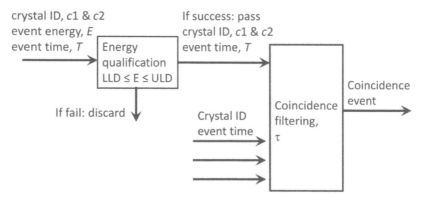

FIGURE 1.5
The signal flow in coincidence detection in PET. τ is the coincidence timing window.

1.3.3.3 Factors Affecting Spatial Resolution

Many PET image reconstruction algorithms assume that, as described earlier, a pair of crystals responds to positrons released on the imaginery LOR associated with them. This assumption is not true in reality and any departures lead to degraded image resolution. Figure 1.6(a) shows the sensitivity profile of a pair of crystals to a point source as it moves along the line halfway between them, by considering the solid angle of coincidence detection. Instead of an delta function like response, the sensitivity function is a triangle whose base equals the crystal width d, therefore having an FWHM equal to $d/2$.

In addition, what is observed in PET is the annihilation of positrons, not their release. Annihilation is most likely when a positron has lost its kinetic energy. Therefore, as shown in Figure 1.6(b), a positron will travel a distance before annihilation, and a pair of crystals sees PET isotopes in a volume larger than volume of annihilation to which it is responsive. The distance traveled in the direction normal to that of the annihilation photons is called *positron range*. This range depends on the isotope and the surrounding medium. For

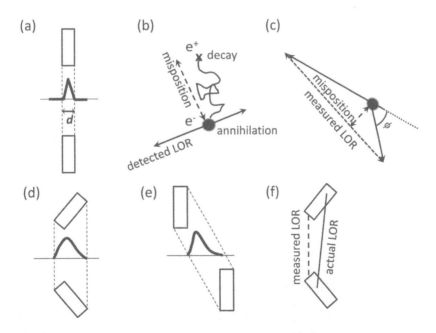

FIGURE 1.6
Spatial resolution in PET is affected by the crystal width (a), positron range (b), photon noncollinearity (c), where ϕ has a zero mean and an FWHM of 0.25 degree, and DOI blurring (d, e). DOI blurring for the case involving two crystals on the same ring is shown in (d) and the case involving two crystals in different rings is shown in (e). DOI blurring can also be examined by considering mispositioning of the LOR when DOI is unknown (f).

F-18, the positron range in water is about 0.54 mm in FWHM. For Rb-82, it is, however, as large as 6.14 mm.

Another factor, depicted in Figure 1.6(c) and called *photon noncolinearity*, is due to the fact that the annihilation photons are not emitted in exact opposite directions. The deviatoin is random; its distribution has a zero mean and an FWHM of about 0.5 degree. As a result, the measured LOR does not pass through the location of annihilation. This effect causes a mispositioning of 0.0044 R in FHWM, where R is the radius of the system.

The fourth factor is related to the positioning errors of block detector due to statistical variations in the coordinates calculated by Equation (1.1). This component depends on the specific design of the block detector and decoding method used.

An empirical rule for the overall spatial resolution, in FWHM, is given by

$$r = k\sqrt{(d/2)^2 + r^2 + (0.0044R)^2 + b^2},$$ (1.2)

where r is the positron range, b is the resolution of the block detector, and $k>1$ describes the effect of image reconstruction. Among them, the dominating factor is the crystal width.

Equation (1.1) considers the resolution at the center of a PET system. Figures 1.6(d) and (e) illustrate situations in a multi-ring system when gamma-rays enter the crystals at an oblique angle. The sensitivity profile can now be considerably wider than the crystal width if the thickness of the crystal is much greater than its width. It also becomes asymmetrical, no longer a triangle as shown in Figure 1.6(a). Another way to look this blurring is to consider mispositioning of the LOR, as depicted in Figure 1.6(f). When a single is detected, by convention the interaction is assumed to be at the front center of the crystal or to be located at the expected position of interaction given the attenuation length of the scintillator. The actual depth into the crystal where interaction occurs, called the *depth-of-interaction* (DOI), is random. When it is different from the assumed depth, the measured LOR, which connects the assumed interaction positions, is misplaced from the actual LOR. Evidently, the mispositioning, which is additional to the $d/2$ blurring in Figure 1.6(a), is more pronounced for crystals having a larger thickness:width ratio. This blurring can be removed or reduced if DOI is measured. It also can be reduced by using crystals having a shorter attenuation length.

A gamma-ray can interact with the detector through multiple interactions if the first interaction is Compton scattering. In theory, the LOR shall be determined based on the first interation. In practice, individual interactions cannot be resolved. For large crystals, this is not an issue because the multiple interactions are likely to occur inside the same crystal. However, when narrow crystals are used the interactions can spread over multiple crystals, and the identified active crystal may not be the first interacting. Blurring caused by this *intercrystal scattering* is significant when crystal "fingers" are used for achieving submillimeter resolution.

1.3.3.4 Considerations for the Scintillator

Performance properties of a PET system are critically affected by the scintillator. Key properties of the scintillator relevant to PET imaging include the attenuation length (at 511 keV), light yield, and decay time. *Attenuaton length*, numerically equals to the inverse of the attenuation coefficient, is the distance into the scintillator where the intensity of a gamma-ray beam drops to 1/e of its value at entry. It is also the average distance a gamma-ray photon travels in the scintillator before interaction. In response to a gamma hit, many scintillators produce a short light pulse exhibiting a single-exponent decay. *Light yield* is the amount of lights in the pulse per gamma-ray energy deposited in the scintillator, often expressed in photons/MeV or as a percentage of the light output of NaI. *Decay time* is the time constant of the decay. Some scintillators can have multiple decay components, characterized by multiple decay times.

Evidently, a small attenuation length is preferred so that short scintillators can be used to effectively detect gamma-rays and mitigate DOI blurring. In this regard, one often favors a material having a high density and/ or a high effective Z number (*high-Z*). Materials having high-Z generally have a large *photofraction*, which is the fractional of gamma-ray interactions that is due to photoelectric effect. Since typically only photopeak events are kept, a large photofraction is preferred. Due to counting statistics, generally a *brighter* scintillator that has a higher light yield will have a better energy resolution, allowing the use of a tighter energy window to reject more scatters without compromising the detection of trues. High-resolution block detectors also require brighter scintillators because they use pulse heights for event positioning (see Equation (1.1)). The accuracy in determining the event time is to first order proportional to the ratio m/δ, where m is the rising slope, and δ is the standard deviation of the noise, of the light pulse. Typically, short decay time also means short rise time; therefore, TOF systems require *fast* scintillators having a short decay time (hence a large m) that is also bright (hence a small δ). Using fast scintillators also reduces pile-ups and dead-time loss.

An ideal PET scintillator therefore shall have a high density, a high effective Z number, a high light yield, and a short decay time. Scintillators that have been considered for PET include NaI, BaF_2, BGO, and L(Y)SO. NaI, the scintillator of choice for gamma cameras and used in early PET systems, is bright but it is slow and has a large attenuation length. BaF_2 has a fast component and was succesfully used in early TOF systems. Unfortunately, the potential gain in the image signal-to-noise ratio by TOF detection was offset by the low detection efficiency of the material. BGO, having high density and high Z value, can provide good detection efficiency, making it the scintillator of choice in the 1990s. Unfortunately, it is not bright nor fast. L(Y)SO is slightly less dense than BGO but is much brighter and faster. As a result, it has become a popular choice for preclinical systems that require high resolution and the present-day clinical TOF PET systems. L(Y)SO has natural radioactivity, which can be a hindrance for imaging weak sources or detecting small changes.

PET spatial resolution has now reached 2–4 mm for clinical systems and approximately 1 mm for preclinical systems. Submillimeter resolution detectors have also been reported. In contrast, the sensitivity is under 1% for most clinical systems (based on NEMA NU-2 standard) and under 5% for many preclinical systems. The coincidence resolving time (CRT) of the latest clinical TOF PET systems is in the range of 200–400 ps (Grant et al 2016; Zhang et al 2018; Carlier 2020). All CRTs given in this chapter are the FHWM values.

1.4 Recent Advances

In this section, we review a number of recent advances in new technologies of radiation detection in SPECT and PET that, we believe, will impact significantly the future ECT instrumentation development. Only selected topics are discussed in this section; more complete and detailed information can be found in the relevant textbooks (Knoll 1999; Wernick and Aarsvold 2004; Cherry et al 2012) and review articles (Madsen 2007; Peterson and Furenlid 2011).

1.4.1 SPECT

1.4.1.1 Scintillators

NaI(Tl) has been the workhorse scintillator widely employed in gamma cameras and SPECT imaging systems, usually in the form of relatively large-area monolithic crystals. "Block detector" designs, similar to those described earlier for PET but consisting of an array of small "pixelated" individual NaI(Tl) crystals of the size of 1–2 mm, have also been developed for gamma-ray and SPECT imaging (Zeniya et al 2006; Xi et al 2010). Pixelated CsI(Tl) and CsI(Na) crystals, that are denser and have higher light yield but slower decay time than those of NaI(Tl), have also been used in building prototype gamma cameras (Truman et al 1994; William et al 2000). CsI(Tl), in the form of microcolumnar crystal arrays, has been employed in building high-resolution gamma cameras and SPECT imaging systems (Tornai et al 2001; Nagarkar et al 2006). $YAlO_3(Ce)$, also known as YAP that is denser and has faster decay time but less light yield than those of NaI(Tl), was utilized in building the YAP-(S)PET small-animal scanner that is capable of both SPECT and PET imaging (DelGuerra et al 2006).

Lanthanum scintillation crystals offer relatively high light output compared to other scintillators (Pani et al 2006), leading to better energy resolution. Both $LaCl_3(Ce)$ and $LaBr_3(Ce)$ have been investigated for uses in gamma-ray and SPECT imaging (van Loef et al 2001; Shah et al 2003; Alzimami et al 2008). Since $LaBr_3(Ce)$ is denser and has a higher light yield between the two, it has been employed in building gamma-ray cameras and SPECT imaging systems for a variety of applications (Russo et al 2009; Yamamoto et al 2010; Roy et al 2014). More recently, a new scintillator, europium-doped strontium iodide ($SrI_2(Eu)$), has received considerable attention for its potential uses in gamma-ray and SPECT imaging (Cherepy et al 2008; Cherepy et al 2009). $SrI_2(Eu)$ offers even more light output than that of $LaBr_3(Ce)$ leading to excellent energy resolution, and is denser than NaI(Tl). Even though its decay time is slower than that of both NaI(Tl) and $LaBr_3(Ce)$, for SPECT imaging, these inferior characteristics are not important. This new scintillation crystal has a great potential to be a candidate to evolve the future landscape of SPECT instrumentation.

More detailed information of advances in scintillation technologies for gamma-ray and SPECT imaging can be found in (van Eijk 2002; Madsen 2007; Peterson and Furenlid 2011).

1.4.1.2 Semiconductor Detectors

Semiconductor (solid-state) detectors, such as silicon (Si), germanium (Ge), cadmium telluride (CdTe), and cadmium-zinc-telluride (CZT), have been extensively investigated for applications in medical imaging including those in nuclear medicine. These semiconductor detectors offer excellent energy resolution when compared to the standard NaI(Tl)-PMT approach. The superior energy resolution results from the relatively large number of electron–hole pairs generated per keV of photon energy deposited from those radioisotopes commonly used in nuclear medicine (Knoll 1999; Cherry et al 2012), leading to relatively low statistical variation in signal response to photon energy. These semiconductor detectors also have the potentials for providing superior spatial resolution. The detector element can be fabricated with very small image pixels, and also the clouds of electrons and holes resulting from photon interactions at the typical radiotracer energies encountered in nuclear medicine are less than a couple of hundred microns after experiencing diffusion and drift in the electric field, leading to a spatial resolution limit on the order of a few hundred microns or better. However, factors such as the requirement for cooling, especially in the case of Ge and Si, relative costs and availability of the semiconductor detector materials, their temperature-sensitive performance parameters, the needs for a large number of signal processing channels and rapid computation, etc., have prevented from the effective and routine utilization of these semiconductor detector technologies in nuclear medicine, especially in SPECT (Scheiber and Cambron 1992; Scheiber 1996; Scheiber 2000; Scheiber and Giakos 2001; Limousin 2003; Sharir et al 2010). But recent advances in these areas have made future incorporation of their advantage of excellent energy and spatial resolution into routine SPECT uses more feasible.

Even though investigation on the use of Si and Ge for gamma-ray imaging have been explored for several decades, their incorporation into routine applications has been somewhat limited, in part because of the requirement of special cooling in order to avoid excessive thermally generated electronic noise, which could be costly and inconvenient for operation. Several gamma-ray or SPECT imaging systems based on Si detectors have been developed in the last decade, which offered very-high spatial resolution, but with relatively small FOV, for relatively low photon energy detection such as that involving I-125 (Peterson et al 2003; Choong et al 2005; Peterson et al 2009; Shokouhi et al 2010). Spatial resolution better than 100 microns can be achieved with specially designed multi-pinhole collimators. High-purity germanium (HPGe), integrated with advanced electronics and compact mechanical cooling system, has also been exploited for building SPECT imaging system (Johnson et al 2011). It was demonstrated that better than 1% in both energy resolution and imaging response uniformity can be achieved. Both these Si and Ge technologies have proved to be useful and feasible for building gamma-ray and SPECT imaging systems for specific applications such as small organ or small animal imaging. It is also worth noting that HPGe has also been employed to build a PET imaging system (Cooper et al 2009).

CdTe and CZT have also been studied extensively to explore their potential uses in gamma-ray and SPECT imaging, especially because, in contrast to Si and Ge, they can be operated in an ordinary room temperature environment. The broad employment of these semiconductor detectors has also been somewhat limited previously, primarily because of their relatively high costs and scarce availability. In the past decade, multiple gamma cameras and SPECT imaging systems have been designed and built with the use of either CdTe or CZT. For example, MediSPECT, a small animal imaging system, employed CdTe pixel detectors and a coded aperture mask collimator to image radioisotopes with low (I-125) or medium (Tc-99m) energy, and achieved a spatial resolution of approximately 1–2 mm (Accorsi et al 2007). A single-head MediSPECT system was later used with a 0.4-mm single pinhole collimator to achieve a spatial resolution of 0.2 mm but within a FOV of only 2 mm (Accorsi et al 2007a). SemiSPECT, another small animal SPECT imaging system based on an array of 8 CZT detectors, achieved an average spatial resolution of 1.45 mm along each axis within the FOV using pinhole collimators of 0.5 mm diameter and offered an overall system sensitivity of approximately 0.1% depending on the specific window setting (Kim 2006). Clinical gamma cameras and SPECT imaging systems based on CZT detectors have also been successfully utilized in routine practice, especially for cardiac imaging (Esteves et al 2009; Duvall et al 2011) and breast imaging (O'Connor et al 2007; Mitchell et al 2013). Other room-temperature compound semiconductors that potentially can be very promising for medical imaging applications, especially SPECT, include HgI_2, PbI_2, and TlBr, in part because of their relatively high density and strong stopping power when compared to those of CdTe or CZT (Peterson and Furenlid 2011).

More information on solid-state detectors and their uses in gamma-ray and SPECT imaging can be found in the relevant chapters in the textbooks (Wernick and Aarsvold 2004; Cherry et al 2012) and selected review articles (Madsen 2007; Peterson and Furenlid 2011; Peterson and Shokouhi 2012).

1.4.1.3 Photodetectors

In addition to scintillation crystals and semiconductor detectors, advances in photodetector technologies for signal readout have also been a driving force in evolving gamma-ray and SPECT imaging instrumentation developments. Position-sensitive PMT (PSPMT) technology, of which 2D position information is provided via two sets of wire anodes that are arranged orthogonally to one another, has been widely employed in gamma cameras and SPECT systems to replace the traditional Anger logic circuits in determining the event position (Kume et al 1986). The PSPMT technology has been integrated with primarily pixelated scintillation crystals including NaI(Tl) (Yasillo et al 1990; Zeniya et al 2006), CsI(Na) (Williams et al 2000), CsI(Tl) (Pani et al 1999), $LaBr_3(Ce)$ (Yamamoto et al 2010), etc.

Photodiodes (PDs), much more compact than PMTs or PSPMTs, can also be used to convert scintillation lights to electrical signals (Choong et al 2002), especially when compact designs are required. However, PDs cannot amplify the signals like the way PMTs do in order to enhance the signals significantly. Avalanche photodiodes (APDs), which operate at higher reverse-bias voltages in a breakdown mode, can achieve signal amplification since the drifting charges are accelerated so that additional electron–hole pairs are created. But even so, the gain of a typical APD is usually only 20–30% of that of the traditional PMTs. Large-area tiled APD arrays can be used to replace PMT arrays in conjunction with the use of monolithic scintillators as in traditional gamma camera (Shah et al 2001). Alternatively, one APD can also be coupled directly to an individual segmented scintillation crystal as the basic detector element for forming the large-area detector array. Position-sensitive APDs (PSAPDs) use charge sharing between additional electrodes on the back surface of the APD to determine event localization with sufficient spatial resolution at a few hundred microns. PSAPDs have been used as the photodetectors in conjunction with CsI(Tl) for use in high-resolution small-animal SPECT systems (Funk et al 2006). It is usually required to cool the PSAPD-based detectors in order to reduce the dark current for maintaining a reasonable level of the SNR.

Charge-coupled devices (CCDs), unlike the event-driven PMTs and APDs, integrate the events over some integration time before sequentially reading out each pixel and outputting the data in a frame transfer mode. CCDs can also be used as the photodetectors in converting scintillation lights to electrical signals. CCDs have high quantum efficiency with their energy resolution affected by dark current and readout electronic noise, which can be reduced by cooling. Some CCDs also employ electron-multiplying approaches

(EMCCDs) to amplify the charge signals serially in order to reduce the readout noise. Since a typical CCD pixel size is about only 20 microns, relatively small compared to the scintillator element (typically CsI(Tl) crystals), demagnification is usually needed in order to extend the detection area and retain the high intrinsic spatial resolution when coupling the scintillator to the CCD. For this de-magnification step, both fiber optic tapers (de Vree et al 2005) and lenses (Nagarkar et al 2006) have been used to build SPECT imaging systems. To compensate for the light loss by using the lens and fiber optic taper, demagnifier (DM) tubes (Meng 2006) and microchannel plate (Miller et al 2008) have also been used to provide optical gains in the high-resolution SPECT imaging systems using CCDs or EMCCDs. These prototype SPECT imaging devices achieved sub-hundred micros spatial resolution for I-125 imaging, and a few hundred microns resolution for Tc-99m imaging.

The latest technology advances in novel photodetectors for gamma-ray or SPECT imaging are those associated with silicon photomultipliers (SiPMs), which are also discussed elsewhere in this book and this chapter. SiPM has been used in conjunction with EMCCD in building high-resolution SPECT imaging systems to provide prior information of events occurred inside the scintillator CsI(Tl) (Heemskerk et al 2010). A compact gamma camera using scintillator YSO(Ce) and an array of SiPMs was built and evaluated, demonstrating an achievable spatial resolution of 1.5 mm (Yamamoto et al 2011). Another recent study showed that SiPMs can be employed successfully in nuclear medicine imaging when coupled with NaI(Tl) for use in SPECT and with LYSO for use in PET, but unsuccessfully when coupled with CsI(Tl) or BGO (Stolin et al 2014). The use of SiPMs in novel design concepts for building new gamma-ray and SPECT imaging systems in the future is expected to grow and requires substantial research efforts to fully explore the potentials of this new technology.

More detailed information regarding photodetectors can be found in the relevant chapters in the textbooks (Wernick and Aarsvold 2004; Cherry et al 2012) and review articles (Madsen 2007; Peterson and Furenlid 2011).

1.4.2 PET

1.4.2.1 DOI Detectors

As discussed earlier, high-resolution and high-sensitivity PET detectors shall use narrow and long crystals having a short attenuation length. However, such crystals are prone to DOI blurring. For higher system sensitivity, one can also increase the overall solid angle of detection by increasing the axial length and/or decreasing the diameter of the system. In addition to increasing scatters and randoms, both design strategies will amplify DOI blurring as more events will enter detectors more obliquely.

Thick detectors capable of DOI measurement are therefore essential for developing high-resolution, high-sensitivity PET systems. DOI can be measured by using dual-ended light readout of a crystal, based on the obser-vation that the relative amounts of light reaching the two ends of the crystal depend on DOI (Moses and Derenzo 1994). For practical reason, single-ended light readout is perferred. It is possible to stack short scinitillator segments, or layers, and devise a scheme to determine the active layer from rear-end measurement. For example, for multilayer block detectors one can devise crystal surface treatment to distribute the scintillation lights to the PDs in such a way that histograms of x and y values calculated by Equation (1.1) for different layers are interleaved (Nishikido et al 2014), therefore, encoding both the active layer and the active crystal within the layer. In a phoswich detector, different layers have different decay times, and a decay-time depen-dent measurement is devised for identification of the active layer (Prout et al 2020). DOI dependent variation in the decay time also can be created for a continuous crystal by phsophor coating of the crystal (Roncali et al 2014).

Monolithic DOI detectors also have been proposed. In this case, the scin-tillator is coupled to a PSPMT, or an array of small PDs, for measuring light distribution at the rear surface of a monolithic scintillator. Ignoring light reflection at the edges, the centroid of the measured light distribution will give the x and y coordinates of the gamma-ray interaction and the width of the distribution the DOI (a shallower event will create a wider distribution). Due to light reflection, the positioning accuracy near the edges will deteriori-ate however. This issue can be mitigated by roughing the surface or apply-ing light absorbing paint. It also can be handled by using more sophisicated positiong algorithms such as maximum-likelihood and nonlinear estimation methods (Joung et al 2002; Ling et al 2007; Li et al 2010). For an array of pix-elated crystals, one can place reflectors at the front surface to redirect scintil-lation lights back into the crystals. The reflectors can be designed to allow estimation of DOI from the spread of the light distribution observed at the rear surface (LaBella et al 2020).

DOI also can be measured by putting PDs along the length of the crys-tal, or an array of PDs on a side surface of a scintillator slab (Levin 2012). DOI detectors exploiting the compactness of SiPM and associated electronics also have been proposed; examples will be given in Section 1.4.2.3. More DOI technologies, including the use of wavelength-shifting fibers and hybrids of the earlier design approaches, are reviewed by Ito et al (2011).

1.4.2.2 TOF Detectors

There is renewed interest for TOF PET since early 2000s (Moses 2003). Initially, clinical systems capable of 500–600 ps CRT was achieved by using L(Y)SO and PMTs (Surti et al 2007; Jakoby et al 2011). Using LaBr$_3$, Kuhn et al reported a 313-ps CRT (Kuhn et al 2006). By increasing the cesium (Ce) dop-ing of LaBr$_3$ to 30%, sub 100 ps CRT was reported (Glodo et al 2005). Other promising scintillators for TOF PET include LuAG, LuYAP, LaCl$_3$, CeBr$_3$, LuI$_3$. LuI$_3$ is of great interest because of its light yield (2.6 times that of NaI), a 4%

energy resolution, and a 125-ps CRT (Moses 2007). Among these scintillators, LSO has a best figure-of-merit when taking into account the tradeoff between detection efficiency and TOF resolution (Conti et al 2009). As already mentioned, the latest clinical TOF systems, all using Lu-based scintillators and SiPMs, have a 240–400 ps CRT. With the recent advances in scintillator, PD and electronics, it may be possible to achieve 10 ps CRT (Lecoq et al 2020).

Increasing the crystal length to increase the detection efficiency is observed to deteriorate TOF resolution (Auffray et al 2013). First, longer crystals have diminished light output and cause a larger spread in the transit time for the scintillation lights before they are detected. Second, the time for the first scintillation light to reach the PD is dependent on the DOI (Toussaint et al 2019); longer crystals have a wider DOI distribution and hence a wider distribution of this first arrival time. By using 2×2 mm^2 LSO, the CRT is found to degrade from 108 ps to 176 ps as the crystal length increases from 3 mm to 20 mm (Gundacker et al 2014). On the other hand, simulation results indicated that the crystal cross-sectional size does not significantly affect CRT. For a system that has a 528-ps CRT and employs $6.75 \times 6.75 \times 25$ mm^3 crystals, the contribution due to the crystal size was about 326 ps (Moses and Ullisch 2006). By correcting for the DOI dependence, Shibuya et al improved the CRT of a four-layer detector slightly from 361ps to 324 ps (Shibuya et al 2008). When time resolution improves, the DOI effect can become more significant (Spanoudaki and Levin 2011).

TOF detectors based on detecting the Cherenkov lights generated in materials such as PbF$_2$ and PWO have also been investigated, and sub 100 ps CRT was shown feasible (Korpar et al 2011; Brunner et al 2014). A limitation of this approach is its poor energy resolution due to the small number of Cherenkov lights produced. Cherenkov-based TOF detection with BGO, which can yield acceptable energy resolution based on scintillation, also has been demonstrated. In the lab, better than 300 ps CRT has been reported (Brunner and Schaart 2017; Gundacker et al 2020). In this case the TOF kernel is not Gaussian.

1.4.2.3 *Silicon Photomulitipliers (SiPMs)*

The advent of SiPM has transformed PET instrumentation (Roncali and Cherry 2011). This solid-state PD, besides being compact and rugged, has a PMT-like gain and TOF-capable fast response, operates at a low voltage, and is insensitive to magnetic fields. Most SiPMs produce analog pulse whose amplitude is proportional to the number of lights hitting the PD (ignoring saturation). Digital SiPMs (dSiPMs), on the other hand, produce digital outputs that contain the number of lights detected and the detection time (Haemisch et al 2012, van Dam et al 2013). Compared to analog SiPMs, dSiPM has compromised detection efficiency as the on-chip electronics reduces its detection sensitive area. Improved designs to address this issue have been proposed (Braga et al 2011). Much attention of SiPM-based PET detector development has been on TOF detection. The principle and response characteristics of

SiPM and its applications in TOF PET are covered in another chapter of this book.

SiPMs are also widely used for developing high-resolution DOI detectors (which can also be TOF capable). Examples include using SiPM arrays to readout pixelated or monolithic scintillator with single-ended readout or dual-ended readout (Schaart et al 2009; Delfino et al 2010; Kang et al 2010; Llosa et al 2010; Song et al 2010; Kwon et al 2011; Kishimoto et al 2013; Nishikido et al 2013; Seifert et al 2013). One can also build DOI detectors by stacking multiple layers of thin, non-DOI detectors that employ single-ended readout (Moehrs et al 2006; Herbert et al 2007; Espana et al 2014). A detector design that encloses all six surfaces of a scintillator block with SiPMs is also reported (Yamaya et al 2011). SiPMs are also exploited for developing detectors for PET/MR imaging (Schulz et al 2011; Hong et al 2012; Thompson et al 2012; Yoon et al 2012; Kim et al 2020), endoscopic detectors for pancreas and prostate imaging (Frisch 2013; Garibaldi et al 2013), and handheld intraoperative imager (Popovic et al 2014).

1.4.2.4 Novel Readout Methods

Due to its small size, the electronic readout of high-density SiPMs has emerged as a challenge. Many groups address this issue with the development of SiPM readout ASICs (Deng et al 2010; Meier et al 2010; Bagliesi 2011; Powolny 2011; Janecek et al 2012; Sacco et al 2013; Stankova et al 2012; Castilla et al 2013; Comerma et al 2013; Goertzen et al 2013). Even with such ASICs, it is still often necessary to use electronic multiplexing to reduce the number of outputs, at the expense of count-rate performance. A natural choice for the electronic multiplexing method is to do something similar to the conventional block detector: while, in the block detector, the scintillation lights are distributed to, usually, four PDs, a resistive network is devised to distribute the electrical outputs of the SiPMS to four outputs (Siegel et al 1996; Song et al 2010; Goertzen et al 2013). Position decoding is done as discussed earlier for the block detector; likewise, any coding errors will degrade the detector resolution. Another method is to obtain row and column sums to yield $2N$ outputs for a $N \times N$ SiPM array (Stratos et al 2013). In this case, SiPM identification is achieved by active row and column, which is easier and more accurate to determine than the previous method. By placing resistive chains between the row and column sums, the number of outputs can be further reduced to four (Wang et al 2012; Stratos et al 2013). Position-sensitive monolithic SiPMs also are developed for addressing the readout challenge (Li et al 2014; Schmall et al 2014; Ferri et al 2015; Li et al 2017; Du et al 2018).

In the discussed methods, each readout channel is connected to the summed capacitance of the SiPMs sharing the channel, leading to increased pulse rise time and deteriorated time resolution. This issue can be addressed by introducing preamplifiers to each SiPMs to decouple their capacitances (Liu and Goertzen 2013), increasing circuit complexity. The idea of compressed sensing also has been proposed for producing only 16 outputs for a 12 × 24 SiPM array (Chinn et al 2012; Dey et al 2013). In strip-line (SL) method, multiple SiPMs are connected to an SL and the signals at the two ends of the SL are taken (Kim et al 2010; Kim et al 2012a). If needed, preamplifiers may be placed at the ends of the SL but they are not needed for individual SiPMs. Analogous to TOF detection, the active SiPM on the SL is determined based on the differential time the signals arriving at the two ends of the SL. This readout is scalable and up to 32 SiPMs on an SL have been successfully demonstrated (Kim et al 2021a). The method also permits a significant separation between the acquisition electronics and detectors – a feature that has been exploited for developnig hybrid PET/MRI and PET/EPRI systems (Kim et al 2020; Kim et al 2021a; Kim et al 2021b). Won et al has reported a similar propagation-delay based method for 2D SiPM arrays (Won et al 2016).

1.4.2.5 Waveform Sampling

As discussed earlier, gamma-ray detection is conventionally done by splitting the signal pulse into a fast time channel, and a slow energy channel in which the pulse is "slowed" down by a shaper and then a small number of samples are obtained at about 40 MHz by using ADC. In principle, both event time and energy can be computed from the same pulse samples by digital signal processing (DSP). Several groups studied this approach by using a sampling rate of 50–100 MHz (Streun et al 2002; Martinez et al 2004; Ziemons et al 2005; Fontaine et al 2006; Olcott et al 2006; Hu et al 2009; Hu et al 2012). They reported a CRT in the range of 2–4 ns, with a best result of 0.7 ns achieved when sampling at 100 MHz and applying linear interpolation to upsample the initial rise of the pulse (Hu et al 2009). With waveform samples, DSP algorithms may be developed to improve time resolution (Kao et al 2011) or to remove RF interferences in PET/MR and PET/EPRI imaging (Kim et al 2021b).

To improve CRT further for TOF detection will require higher sampling rates, and in lab high-end digital oscilloscopes can be used. Indeed, the best reported TOF results are often obtained this way. Inexpensive low-power sampling chips based on switched-capacitor arrays can provide a more practical solution. A popular choice is the DRS4 chip that provides eight channels of 0.7–5 GHz sampling with 1024 sampling cells per channel. Unlike ADC, this type of devices provides high-speed sampling for a short duration (e.g., about 200 ns for DRS4 at 5 GHz) but reading out the samples constitute a considerable dead time (e.g., about 30 μs for DRS4). Fortunately, for PET this dead time is acceptable or the issue can be addressed by smart triggering schemes (Ashmanskas et al 2011). DRS4 have been employed for developing TOF detectors (Kim et al 2010; Ashmanskas et al 2011; Kim et al 2011; Kim et al 2012a; Kim et al 2013; Ronzhin et al 2013). For LYSO, Kim et al has shown that the sampling rate can be reduced to 2 GHz, and the sampling duration to 50 ns, without noticeably compromising time and energy resolutions (Kim et al 2012b); and that time calibration is critical (Kim et al 2014). Sampling chips capable of greater than 10 GHz sampling have been developed (Oberla et al 2014).

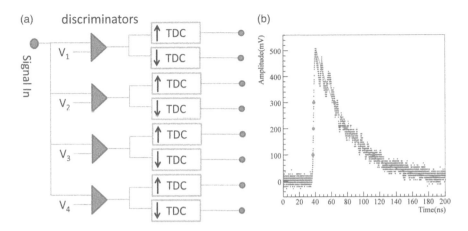

FIGURE 1.7
(a) The MVT method samples a pulse by determining the time it crosses a number of voltage thresholds, using discriminators and TDCs. (b) A sample pulse (noisy curve), the MVT samples (dots), and the fitted pulse from samples (smooth curve).

An alternative sampling method is the multi-voltage threshold (MVT) proposed in (Xie et al 2005; Kim et al 2009). This method, as illustrated in Figure 1.7, samples a pulse with respect to a few voltage thresholds; information about the pulse is then obtained by fitting the samples to a formula describing the pulse shape. The MVT method can always sample the fast rising edge of the pulse if the thresholds are properly defined. This is not the case with DRS4 (and ADC) if the sampling rate is too low. An FPGA-only implementation of MVT has been reported (Xi et al 2013) and it has been successfully applied to obtain a CRT of about 300 ps for LYSO-based detectors (Xie et al 2013).

1.5 Future Trends and Summary

Based on the advances made in recent years that are discussed in previous section regarding relevant technologies for SPECT and PET, vision for future directions and significant trands of this two molecular imaging modalities are discussed briefly in this section.

1.5.1 Modular and Reconfigurable Detectors

The availability of compact semiconductor detectors such as CdTe and CZT, as well as SiPMs that can replace the bulky PMTs, will enable novel and flexible designs of modular detector components that can be used as basic building blocks to assemble imaging systems very flexibly to meet the specific clinical or research needs. It is also anticipated that these compact detector components can be reconfigured conveniently whenever needed in order to provide the optimal imaging configuration and scanning trajectory to maximize the resulting image quality as well as the ultimal clinical or research outcomes. These modular and reconfigurable detectors can facilitate the advances of many application-specific imaging devices discussed in section 1.5.2.

1.5.2 Application-Specific Imaging Systems

Most current SPECT and PET systems are desinged for use as general-purpose equipment to cover the full-range of potential clinical and research needs. Although very useful, these general-purpose imaging systems are usually not optimal for certain specific applications. Modular and reconfigurable detectors made possible by the recent advances of relevant technologies as discussed in section 1.5.1 can facilitate the development of application-specific imaging devices that are optimized specially for the targeted applications. Application-specific gamma camera, SPECT, and PET imaging systems have been developed to provide high sensitivity and resolution for imaging of the breast (Thompson et al 1994; Doshi et al 2000; Doshi et al 2001; Levine et al 2003; Moses 2004; Abreu et al 2006; Karellas and Vedantham 2008; Raylman et al 2008; Bowen et al 2012; Koolen et al 2012; Moliner et al 2012; Miyake et al 2014), prostate (Huber et al 2001; Huber et al 2005; Majewski et al 2011a) and brain (Watanabe et al 2002; Wienhard et al 2002; Karp et al 2003; van Velden et al 2009; Yamamoto et al 2011; Majewski et al 2011b) have been proposed or developed.

Another example is special imaging device specifically designed for use under targeted study conditions. For example, typically, animal are imaged under anesthesia. For brain studies, anesthesia, however, alters the brian activity. There are also experiments that require the animal to be awake and perform tasks. High-resolution, wearable "RatCAP" imagers for imaging the brain of an awake mouse has been developed (Schlyer et al 2007; Vaska et al 2007; Schulz and Vaska 2011). SiPM-based endoscopic PET detectors and light-weigh handheld imaging probes are also been proposed as discussed earlier.

1.5.3 Multi-Modality Imaging Systems

Dual-modailty imaging systems such as PET/CT and SPECT/CT are now considered standard clinical equipment for routine uses. As mentioned earlier, there are also active efforts on developing hybrid PET/MRI and SPECT/MRI systems. Most of these efforts are concerned with the development of PET or SPECT insert detectors. Insert detectors for whole-body PET systems have also been proposed for enhancing the resolution (Wu et al 2008), for improving breast imaging (Mathews et al 2013), and for zoom-in imaging to enhance the detection of small lesions (Zhou and Qi 2009; Zhou and Qi 2011).

1.5.4 Ultrahigh-Sensitivity SPECT and PET

Many potential clinical and research applications of SPECT and PET are severely limited by the relatively poor system semsitivity of these two imaging modalities. This is perhaps one of the most significant challenges for developing future SPECT and PET systems. Recently, some interesting developments iniatiated this new line of innovasations. For example, in SPECT, novel collimator designs using the innovative concept of inverted compound-eye that incorporates hundreds of channels in multi-pinhole collimator for each of a couple of dozen detector modules, resulting in an increase in system sensitivity of 30-fold and potentially more (Lai and Meng, 2018). The concept of "total-body" long-axial FOV PET imaging systems has resurfaced in the last few years, offering a potential factor of 40 in inceased system sensitivity (Poon et al 2012; Cherry et al 2017; Cherry et al 2018), which has led to a successful commercial system that is 200-cm long with CT (Badawi et al 2019), and a broad interest and extensive efforts in the community to advance this ultrahigh-sensitivity PET technology (Badawi et al 2021).

1.5.5 SPECT and PET in Image-Guided Therapy and Theranostics

Recently, the value of PET to provide range verification and dose monitoring in proton and heavy-ion therapy has been demonstrated (Pshenichnov et al 2006; Attanasi et al 2011; Parodi 2012; Aiello et al 2013; Jan et al 2013; Zhu and El Fakhri 2013), and in-beam PET systems proposed (Enghardt et al 2004; Attanasi et al 2008; Parodi et al 2008; Vecchio et al 2009; Shakirin et al 2011; Zhu et al 2011; Shao et al 2014; Sportelli et al 2014). Stationary partial-ring and dual-head configurations are favorable for such systems because they provide access to the patient for the treatment beams (Shakirin et al 2007; An et al 2013). Image artifacts that are typically generated with these limited-view systems are removed if they are TOF systems. Similarly, dedicated TOF PET breast systems employing partial rings to provide access for biopsy needles have been proposed (Surti and Karp 2008; Chen et al 2011). The Open PET systems are also very promising for supporting image-guided applications (Yamaya et al 2008; Yoshida et al 2011 Tashima et al 2012; Yoshida et al 2013).

In the area of targeted radionuclide therapy (TRT) that has seen rapid advancing in recent years, the radionuclides used in these internal radotherapy approaches, such as lutitium-177 (Lu-177) in β-particle therapy (Wester and Schottelius 2019) and actinium-255 (Ac-225) in α-particle therapy (Nelson et al 2021), often also emit gamma-rays, which can be utilized in SPECT imaging for treatment planning, improvement, assessment, etc. Some TRT radioisotopes, such as copper-67 (Cu-67) and scandium-47 (Sc-47), both used in β-particle therapy, also have pairing positron-emitting isotopes (Cu-64 and Sc-43/44, respectively), which can be employed in PET imaging for planning, imporving, and evaluatiing the internal radiotherapy

(Cullinane et al 2020; Muller et al 2014). These "theranostic" radioisotopes are expected to receive considerable attention at the forefront in cancer research.

1.5.6 Summary

The advances in scintillators, semiconductor detectors, photodetectors such as SiPM, and reaodut electronics in the past decades have led to significant progress in SPECT and PET in terms of performance as well as applications. The advances are likely to finally enable the developemnt of high resolution, high efficiency DOI- and TOF-capable detectors. The new detectors are also likely to be much more rugged and versatile. More application-specific SPECT and PET imaging systems, as well as SPECT/MRI and PET/MRI systems can be expected. The use of SPECT and PET in imaging-guided therapy and theranostics can be routine applications soon as well. The development of long-axial FOV systems to greatly increase the sensitivity of SPECT and PET imaging to enable the detection of abnormalities at the earliest stage possible is also of great interest and significance. These advances in SPECT and PET instrumentation, however, need to be accompanied by parallel advances in image reconstruction methods and imaging tracer probes.

References

Abreu, M.C., J.D. Aguiar, F.G. Almeida, et al. 2006. "Design and Evaluation of the Clear-PEM Scanner for Positron Emission Mammography." *IEEE Trans. Nucl. Sci.* 53:71–77.

Accorsi, R., M. Autiero, L. Celentano, et al. 2007. "MediSPECT: Single Photon Emission Computed Tomography System for Small Field of View Small Animal Imaging Based on a CdTe Hybrid Pixel Detector." *Nucl. Instr. Meth. A.* 571:44–47.

Accorsi, R., A.S. Curion, P. Frallicciardi, et al. 2007a. "Preliminary Evaluation of the Tomographic Performance of the mediSPECT Small Animal Imaging System." *Nucl. Instr. Meth. A.* 571:415–418.

Aiello, M., F. Attanasi, N. Belcari, et al. 2013. "A Dose Determination Procedure by PET Monitoring in Proton Therapy: Monte Carlo Validation." *IEEE Trans. Nucl. Sci.* 60:3298–3304.

Alzimami, K.S., N.M. Spyrou, and S.A. Sassi. 2008. "Investigation of LaBr$_3$:Ce and LaCl$_3$:Ce Scintillators for SPECT Imaging." ISBI 2008, the 5th IEEE International Symposium on Biomedical Imaging: From Nano to Macro. 14-17 May 2008.

An, S.J., C.H. Beak, K. Lee, et al. 2013. "A Simulation Study of a C-Shaped In-Beam PET System for Dose Verification in Carbon Ion Therapy." *Nucl. Instr. Meth. A.* 698:37–43.

Anger, H.O. 1958. "Scintillation Camera." *Rev. Sci. Instr.* 29:27–33.

Anger, H.O. 1964. "Scintillation Camera with Multichannel Collimators." *J. Nucl. Med.* 5:515–531.

Anger, H.O. 1969. "Multiplane Tomographic Gamma-Ray Scanner." *Medical Radioisotope Scintigraphy.* Vienna: International Atomic Energy Agency.

Anger, H.O., and D.J. Rosental. 1959. *"Scintillation Camera and Position Camera."* Medical Radioisotope Scanning. Vienna: IAEA and WHO.

Ashmanskas, W.J., B.C. LeGeyt, F.M. Newcomer, et al. 2011. "Waveform-Sampling Electronics for Time-of-Flight PET Scanner." *The 2011 IEEE Nucl. Sci. Symp. and Med. Imag. Conf. Record (NSS/MIC)* 3347–3350.

Attanasi, F., N. Belcari, M. Camarda, et al. 2008. "Preliminary Results of an In-Beam PET Prototype for Proton Therapy." *Nucl. Instr. Meth. A* 591:296–299.

Attanasi, F., A. Knopf, K. Parodi, et al. 2011. "Extension and Validation of an Analytical Model for in vivo PET Verification of Proton Therapy – A Phantom and Clinical Study." *Phys. Med. Biol.* 56:5079–5098.

Auffray, E., B. Frisch, F. Geraci, et al. 2013. "A Comprehensive & Systematic Study of Coincidence Time Resolution and Light Yield Using Scintillators of Different Size and Wrapping." *IEEE Trans. Nucl. Sci.* 60:3163–3172.

Badawi, R.D., H. Shi, P. Hu, S. Chen, T. Xu, P.M. Price, et al. 2019. "First Human Imaging Studies with the EXPLORER Total-body PET Scanner." *J. Nucl. Med.* 60(3):299–303.

Badawi, R.D., J.S. Karp, L. Nardo, et al. 2021. "Special Issue: Total Body PET Imaging." *PET Cliniics.* 15(1), the entire Special Issue.

Bagliesi, M.G., C. Avanzini, G. Bigongiari, et al. 2011. "A Custom Front-End ASIC for the Readout and Timing of 64 SiPM Photosensors." *Nucl. Phys. B Proc. Suppl.* 215:344–348.

Bowen, S.L., A. Ferrero, and R.D. Badawi. 2012. "Quantification with a Dedicated Breast PET/CT Scanner." *Med. Phys.* 39:2694–2707.

Bowley, A.R., C.G. Taylor, D.A. Causer, et al. 1973. "Radioisotope Scanner for Rectilinear, Arc, Transverse Section and Longitudinal Section Scanning – (Ass-Aberdeen Section Scanner)." *Brit. J. Radiol.* 46:262–271.

Braga, L.H.C., L. Pancheri, L. Gasparini, et al. 2011. "A CMOS Mini-SiPM Detector with In-Pixel Data Compression for PET Applications." *The 2011 IEEE Nucl. Sci. Symp. Med. Imag. Conf. Record (NSS/MIC).* 548–552.

Brooks, R.A., V.J. Sank, G. Dichiro, et al. 1980. "Design of a High-Resolution Positron Emission Tomograph – Neuro-PET." *J. Comput. Assist. Tomogr.* 4:5–13.

Brownell, G.L., and C.A. Burnham. 1973. "MGH Positron Camera." In *Tomographic Imaging in Nuclear Medicine,* edited by Freedman G.S. New York, NY: Society of Nuclear Medicine.

Brownell, G.L., and W.H. Sweet. 1953. "Localization of Brain Tumors with Positron Emitters." *Nucleonics.* 11:40–45.

Brownell, G.L., C.A. Burnham, D.A. Chesler, et al. 1977. "Transverse Section Imaging of Radionuclide Distributions in Heart, Lung and Brain." In *Reconstruction Tomography in Diagnostic Radiology and Nuclear Medicine,* edited by M.M. Ter-Pogossian, M.E. Phelps, G.L. Brownell, J.R. Cox, D.O. Davis, R.G. Evens. Baltimore, MD: University Park Press.

Brunner, S.E., L. Gruber, J. Marton, et al. 2014. "Studies on the Cherenkov Effect for Improved Time Resolution of TOF-PET." *IEEE Trans. Nucl. Sci.* 61:443–447.

Brunner, S.E., and D.R. Schaart. 2017. "BGO as a Hybrid Scintillator/Cherenkov Radiator for Cost-Effective Time-of-Flight PET." *Phys. Med. Biol.* 62:4421–4439.

Budinger, T.F., and G.T. Gullberg. 1974. "Three-Dimensional Reconstruction in Nuclear-Medicine Emission Imaging." *IEEE Trans. Nucl. Sci.* 21:2–20.

Budinger, T.F., S.E. Derenzo, G.T. Gullberg, et al. 1977. "Emission Computer-Assisted Tomography with Single-Photon and Positron-Annihilation Photon Emitters." *J. Comput. Assist. Tomogr.* 1:131–145.

Carlier, T., L. Ferrer, M. Conti, et al. 2020. "From a PMT-Based to a SiPM-Based PET System: A Study to Define Matched Acquisition/Reconstruction Parameters and NEMA Performance of the Biograph Vision 450." *EJNMMI Phys.* 7:55.

Castilla, J., J.M. Cela, A. Comerma, et al. 2013. "Evaluation of the FlexToT ASIC on the Readout of SiPM Matrices and Scintillators for PET." *The 2013 IEEE Nucl. Sci. Symp. Med. Imag. Conf. Record (NSS/MIC).* 1–4.

Chen, Y., K. Saha, and S.J. Glick. 2011. "Investigating Performance of Limited Angle Dedicated Breast TOF PET." *The 2011 IEEE Nucl. Sci. Symp. and Med. Imag. Conf. Record (NSS/MIC).* 2346–2349.

Cherepy, N.J., G. Hull, A.D. Drobshoff, et al. 2008. "Strontium and Barium Iodide High Light Yield Scintillators." *Appl. Phys. Lett.* 92:083508. http://dx.doi.org/10.1063/1.2885728

Cherepy, N.J., S.A. Payne, S.J. Asztalos, et al. 2009. "Scintillators with Potential to Supersede Lanthanum Bromide." *IEEE Trans. Nucl. Sci.* 56:873–880.

Cherry, S.R., J.A. Sorenson, and M.E. Phelps. 2012. *"Physics in Nuclear Medicine."* Philadelphia: Elsevier/Saunders.

Cherry S.R., R.D. Badawi, J.S. Karp, W.W. Moses, P. Price, and T. Jones. 2017. "Total-body Imaging: Transforming the Role of Positron Emission Tomography." *Sci. Transl. Med.* 9:eaaf6169.

Cherry, S.R., T. Jones, J.S. Karp, J. Qi, W.W. Moses, and R.D. Badawi, 2018. "Total-body PET: Maximizing Sensitivity to Create New Opportunities for Clinical Research and Patient Care." *J. Nucl. Med.* 59:3–12.

Chinn, G., P.D. Olcott, and C.S. Levin. 2012. "Improved Compressed Sensing Multiplexing for PET Detector Readout." *The 2012 IEEE Nucl. Sci. Symp. Med. Imag. Conf. Record (NSS/MIC).* 2472–2474.

Cho, Z.H., and M.R. Farukhi. 1977. "Bismuth Germanate as a Potential Scintillation Detector in Positron Cameras." *J. Nucl. Med.* 18:840–844.

Choong, W.S., G.J. Gruber, W.W. Moses, et al. 2002. "A Compact 16-Module Camera Using 64-Pixel CsI(Tl)/Si P-I-N Photodiode Imaging Modules." *IEEE Trans. Nucl. Sci.* 49:2228–2235.

Choong, W.S., W.W. Moses, C.S. Tindall, et al. 2005. "Design for A High-Resolution Small-Animal SPECT System Using Pixellated Si(Li) Detectors for in vivo I-125 Imaging." *IEEE Trans. Nucl. Sci.* 52:174–180.

Comerma, A., D. Gascon, L. Garrido, et al. 2013. "Front End ASIC Design for SiPM Readout." *JINST.* 8:C01048, doi 10.1088/1748-0221/8/01/C01048.

Conti, M., L. Eriksson, H. Rothfuss, et al. 2009. "Comparison of Fast Scintillators with TOF PET Potential." *IEEE Trans. Nucl. Sci.* 56:926–933.

Cooper, R.J., A.J. Boston, H.C. Boston, et al. 2009. "Positron Emission Tomography Imaging with the Smart PET System." *Nucl. Instr. Meth. A.* 606:523–532.

Crandall, P.H., and B. Cassen. 1966. "High Speed Section Scanning of Brain." *Arch. Neurol.* 15(2):163.

Cullinane, C., C.M. Jeffery, P.D. Roselt, E.M. van Dam, S. Jackson, K. Kuan, J. Jackson, D. Binns, J. van Zuylekom, M.J. Harris, R.J. Hicks, and P.S. Donnelly, 2020. "Peptide Receptor Radionuclide Therapy with ^{67}Cu-CuSarTATE Is Highly Efficacious against a Somatostatin-Positive Neuroendocrine Tumor Model." *J. Nucl. Med.* 61:1800–1805.

Delfino, E.P., S. Majewski, R.R. Raylman, et al. 2010. "Towards 1 mm PET Resolution Using DOI Modules Based on Dual-Sided SiPM Readout." *The 2010 IEEE Nucl. Sci. Symp. Med. Imag. Conf. Record (NSS/MIC)*. 3442–3449.

DelGuerra, A., A. Bartoli, N. Belcari, et al. 2006. "Performance Evaluation of the Fully Engineered YAP-(S)PET Scanner for Small Animal Imaging." *IEEE Trans. Nucl. Sci.* 53:1078–1083.

Deng, Z, A.K. Lan, X. Sun, et al. 2010. "Development of an 8-Channel Time Based Readout ASIC for PET Applications." *The 2010 IEEE Nucl. Sci. Symp. Med. Imag. Conf. Record (NSS/MIC)*. 1684–1689.

Derenzo, S.E. 1977. "Positron Ring Cameras for Emission-Computed Tomography." *IEEE Trans. Nucl. Sci.* 24:881–885.

Derenzo, S.E., H. Zaklad, and T.F. Budinger. 1975. "Analytical Study of a High-Resolution Positron Ring Detector System for Transaxial Reconstruction Tomography." *J. Nucl. Med.* 16:1166–1173.

Derenzo, S.E., T.F. Budinger, J.L. Cahoon, et al. 1977. "High-Resolution Computed Tomography of Positron Emitters." *IEEE Trans. Nucl. Sci.* 24:544–558.

Derenzo, S.E., T.F. Budinger, R.H. Huesman, et al. 1981. "Imaging Properties of a Positron Tomograph with 280-Bgo-Crystals." *IEEE Trans. Nucl. Sci.* 28:81–89.

Dey, S., E. Myers, T.K. Lewellen, et al. 2013. "A Row-Column Summing Readout Architecture for SiPM Based PET Imaging Systems." *The 2013 IEEE Nucl. Sci. Symp. Med. Imag. Conf. Record (NSS/MIC)*. 1–5.

Doshi, N.K., Y.P. Shao, R.W. Silverman, et al. 2000. "Design and Evaluation of an LSO PET Detector for Breast Cancer Imaging." *Med. Phys.* 27:1535–1543.

Doshi, N.K., R.W. Silverman, Y. Shao, et al. 2001. "maxPET: A Dedicated Mammary and Axillary Region PET Imaging System for Breast Cancer." *IEEE Trans. Nucl. Sci.* 48:811–815.

Du, J., et al. 2018. "Performance of a High-resolution Depth-encoding Pet Detector Module Using Linearly-graded SiPM Arrays." *Phys. Med. Biol.* 63:035035.

Duvall, W.L., L.B. Croft, E.S. Ginsberg, et al. 2011. "Reduced Isotope Dose and Imaging Time with a High-Efficiency CZT SPECT Camera." *J. Nucl. Cardiol.* 18:847–857.

Enghardt, W., P. Crespo, F. Fiedler, et al. 2004. "Charged Hadron Tumour Therapy Monitoring by Means of PET." *Nucl. Instr. Meth. A.* 525:284–288.

Espana, S., R. Marcinkowski, V. Keereman, et al. 2014. "DigiPET: Sub-Millimeter Spatial Resolution Small-Animal PET Imaging Using Thin Monolithic Scintillators." *Phys. Med. Biol.* 59:3405–3420.

Esteves, F.P., P. Raggi, R.D. Folks, et al. 2009. "Novel Solid-State-Detector Dedicated Cardiac Camera for Fast Myocardial Perfusion Imaging: Multicenter Comparison with Standard Dual Detector Cameras." *J. Nucl. Cardiol.* 16:927–934.

Ferri, A., et al. 2015. "Characterization of Linearly Graded Position-sensitive Silicon Photomultipliers." *IEEE Trans. Nucl. Sci.* 62:688–693.

Fontaine, R., M.A. Tetrault, F. Belanger, et al. 2006. "Real Time Digital Signal Processing Implementation for an APD-Based PET Scanner with Phoswich Detectors." *IEEE Trans. Nucl. Sci.* 53:784–788.

Frisch, B., and EndoTOFPET-US Collaboration. 2013. "Combining Endoscopic Ultrasound with Time-of-Flight PET: The EndoTOFPET-US Project." *Nucl. Instr. Meth. A.* 732:577–580.

Funk, T., P. Despres, W.C. Barber, et al. 2006. "A Multipinhole Small Animal SPECT System with Submillimeter Spatial Resolution." *Med. Phys.* 33:1259–1268.

Garibaldi, F., S. Capuani, S. Colilli, et al. 2013. "TOPEM: A PET-TOF Endorectal Probe, Compatible with MRI for Diagnosis and Follow Up of Prostate Cancer." *Nucl. Instr. Meth. A.* 702:13–15.

Gariod, R., R. Allemand, E. Cormoreche, et al. 1982. "The LETI Positron Tomography Architecture and Time-of-Flight Improvement." In *Proc. of Workshop on Time-of-Flight Tomography.* St. Louis, MO: IEEE Computer Society Press, Silver Spring, Maryland.

Genna, S., and A.P. Smith. 1988. "The Development of ASPECT: An Annular Single-Crystal Brain Camera for High-Efficiency SPECT." *IEEE Trans. Nucl. Sci.* 35:654–658.

Glodo, J., W.W. Moses, W.M. Higgins, et al. 2005. "Effects of Ce Concentration on Scintillation Properties of $LaBr_3$:Ce." *IEEE Trans. Nucl. Sci.* 52:1805–1808.

Goertzen, A.L., X.Z. Zhang, M.M. McClarty, et al. 2013. "Design and Performance of a Resistor Multiplexing Readout Circuit for a SiPM Detector." *IEEE Trans. Nucl. Sci.* 60:1541–1549.

Grant, A.M., et al. 2016. "NEMA NU 2-2012 Performance Studies for the SiPM-Based ToF-PET Component of the GE SIGNA PET/MR System." *Med. Phys.* 43:2334–2343.

Gundacker, S., A. Knapitsch, E. Auffray, et al. 2014. "Time Resolution Deterioration with Increasing Crystal length in A TOF-PET System." *Nucl. Instr. Meth. A.* 737:92–100.

Gundacker, S., et al. 2020. "Experimental Time Resolution Limits of Modern SiPMs and TOF-PET Detectors Exploring Different Scintillators and Cherenkov Emission." *Phys. Med. Biol.* 65:025001.

Haemisch, Y., T. Frach, C. Degenhardt, et al. 2012. "Fully Digital Arrays of Silicon Photomultipliers (dSiPM) – A Scalable Alternative to Vacuum Photomultiplier Tubes (PMT)." *Phys. Procedia.* 37:1546–1560.

Harper, P.V., R.N. Beck, D.E. Charlest, et al. 1965. "The Three Dimensional Mapping and Display of Radioisotope Distributions." *J. Nucl. Med.* 6:332.

Heemskerk, J.W.T., M.A.N. Korevaar, J. Huizenga, et al. 2010. "An Enhanced High-Resolution EMCCD-Based Gamma Camera Using SiPM Side Detection." *Phys. Med. Biol.* 55:6773–6784.

Herbert, D.J., S. Moehrs, N. D'Ascenzo, et al. 2007. "The Silicon Photomultiplier for Application to High-Resolution Positron Emission Tomography." *Nucl. Instr. Meth. A.* 573:84–87.

Hoffman, E.J., A.R. Ricci, L.M.A.M. Vanderstee, et al. 1983. "Ecat-III – Basic Design Considerations." *IEEE Trans. Nucl. Sci.* 30:729–733.

Hong, S.J., H.G. Kang, G.B. Ko, et al. 2012. "SiPM-PET with a Short Optical Fiber Bundle for Simultaneous PET-MR Imaging." *Phys. Med. Biol.* 57:3869–3883.

Hu, W., Y. Choi, K. Hong, et al. 2012. "Free-Running ADC- and FPGA-Based Signal Processing Method for Brain PET Using GAPD Arrays." *Nucl. Instr. Meth. A.* 664:370–375.

Hu, W., Y. Choi, J.H. Jung, et al. 2009. "A Simple and Improved Digital Timing Method for Positron Emission Tomography." *The 2009 IEEE Nucl. Sci. Symp. Med. Imag. Conf. Record (NSS/MIC).* 3893–3896.

Huber, J.S., W.S. Choong, W.W. Moses, et al. 2005. "Characterization of a PET Camera Optimized for Prostate Imaging." *The 2005 IEEE Nucl. Sci. Symp. Med. Imag. Conf. Record (NSS/MIC).* 1556–1559.

Huber, J.S., S.E. Derenzo, J. Qi, et al. 2001. "Conceptual Design of a Compact Positron Tomograph for Prostate Imaging." *IEEE Trans. Nucl. Sci.* 48:1506–1511.

Hutton, B.F. 2014. "The Origins of SPECT and SPECT/CT." *Eur. J. Nucl. Med. Mol. Imag.* 41(Suppl 1):S3–16.

Ito, M, S.J. Hong, and J.S. Lee. 2011. "Positron Emission Tomography (PET) Detectors with Depth-of-Interaction (DOI) Capability." *Biomed. Engin, Lett.* 1:70–81.

Jakoby, B.W., Y. Bercier, M. Conti, et al. 2011. "Physical and Clinical Performance of the mCT Time-of-Flight PET/CT Scanner." *Phys. Med. Biol.* 56:2375–2389.

Jan, S., T. Frisson, and D. Sarrut. 2013. "GATE Simulation of C-12 Hadrontherapy Treatment Combined with a PET Imaging System for Dose Monitoring: A Feasibility Study." *IEEE Trans. Nucl. Sci.* 60:423–429.

Janecek, M., J.P. Walder, P.J. McVittie, et al. 2012. "A High-Speed Multi-Channel Readout for SSPM Arrays." *IEEE Trans. Nucl. Sci.* 59:13–18.

Jaszczak, R.J. 2006. "The Early Years of Single Photon Emission Computed Tomography (SPECT): An Anthology of Selected Reminiscences." *Phys. Med. Biol.* 51:R99–R115.

Jaszczak, R.J., L.T. Chang, N.A. Stein, et al. 1979. "Whole-Body Single-Photon Emission Computed-Tomography Using Dual, Large-Field-of-View Scintillation Cameras." *Phys. Med. Biol.* 24:1123–1143.

Jaszczak, R.J., P.H. Murphy, D. Huard, et al. 1977. "Radionuclide Emission Computed Tomography of Head with Tc-99m and a Scintillation Camera." *J. Nucl. Med.* 18:373–380.

Johnson, L.C., D.L. Campbell, E.L. Hull, et al. 2011. "Characterization of a High-Purity Germanium Detector for Small-Animal SPECT." *Phys. Med. Biol.* 56:5877–5888.

Joung, J., R.S. Miyaoka, and T.K. Lewellen. 2002. "cMiCE: A High Resolution Animal PET Using Continuous LSO with a Statistics Based Positioning Scheme." *Nucl. Instr. Meth. A.* 489:584–598.

Kang, J., Y. Choi, K.J. Hong, et al. 2010. "Dual-Ended Readout PET Detector Module Based on GAPD Having Large-Area Microcells." *The 2010 IEEE Nucl. Sci. Symp. Med. Imag. Conf. Record (NSS/MIC).* 3205–3209.

Kao, C.-M., H. Kim, and C.-T. Chen. 2011. "Event-Time Determination by Waveform Analysis for Time-of-Flight Positron Emission Tomography." *The 2011 IEEE Nucl. Sci. Symp. Med. Imag. Conf. Record (NSS/MIC).* 3874–3879.

Karellas, A., and S. Vedantham. 2008. "Breast Cancer Imaging: A Perspective for the Next Decade." *Med. Phys.* 35:4878–4897.

Karp, J.S., S. Surti, M.E. Daube-Witherspoon, et al. 2003. "Performance of a Brain PET Camera Based on Anger-Logic Gadolinium Oxyorthosilicate Detectors." *J. Nucl. Med.* 44:1340–1349.

Keyes, J.W., N. Orlandea, W.J. Heetderks, et al. 1977. "Humongotron – Scintillation-Camera Transaxial Tomograph." *J. Nucl. Med.* 18:381–387.

Kim, H., C.-T. Chen, H.-T. Chen, et al. 2013. "A TOF PET Detector Development Using Waveform Sampling and Strip-Line Based Data Acquisition." *The 2013 IEEE Nucl. Sci. Symp. Med. Imag. Conf. Record (NSS/MIC).* 1–4.

Kim, H., C.-T. Chen, N. Eclov, et al. 2014. "A New Time Calibration Method for Switched-Capacitor-Array-Based Waveform Samplers." *Nucl. Instr. Meth. A.* 767:67–74.

Kim, H., C.-T. Chen, A. Ronzhin, et al. 2012a. "A Silicon Photomultiplier Signal Readout Using Transmission-Line and Waveform Sampling for Positron Emission Tomography." *The 2012 IEEE Nucl. Sci. Symp. Med. Imag. Conf. Record (NSS/MIC).* 1466–2468.

Kim, H., C.-T. Chen, A. Ronzhin, et al. 2012b. "A Study on the Optimal Sampling Speed of DRS4-Based Waveform Digitizer for Time-of-Flight Positron Emission Tomography Application." *The 2012 IEEE Nucl. Sci. Symp. Med. Imag. Conf. Record (NSS/MIC).* 2469–2471.

Kim, H., H. Frisch, C.-T. Chen, et al. 2010. "A Design of a PET Detector Using Micro-Channel Plate Photomultipliers with Transmission-Line Readout." *Nucl. Instr. Meth. A.* 622:628–636.

Kim H., L.R. Furenlid, M.J. Crawford, et al. 2006. "SemiSPECT: A Small-Animal SPECT Imaging Based on Eight CZT Detector Arrays." *Med. Phys.* 3:465–474.

Kim, H., C.-M. Kao, S. Kim, et al. 2011. "A Development of Waveform Sampling Readout Board for PET Using DRS4." *The 2011 IEEE Nucl. Sci. Symp. Med. Imag. Conf. Record (NSS/MIC).* 2393–2396.

Kim, H., C.-M. Kao, Q. Xie, et al. 2009. "A Multi-Threshold Sampling Method for TOF-PET Signal Processing." *Nucl. Instr. Meth. A.* 602:618–621.

Kim, H., et al. 2020. "Design, Evaluation and Initial Imaging Results of a PET Insert Based on Strip-Line Readout for Simultaneous PET/MRI." *Nucl. Instrum. Meth. A.* 959:163575.

Kim, H., et al. 2021a. "Multiplexing Readout for Time-of-flight (TOF) PET Detectors Using Striplines." *IEEE Trans. Radiat. Plasma Med. Sci.,* doi: 10.1109/TRPMS.2021.3051364.

Kim, H., et al. 2021b. "Development of a PET/EPRI Combined Imaging System for Assessing Tumor Hypoxia." *J. Instrum.* 16: P03031.

Kishimoto, A., J. Kataoka, T. Kato, et al. 2013. "Development of a Dual-Sided Readout DOI-PET Module Using Large-Area Monolithic MPPC-Arrays." *IEEE Trans. Nucl. Sci.* 60:38–43.

Knoll, GF. 1999. *"Radiation Detection and Measurement."* 4th ed. Hoboken, NJ: John Wiley & Son.

Koolen, B.B., W.V. Vogel, M.J. Vrancken, et al. 2012. "Molecular Imaging in Breast Cancer: From Whole-Body PET/CT to Dedicated Breast PET." *J. Oncol.* 2012:438647. doi:10.1155/2012/438647.

Korpar, S., R. Dolenec, P. Krizan, et al. 2011. "Study of TOF PET Using Cherenkov Light." *Nucl. Instr. Meth. A.* 654:532–538.

Kuhl, D.E., and R.Q. Edwards. 1963. "Image Separation Radioisotope Scanning." *Radiology.* 80:653–662.

Kuhl, D.E., and R.Q. Edwards. 1964. "Cylindrical and Section Radioisotope Scanning of the Liver and Brain." *Radiology.* 83:926–936.

Kuhl, D.E., R.Q. Edwards, A.R. Ricci, et al. 1976. "Mark-4 System for Radionuclide Computed Tomography of Brain." *Radiology.* 121:405–413.

Kuhn, A., S. Surti, J.S. Karp, et al. 2006. "Performance Assessment of Pixelated LaBr$_3$ Detector Modules for Time-of-Flight PET." *IEEE Trans. Nucl. Sci.* 53:1090–1095.

Kume, H., S. Muramatsu, and M. Iida. 1986. "Position-Sensitive Photomultiplier Tubes for Scintillation Imaging." *IEEE Trans. Nucl. Sci.* 33:359–363.

Kwon, S.I., J.S. Lee, H.S. Yoon, et al. 2011. "Development of Small-Animal PET Prototype Using Silicon Photomultiplier (SiPM): Initial Results of Phantom and Animal Imaging Studies." *J. Nucl. Med.* 52:572–579.

LaBella, A., et al. 2020. "High-Resolution Depth-Encoding PET Detector Module with Prismatoid Light-Guide Array." *J. Nucl. Med.* 61:1528–1533.

Lai, X., and L.-J. Meng, 2018. "Simulation Study of the Second-Generation MR-Compatible SPECT System Based on the Inverted Compound-Eye Gamma Camera Design." *Phys. Med. Biol.* 63:045008.

Larsson, S.A. 1980. "Gamma Camera Emission Tomography. Development and Properties of a Multi-Sectional Emission Computed Tomography System." *Acta Radiol Suppl.* 363:1–75.

Lecoq, P., et al. 2020. "Roadmap toward the 10 ps time-of-flight PET challenge." *Phys. Med. Biol.*65:21RM01.

Levin, C.S. 2012. "Promising New Photon Detection Concepts for High-Resolution Clinical and Preclinical PET." *J. Nucl. Med.* 53:167–170.

Levine, E.A., R.I. Freimanis, N.D. Perrier, et al. 2003. "Positron Emission Mammography: Initial Clinical Results." *Annal. Surg. Oncol.* 10:86–91.

Li, B., et al. 2017. "Time-Resolving Characteristics of Pixel- and Charge-Division-Type Position Sensitive SiPMs with Epitaxial Euenching Resistors." *IEEE Trans. Electron. Devices.* 64:2239–2243.

Li, C., et al. 2014. "Position Sensitive Silicon Photomultiplier with Intrinsic Continuous Cap Resistive Layer." *IEEE Trans. Electron. Devices.* 61:3229–3232.

Li, Z., M. Wedrowski, P. Bruyndonckx, et al. 2010. "Nonlinear Least-Squares Modeling of 3D Interaction Position in a Monolithic Scintillator Block." *Phys. Med. Biol.* 55:6515–6532.

Lim, C.B., D. Chu, L. Kaufman, et al. 1975. "Initial Characterization of a Multi-Wire Proportional Chamber Positron Camera." *IEEE Trans. Nucl. Sci.* 22:388–394.

Limousin, O. 2003. "New Trends in CdTe and CdZnTe Detectors for X- and Gamma-Ray Applications." *Nucl. Instr. and Meth. A.* 504:24–37.

Ling, T., T.K. Lewellen, and R.S. Miyaoka. 2007. "Depth of Interaction Decoding of a Continuous Crystal Cetector Module." *Phys. Med. Biol.* 52:2213–2228.

Liu, C.Y, and A.L. Goertzen. 2014. "Multiplexing Approaches for a 12 × 4 Array of Silicon Photomultipliers." *IEEE Trans. Nucl. Sci.* 61(1):35–43.

Llosa, G., J. Barrio, C. Lacasta, et al. 2010. "Characterization of A PET Detector Head Based on Continuous LYSO Crystals and Monolithic, 64-Pixel Silicon Photomultiplier Matrices." *Phys. Med. Biol.* 55:7299–7315.

Madsen, M.T. 2007. "Recent Advances in SPECT Imaging." *J. Nucl. Med.* 48:661–673.

Majewski, S., J. Proffitt, J. Brefczynski-Lewis, et al. 2011b. "HelmetPET: A Silicon Photomultiplier Based Wearable Brain Imager." *The 2011 IEEE Nucl. Sci. Symp. Med. Imag. Conf. Record (NSS/MIC).* 4030–4034.

Majewski, S., A. Stolin, P. Martone, et al. 2011a. "Dedicated Mobile PET Prostate Imager." *J. Nucl. Med.* 52S:1945.

Martinez, J.D., J.M. Benlloch, J. Cerda, et al. 2004. "High-Speed Sata Acquisition and Digital Signal Processing System for PET Imaging Techniques Applied to Mammography." *IEEE Trans. Nucl. Sci.* 51:407–412.

Mathews, A.J., S. Komarov, H.Y. Wu, et al. 2013. "Improving PET Imaging for Breast Cancer Using Virtual Pinhole PET Half-Ring Insert." *Phys. Med. Biol.* 58:6407–6427.

Meier, D., S. Mikkelsen, J. Talebi, et al. 2010. "An ASIC for SiPM/MPPC Readout." *The 2010 IEEE Nucl. Sci. Symp. Med. Imag. Conf. Record (NSS/MIC).* 1653–1657.

Meng, L.J. 2006. "An Intensified EMCCD Camera for Low Energy Gamma Ray Imaging Applications." *IEEE Trans. Nucl. Sci.* 53:2376–2384.

Miller, B.W., H.H. Barrett, L.R. Furenlid, et al. 2008. "Recent Advances in BazookaSPECT: Real-Time Data Processing and the Development of a Gamma-Ray Microscope." *Nucl. Instr. Meth. A.* 591:272–275.

Mitchell, D., C.B. Hruska, J.C. Boughey, et al. 2013. "Tc-99m-Sestamibi Using a Direct Conversion Molecular Breast Imaging System to Assess Tumor Response to Neoadjuvant Chemotherapy in Women with Locally Advanced Breast Cancer." *Clin. Nucl. Med.* 38:949–956.

Miyake, K.K., K. Matsumoto, M. Inoue, et al. 2014. "Performance Evaluation of a New Dedicated Breast PET Scanner Using NEMA NU4-2008 Standards." *J. Nucl. Med.* 55:1198–1203.

Moehrs, S., A. Del Guerra, D.J. Herbert, et al. 2006. "A Detector Head Design for Small-Animal PET with Silicon Photomultipliers (SiPM)." *Phys. Med. Biol.* 51:1113–1127.

Moliner, L., A.J. Gonzalez, A. Soriano, et al. 2012. "Design and Evaluation of the MAMMI Dedicated Breast PET." *Med. Phys.* 39:5393–5404.

Moses, W.W. 2003. "Time of Flight in PET Revisited." *IEEE Trans. Nucl. Sci.* 50:1325–1330.

Moses, W.W. 2004. "Positron Emission Mammography Imaging." *Nucl. Instr. Meth. A.* 525:249–252.

Moses, W.W. 2007. "Recent Advances and Future Advances in Time-of-Flight PET." *Nucl. Instr. Meth. A.* 580:919–924.

Moses, W.W., and S.E. Derenzo. 1994. "Design Studies for a PET Detector Module Using a Pin Photodiode to Measure Depth of Interaction." *IEEE Trans. Nucl. Sci.* 41:1441–1445.

Moses, W.W., and M. Ullisch. 2006. "Factors Influencing Timing Resolution in a Commercial LSO PET Camera." *IEEE Trans. Nucl. Sci.* 53(1):78–85.

Muehllehner, G. 1970. "Rotating Collimator Tomography." *J. Nucl. Med.* 11:347.

Muehllehner, G. 1971. "A Tomographic Scintillation Camera." *Phys. Med. Biol.* 16:87.

Muehllehner, G., F. Atkins, and P.V. Harper. 1977. "Positron Camera with Longitudinal and Transverse Tomographic Ability." In *Medical Radionuclide Imaging*. Vienna: IAEA.

Muehllehner, G., and J.S. Karp. 2006. "Positron Emission Tomography." *Phys. Med. Biol.* 51:R117–R137.

Mullani, N.A., D.C. Ficke, and M.M. Terpogossian. 1980. "Cesium Fluoride – New Detector for Positron Emission Tomography." *IEEE Trans. Nucl. Sci.* 27:572–575.

Mullani, N.A., C.S. Higgins, J.T. Hood, et al. 1978. "ETTt-IV – Design Analysis and Performance-Characteristics." *IEEE Trans. Nucl. Sci.* 25:180–183.

Mullani, N.A., J. Markham, and M.M. Terpogossian. 1980. "Feasibility of Time-of-Flight Reconstruction in Positron Emission Tomography." *J. Nucl. Med.* 21:1095–1097.

Mullani, N.A., W.H. Wong, R.K. Hartz, et al. 1982. "Design of TOFPET: A High Resolution Time-of-Flight Positron Camera." In *Proc. of Workshop Time-of-Flight Tomography*. St. Louis, MO: IEEE Computer Society Press, Silver Spring, Maryland.

Muller, C., M. Bunka, S. Haller, U. Koster, V. Groehn, P. Bernhardt, N. van der Meulen, A. Turler, R. Schibli, 2014. "Promising Prospects for 44Sc-/47Sc-Based Theragnostics: Application of 47Sc for Radionuclide Tumor Therapy in Mice." *J Nucl Med.* 55(10):1658–1664.

Murphy, P., J. Burdine, M. Moore, et al. 1978. "Single Photon Emission Computed Tomography (ECT) of Body." *J. Nucl. Med.* 19:683–683.

Nagarkar, V.V., I. Shestakova, V. Gaysinskiy, et al. 2006. "A CCD-Based Detector for SPECT." *IEEE Trans. Nucl. Sci.* 53:54–58.

Nelson, B.J.B., J.D. Andersson, and F. Wuest. 2021. "Targeted Alpha Therapy: Progress in Radionuclide Production, Radiochemistry, and Applications." *Pharmaceutics.* 13:49. https://doi.org/10.3390/ pharmaceutics13010049

Nishikido, F., N. Inadama, E. Yoshida, et al. 2013. "Four-Layer DOI PET Detectors Using a Multi-Pixel Photon Counter Array and the Light Sharing Method." *Nucl. Instr. Meth. A.* 729:755–761.

Nishikido, F., et al. 2014. "Feasibility of a Brain-Dedicated PET-MRI System Using Four-Layer DOI Detectors Integrated with an RF Head Coil." *Nucl. Instrum. Meth. A.* 756:6–13.

Nutt, R. 2002. "1999 ICP Distinguished Scientist Award. The History of Positron Emission Tomography." *Mol Imaging Biol.* 4(1):11–26.

Oberla, E., et al. 2014. "A 15GSa/s, 1.5GHz Bandwidth Waveform Digitizing ASIC." *Nucl. Instrum. Meth. A.* 735:452–461.

O'Connor, M.K., S.W. Phillips, C.B. Hruska, et al. 2007. "Molecular Breast Imaging: Advantages and Limitations of a Scintimammographic Technique in Patients with Small Breast Tumors." *Breast J.* 13:3–11.

Olcott, P.D., A. Fallu-Labruyere, F. Habte, et al. 2006. "A High Speed Fully Digital Data Acquisition System for Positron Emission Tomography." *The 2006 IEEE Nucl. Sci. Symp. Med. Imag. Conf. Record (NSS/MIC).* 1909–1911.

Pani, R., P. Bennati, M. Betti, et al. 2006. "Lanthanum Scintillation Crystals for Gamma Ray Imaging." *Nucl. Instr. Meth. A.* 567:294–297.

Pani, R., A. Soluri, R. Scafe, et al. 1999. "Multi-PSPMT Scintillation Camera." *IEEE Trans. Nucl. Sci.* 46:702–708.

Parodi, K. 2012. "PET Monitoring of Hadrontherapy." *Nucl. Med. Rev.* 15:C37–C42.

Parodi, K., T. Bortfeld, and T. Haberer. 2008. "Comparison between In-Beam and Offline Positron Emission Tomography Imaging of Proton and Carbon Ion Therapeutic Irradiation at Synchrotron- and Cyclotron-Based Facilities." *Int'l J. Rad. Oncol. Biol. Phys.* 71:945–956.

Patton, J., A.B. Brill, J. Erickson, et al. 1969. "A New Approach to Mapping 3-Dimensional Radionuclide Distributions." *J. Nucl. Med.* 10:363.

Peterson, T.E., and L.R. Furenlid. 2011. "SPECT Detectors: The Anger Camera and Beyond." *Phys. Med. Biol.* 56:R145–R182.

Peterson, T.E., and S. Shokouhi. 2012. "Advances in Preclinical SPECT Instrumentation." *J. Nucl. Med.* 53:841–844.

Peterson, T.E., S. Shokouhi, L.R. Furenlid, et al. 2009. "Multi-Pinhole SPECT Imaging with Silicon Strip Detectors." *IEEE Trans. Nucl. Sci.* 56:646–652.

Peterson, T.E., D.W. Wilson, and H.H. Barrett. 2003. "Application of Silicon Strip Detectors to Small-Animal Imaging." *Nucl. Instr. and Meth. A.* 505:608–611.

Phelps, M.E., E.J. Hoffman, S.C. Huang, et al. 1978. "ECAT – New Computerized Tomographic Imaging-System for Positron-Emitting Radiopharmaceuticals." *J. Nucl. Med.* 19:635–647.

Phelps, M.E., E.J. Hoffman, N.A. Mullani, et al. 1975. "Application of Annihilation Coincidence Detection to Transaxial Reconstruction Tomography." *J. Nucl. Med.* 16:210–224.

Poon J.K., Dahlbom M.L., Moses WW, et al. 2012. "Optimal Whole-Body PET Scanner Configurations for Different Volumes of LSO Scintillator: A Simulation Study." *Phys Med Biol.* 57:4077–4094.

Popovic, K., J.E. McKisson, B. Kross, et al. 2014. "Development and Characterization of a Round Hand-Held Silicon Photomultiplier Based Gamma Camera for Intraoperative Imaging." *IEEE Trans. Nucl. Sci.* 61:1084–1091.

Powolny, F., E. Auffray, S.E. Brunner, et al. 2011. "Time-Based Readout of a Silicon Photomultiplier (SiPM) for Time of Flight Positron Emission Tomography (TOF-PET)." *IEEE Trans. Nucl. Sci.* 58:597–604.

Prout, D.L., et al. 2020. "A Digital Phoswich Detector Using Time-Over-Threshold for Depth of Interaction in PET." *Phys. Med. Biol.* 65:245017.

Pshenichnov, I., I. Mishustin, and W. Greiner. 2006. "Distributions of Positron-Emitting Nuclei in Proton and Carbon-Ion Therapy Studied with GEANT4." *Phys. Med. Biol.* 51:6099–6112.

Rankowitz S., J.S. Robertson, W.A. Higinbotham, et al. 1962. "Positron Scanner for Locating Brain Tumors." *BNL 6049 and 1962 IRE Int. Convent. Rec.* 10:49.

Raylman, R.R., S. Majewski, M.F. Smith, et al. 2008. "The Positron Emission Mammography/Tomography Breast Imaging and Biopsy System (PEM/PET): Design, Construction and Phantom-Based Measurements." *Phys. Med. Biol.* 53:637–653.

Robertson, J.S., and A.M. Niell. 1962. "Use of a Digital Computer in the Development of a Positron Scanning Procedure." In: *The 4th IBM Medical Symposium.* New York, NY: Endicott.

Rogers, W.L., N.H. Clinthorne, L. Shao, et al. 1988. "Sprint-Ii – A 2nd Generation Single Photon Ring Tomograph." *IEEE Trans. Med. Imaging.* 7:291–297.

Roncali, E., and S.R. Cherry. 2011. "Application of Silicon Photomultipliers to Positron Emission Tomography." *Annals Biomed. Engin.* 39:1358–1377.

Roncali, E., V. Viswanath, and S.R. Cherry. 2014. "Design Considerations for DOI-encoding PET Detectors Using Phosphor-Coated Crystals." *IEEE Trans. Nucl. Sci.* 61:67–73.

Ronzhin, A., M. Albrow, S. Los, M. Martens, et al. 2013. "A SiPM-Based TOF-PET Detector with High Speed Digital DRS4 Readout." *Nucl. Instr. Meth. A.* 703:109–113.

Roy, T., J. Ratheesh, and A. Sinha. 2014. "Three-Dimensional SPECT Imaging with $LaBr_3$:Ce Scintillator for Characterization of Nuclear Waste." *Nucl. Instr. Meth. A.* 735:1–6.

Russo, P., G. Mettivier, R. Pani, et al. 2009. "Imaging Performance Comparison between a $LaBr_3$:Ce Scintillator Based and a CdTe Semiconductor Based Photon Counting Compact Gamma Camera." *Med. Phys.* 36:1298–1317.

Sacco, I., P. Fischer, M. Ritzert, et al. 2013. "A Low Power Front-End Architecture for SiPM Readout with Integrated ADC and Multiplexed Readout." *JINST.* 8:C01023. doi:10.1088/1748-0221/8/01/C01023.

Schaart, D.R., H.T. van Dam, S. Seifert, et al. 2009. "A Novel, SiPM-Array-Based, Monolithic Scintillator Detector for PET." *Phys. Med. Biol.* 54:3501–3512.

Scheiber, C. 1996. "New Developments in Clinical Applications of CdTe and CdZnTe Detectors." *Nucl. Instr. Meth. A.* 380:385–391.

Scheiber, C. 2000. "CdTe and CdZnTe Detectors in Nuclear Medicine." *Nucl. Instr. Meth. A.* 448:513–524.

Scheiber, C., and J. Chambron. 1992. "CdTe Detectors in Medicine – A Review of Current Applications and Future Perspectives." *Nucl. Instr. Meth. A.* 322:604–614.

Scheiber, C., and G.C. Giakos. 2001. "Medical Applications of CdTe and CdZnTe Detectors." *Nucl. Instr. and Meth. A.* 458:12–25.

Schlyer, D., P. Vaska, D. Tomasi, et al. 2007. "A Simultaneous PET/MRI Scanner Based on RatCAP in Small Animals." *The 2007 IEEE Nucl. Sci. Symp. Med. Imag. Conf. Record (NSS/MIC).* 3256–3259.

Schmall, J.P., et al. 2014. "A Study of Position-Sensitive Solid-State Photomultiplier Signal Properties." *IEEE Trans. Nucl. Sci.* 61:1074–1083.

Schulz, D., and P. Vaska. 2011. "Integrating PET with Behavioral Neuroscience Using RatCAP Tomography." *Rev. Neurosciences.* 22:647–655.

Schulz, V., B. Weissler, P. Gebhardt, et al. 2011. "SiPM Based Preclinical PET/MR Insert for a Human 3T MR: First Imaging Experiments." *The 2011 IEEE Nucl. Sci. Symp. Med. Imag. Conf. Record (NSS/MIC).* 4467–4469.

Seifert, S., G. van der Lei, H.T. van Dam, et al. 2013. "First Characterization of a Digital SiPM Based Time-of-Flight PET Detector with 1 mm Spatial Resolution." *Phys. Med. Biol.* 58:3061–3074.

Shah, K.S., R. Farrell, R. Grazioso, et al. 2001. "Large-Area APDs and Monolithic APD Arrays." *IEEE Trans. Nucl. Sci.* 48:2352–2356.

Shah, K.S., J. Glodo, M. Klugerman, et al. 2003. "LaBr$_3$:Ce Scintillators for Gamma-Ray Spectroscopy." *IEEE Trans. Nucl. Sci.* 50:2410–2413.

Shakirin, G., H. Braess, F. Fiedler, et al. 2011. "Implementation and Workflow for PET Monitoring of Therapeutic ion Irradiation: A Comparison of In-Beam, In-Room, and Off-Line Techniques." *Phys. Med. Biol.* 56:1281–1298.

Shakirin, G., P. Crespo, H. Braess, et al. 2007. "Influence of the Time of Flight Information on the Reconstruction of In-Beam PET Data." *The 2007 IEEE Nucl. Sci. Symp. Med. Imag. Conf. Record (NSS/MIC).* 4395–4396.

Shao, Y.P., X.S. Sun, K. Lou, et al. 2014. "In-Beam PET Imaging for On-Line Adaptive Proton Therapy: An Initial Phantom Study." *Phys. Med. Biol.* 59:3373–3388.

Sharir, T., P.J. Slomka, and D.S. Berman. 2010. "Solid-State SPECT Technology: Fast and Furious." *J. Nucl. Cardiol.* 17:890–896.

Shibuya, K., F. Nishikido, T. Tsuda, et al. 2008. "Timing Resolution Improvement Using DOI Information in a Four-Layer Scintillation Detector for TOF-PET." *Nucl. Instr. Meth. A.* 593:572–577.

Shokouhi, S., D.W. Wilson, S.D. Metzler, et al. 2010. "Evaluation of Image Reconstruction for Mouse Brain Imaging with Synthetic Collimation from Highly Multiplexed SiliSPECT Projections." *Phys. Med. Biol.* 55:5151–5168.

Siegel, S., R.W. Silverman, Y.P. Shao, et al. 1996. "Simple Charge Division Readouts for Imaging Scintillator Arrays Using a Multi-Channel PMT." *IEEE Trans. Nucl. Sci.* 43:1634–1641.

Song, T.Y., H.Y. Wu, S. Komarov, et al. 2010. "A Sub-Millimeter Resolution PET Detector Module Using a Multi-Pixel Photon Counter Array." *Phys. Med. Biol.* 55:2573–2587.

Spanoudaki, V.C., and C.S. Levin. 2011. "Investigating the Temporal Resolution Limits of Scintillation Detection from Pixelated Elements: Comparison between Experiment and Simulation." *Phys. Med. Biol.* 56:735–756.

Sportelli, G., N. Belcari, N. Camarlinghi, et al. 2014. "First Full-Beam PET Acquisitions in Proton Therapy with a Modular Dual-Head Dedicated System." *Phys. Med. Biol.* 59:43–60.

Stankova, V., J. Barrio, J.E. Gillam, et al. 2012. "Multichannel DAQ System for SiPM Matrices." *The 2012 IEEE Nucl. Sci. Symp. Med. Imag. Conf. Record (NSS/MIC).* 1069–1071.

Stolin, A., S. Majewski, G. Jaliparthi, et al. 2014. "Evaluation of Imaging Modules Based on SensL Array SB-8 for Nuclear Medicine Applications." *IEEE Trans. Nucl. Sci.* 61:2433–2438.

Stratos, D., G. Maria, F. Eleftherios, et al. 2013. "Comparison of Three Resistor Network Division Circuits for the Readout of 4 × 4 Pixel SiPM Arrays." *Nucl. Instr. Meth. A.* 702:121–125.

Streun, M., G. Brandenburg, H. Larue, et al. 2002. "Coincidence Detection by Digital Processing of Free-Running Sampled Pulses." *Nucl. Instr. Meth. A.* 487(3):530–534.

Surti, S., and J.S. Karp. 2008. "Design Considerations for a Limited Angle, Dedicated Breast, TOF PET Scanner." *Phys. Med. Biol.* 53:2911–2921.

Surti, S., A. Kuhn, M.E. Werner, et al. 2007. "Performance of Philips Gemini TF PET/CT Scanner with Special Consideration for Its Time-of-Flight Imaging Capabilities." *J. Nucl. Med.* 48:471–480.

Tanaka, E., H. Toyama, and H. Murayama. 1984. "Convolutional Image-Reconstruction for Quantitative Single Photon-Emission Computed-Tomography." *Phys. Med. Biol.* 29:1489–1500.

Tashima, H., T. Yamaya, E. Yoshida, et al. 2012. "A Single-Ring OpenPET Enabling PET Imaging during Radiotherapy." *Phys. Med. Biol.* 57:4705–4718.

Ter-Pogossian, M.M. 1977. "Basic Principles of Computed Axial Tomography." *Semi. Nucl. Med.* 7:109–127.

Ter-Pogossian, M.M., D.C. Ficke, J.T. Hood, Sr., et al. 1982. "PETT VI: A Positron Emission Tomograph Utilizing Cesium Fluoride Scintillation Detectors." *J. Comput. Assist. Tomogr.* 6:125–133.

Ter-Pogossian, M.M., N.A. Mullani, D.C. Ficke, et al. 1981. "Photon Time-of-Flight-Assisted Positron Emission Tomography." *J. Comput. Assist. Tomogr.* 5:227–239.

Ter-Pogossian, M.M., M.E. Phelps, E.J. Hoffman, et al. 1975. "A Positron-Emission Transaxial Tomograph for Nuclear Imaging (PETT)." *Radiology.* 114:89–98.

Thompson, C.J., A.L. Goertzen, E.J. Berg, et al. 2012. "Evaluation of High Density Pixellated Crystal Blocks with SiPM Readout As Candidates for PET/MR Detectors in a Small Animal PET Insert." *IEEE Trans. Nucl. Sci.* 59:1791–1797.

Thompson, C.J., K. Murthy, I.N. Weinberg, et al. 1994. "Feasibility Study for Positron Emission Mammography." *Med. Phys.* 21:529–538.

Thompson, C.J., Y.L. Yamamoto, and E. Meyer. 1979. "Positome 2. High-Efficiency Positron Imaging Device for Dynamic Brain Studies." *IEEE Trans. Nucl. Sci.* 26:583–589.

Tornai, M.P., C.N. Archer, A.G. Weisenberger, et al. 2001. "Investigation of Microcolumnar Scintillators on an Optical Fiber Coupled Compact Imaging System." *IEEE Trans. Nucl. Sci.* 48:637–644.

Toussaint M., et al. 2019. "Analytic Model of DOI-Induced Time Bias in Ultra-Fast Scintillation Detectors for TOF-PET." *Phys. Med. Biol.* 64:065009.

Truman, A., A.J. Bird, D. Ramsden, et al. 1994. "Pixellated CsI(Tl) Arrays with Position-Sensitive PMT Readout." *Nucl. Instr. Meth. A.* 353:375–378.

van Dam, H.T., G. Borghi, S. Seifert, et al. 2013. "Sub-200 ps CRT in Monolithic Scintillator PET Detectors Using Digital SiPM Arrays and Maximum Likelihood Interaction Time Estimation." *Phys. Med. Biol.* 58:3243–3257.

van Eijk, C.W.E. 2002. "Inorganic Scintillators in Medical Imaging." *Phys. Med. Biol.* 47:R85–R106.

van Loef, E.V.D., P. Dorenbos, K. Kramer, et al. 2001. "Scintillation Properties of LaCl$_3$:Ce3+ Crystals: Fast, Efficient, and High-Energy Resolution Scintillators." *IEEE Trans. Nucl. Sci.* 48:341–345.

van Velden, F.H.P., R.W. Kloet, B.N.M. van Berckel, et al. 2009. "HRRT versus HR Plus Human Brain PET Studies: An Interscanner Test-Retest Study." *J. Nucl. Med.* 50:693–702.

Vaska, P., C. Woody, D. Schlyer, et al. 2007. "The Design and Performance of the 2nd-Generation RatCAP Awake Rat Brain PET System." *The 2007 IEEE Nucl. Sci. Symp. Med. Imag. Conf. Record (NSS/MIC).* 4181–4184.

Vecchio, S., F. Attanasi, N. Belcari, et al. 2009. "A PET Prototype for 'In-Beam' Monitoring of Proton Therapy." *IEEE Trans. Nucl. Sci.* 56:51–56.

de Vree, G.A., A.H. Westra, I. Moody, et al. 2005. "Photon-Counting Gamma Camera Based on an Electron-Multiplying CCD." *IEEE Trans. Nucl. Sci.* 52:580–588.

Wang, Y., Z. Zhang, D. Li., et al. 2012. "Design and Performance Evaluation of a Compact, Large-Area PET Detector Module Based on Silicon Photomultipliers." *Nucl. Instr. Meth. A.* 670:49–54.

Watanabe, M., K. Shimizu, T. Omura, et al. 2002. "A New High-Resolution PET Scanner Dedicated to Brain Research." *IEEE Trans. Nucl. Sci.* 49:634–639.

Wernick, M., and J. Aarsvold. 2004. *"Emission Tomography: The Fundamentals of PET and SPECT."* Amsterdam; Boston, MA: Elsevier Academic Press.

Wester, H.J., and M. Schottelius. 2019. "PSMA-Targeted Radiopharmaceuticals for Imaging and Therapy." *Semin. Nucl. Med.* 49:302–312.

Wienhard, K., M. Schmand, M.E. Casey, et al. 2002. "The ECAT HRRT: Performance and First Clinical Application of the New High Resolution Research Tomograph." *IEEE Trans. Nucl. Sci.* 49:104–110.

Williams, M.B., A.R. Goode, et al. 2000. "Performance of A PSPMT Based Detector for Scintimammography." *Phys. Med. Biol.* 45:781–800.

Won, J.Y., et al. 2016. "Delay Grid Multiplexing: Simple Time-Based Multiplexing and Readout Method for Silicon Photomultipliers." *Phys. Med. Biol.* 61:7113–7135.

Wrenn, F.R., M.L. Good, and P. Handler. 1951. "The Use of Positron-Emitting Radioisotopes for the Localization of Brain Tumors." *Science.* 113:525–527.

Wu, H.Y., D. Pal, T.Y. Song, et al. 2008. "Micro Insert: A Prototype Full-Ring PET Device for Improving the Image Resolution of a Small-Animal PET Scanner." *J. Nucl. Med.* 49:1668–1676.

Xi, D.M., C.-M. Kao, W. Liu, et al. 2013. "FPGA-Only MVT Digitizer for TOF PET." *IEEE Trans. Nucl. Sci.* 60:3253–3261.

Xi, W.Z., J. Seidel, J.W. Kakareka, et al. 2010. "MONICA: A Compact, Portable Dual Gamma Camera System for Mouse Whole-Body Imaging." *Nucl. Med. Biol.* 37:245–253.

Xie, Q., Y.B. Chen, J. Zhu, et al. 2013. "Implementation of LYSO/PSPMT Block Detector with All Digital DAQ System." *IEEE Trans. Nucl. Sci.* 60:1487–1494.

Xie, Q., C.-M. Kao, Z. Hsiau, et al. 2005. "A New Approach for Pulse Processing in Positron Emission Tomography." *IEEE Trans. Nucl. Sci.* 52:988–995.

Yamamoto, S., M. Honda, T. Oohashi, et al. 2011. "Development of a Brain PET System, PET-Hat: A Wearable PET System for Brain Research." *IEEE Trans. Nucl. Sci.* 58:668–673.

Yamamoto, S., M. Imaizumi, E. Shimosegawa, et al. 2010. "Development of a Compact and High Spatial Resolution Gamma Camera System Using $LaBr_3(Ce)$." *Nucl. Instr. Meth. A.* 622:261–269.

Yamaya, T., T. Inaniwa, S. Minohara, et al. 2008. "A Proposal of an Open PET Geometry." *Phys. Med. Biol.* 53:757–773.

Yamaya, T., T. Mitsuhashi, T. Matsumoto, et al. 2011. "A SiPM-Based Isotropic-3D PET Detector X'tal Cube with a Three-Dimensional Array of 1 mm^3 Crystals." *Phys. Med. Biol.* 56:6793–6807.

Yasillo, N.J., R.N. Beck, and M. Cooper. 1990. "Design Considerations for a Single Tube Gamma-Camera." *IEEE Trans. Nucl. Sci.* 37:609–615.

Yoon, H.S., G.B. Ko, S. Il Kwon, et al. 2012. "Initial Results of Simultaneous PET/MRI Experiments with an MRI-Compatible Silicon Photomultiplier PET Scanner." *J. Nucl. Med.* 53:608–614.

Yoshida, E., S. Kinouchi, H. Tashima, et al. 2011. "System Design of a Small OpenPET Prototype with 4-Layer DOI Detectors." *Radiol. Phys. Technol.* 5:92–97.

Yoshida, E., H. Tashima, H. Wakizaka, et al. 2013. "Development of a Single-Ring OpenPET Prototype." *Nucl. Instr. Meth. A.* 729:800–808.

Zeniya, T., H. Watabe, T. Aoi, et al. 2006. "Use of a Compact Pixellated Gamma Camera for Small Animal Pinhole SPECT Imaging." *Annals Nucl. Med.* 20:409–416.

Zhang, J., et al. 2018. "Performance Evaluation of the Next Generation Solid-State Digital Photon Counting PET/CT System." *EJNMMI.* 8:97.

Zhou, J., and J.Y. Qi. 2009. "Theoretical Analysis and Simulation Study of a High-Resolution Zoom-In PET System." *Phys. Med. Biol.* 54:5193–5208.

Zhou, J.A., and J.Y. Qi. 2011. "Adaptive Imaging for Lesion Detection Using a Zoom-in PET System." *TEEE Trans. Med. Imaging.* 30:119–130.

Zhu, X., and G. El Fakhri. 2013. "Proton Therapy Verification with PET Imaging." *Theranostics.* 3:731–740.

Zhu, X.P., S. Espana, J. Daartz, et al. 2011. "Monitoring Proton Radiation Therapy with In-Room PET Imaging." *Phys. Med. Biol.* 56:4041–4057.

Ziemons, K., E. Auffray, R. Barbier, et al. 2005. "The ClearPET (TM) Project: Development of a 2nd Generation High-Performance Small Animal PET Scanner." *Nucl. Instr. Meth. A.* 537:307–311.

2

Imaging Technologies and Potential Clinical Applications of Photon Counting X-Ray Computed Tomography

Katsuyuki Taguchi

CONTENTS

2.1 Imaging Technologies

We outline the overall strategies and the current status and our perspective on the imaging technologies that will be necessary to enable photon counting detector-computed tomography (PCD-CT) systems.

DOI: 10.1201/9781003218364-2

2.1.1 Overall Strategy

When an x-ray photon hits a PCD, it generates a pulse. If the PCD detection system is not fast enough, consecutive pulses generated by quasi-coincident photons will be integrated and will produce only one count recorded at the wrong energy. This is called *pulse pileup*. And with the loss of counts, the recorded spectrum will be distorted. One can decrease pulse pileups by making PCD pixels smaller and faster. A smaller PCD will receive fewer photons than will a larger PCD at the same x-ray intensity, resulting in fewer coincidences.

A distorted spectrum is also caused by spectral response effect (SRE) which includes the depth of the interaction effect and splitting energy due to charge sharing, K-escape, and Compton scattering. The SRE occurs even with very weak x-ray beams and is thus potentially more problematic than pulse pileups (which occur only near the object contour where x-rays are intense). SRE cannot be ignored, because we need to use energy/spectral information to allow a lower dose and many new clinical applications we will discuss later. A PCD with no spectral information and geometrical efficiency of 80% would allow for a dose reduction of 30–40% (depending on the quality of the "current" energy-integrating detectors [EIDs]), but this is not sufficient to reach desirable low-dose levels. Furthermore, ignoring SRE and using the uncorrected output of energy windows would result in shading artifacts and biases in images. Thus, SRE must be compensated for. One can decrease the SRE by making PCD pixels larger (to avoid splitting energy) and slower (to integrate all of the split energies within each pixel).

Notice that pulse pileup and SRE have opposite solutions, thus, no PCD can address both of the problems simultaneously. It is desirable to develop imaging technologies that could compensate for SRE and pulse pileups during the image reconstruction process, similar to how attenuation and scatter are compensated for in single-photon emission computed tomography (SPECT) and positron emission tomography (PET). In addition to improving the detector technologies, we believe it is necessary to advance and integrate imaging methods in the following four areas to make PCD-CT systems viable for imaging:

1. X-ray beam-shaping filters to optimize the intensity and spectrum of x-rays.
2. Calibration and compensation methods for the degradation effects of PCDs.
3. Models of the PCD's degradation effects.
4. Image reconstruction to provide accurate images from PCD data.

2.1.2 X-Ray Beam-Shaping Filters

New x-ray beam-shaping filters are needed for optimizing the x-ray flux and patient dose. Such a filter may consist of two components, a stationary part and a dynamic part, or a dynamic part only. With the two-component design, the stationary part "shapes" the intensity and spectrum of the x-ray beam across the entire field of view and the dynamic part specifically shapes the x-ray beam near the edge of the object being imaged.

The stationary part is similar to a conventional attenuating filter used in CT systems, which is often called a bow-tie filter because it is thin in the middle and thick at each end. The purpose of the shaping filter is to equalize the x-ray intensity at the detector and to reduce the dose to the patient periphery. It is essential to decrease the intensity of x-rays that go through near or outside edges of objects for PCDs, because the unattenuated (or less attenuated) x-ray flux would otherwise be very intense. Further, for PCDs, the spectrum incident on the object needs to be shaped to maximize the spectral information acquired from the object.

A single stationary filter alone would not be sufficient, because the fan-angles in projections that correspond to the object's edge change as the gantry rotates around the object and different portions of the object are scanned. It will be required to have additional filtrations or collimations which dynamically track the edge for each projection. With such dynamic tracking filters, the maximum count rate requirement for the PCD could be reduced significantly. For example, the count rate of the unattenuated x-ray beam with 120 kVp may be 10^9 cps/mm² while it will be reduced to 10^8 cps/mm² for the x-ray beam exiting the stationary bowtie filter, and further reduced to 10^5–10^8 cps/mm² with a dynamic bowtie filter and presence of the object (Figure 2.1).

Dynamic filters with no stationary components have already been studied. One design split the stationary bowtie filter into two parts in the middle of the fan beam, and each part moves independently along the fan angles to adjust the intensity of the x-ray beams. Another design has a set of triangular wedges

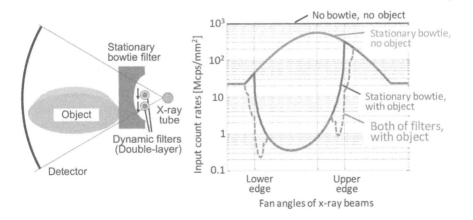

FIGURE 2.1
Calculated true count rates in the lateral view of an elliptic water phantom when it is 5 cm off-center. Dynamic and stationary bowtie filters (left) decrease the count rates near the edges of the object (right, red curve), compared to the results without the dynamic filters (blue curve). Figures are from Ref. (1).

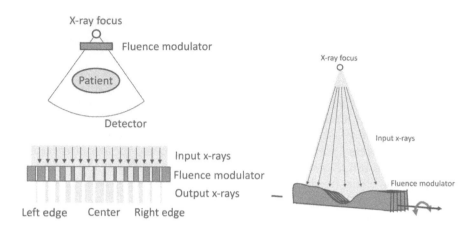

FIGURE 2.2
X-ray fluence/intensity modulators. X-rays that go to the attenuation materials of the fluence modulator are attenuated almost completely, whereas those that go to the opening of the fluence modulator will pass through the module.

and each wedge moves independently longitudinally.[2] A third design has a hollow ellipse, which rotates in the direction opposite to the gantry rotation.[3]

Recently, a fourth design that modifies the fluence (intensity) of x-ray beams without changing the spectrum of x-rays[4] has been drawing attention in the community (Figure 2.2). There are several design types for the x-ray fluence (or intensity) modulator, but all of them are based on the same concept: the device consists of two part, one part that blocks the x-rays entirely (e.g., tungsten) and the other part that passes through the x-rays with no (or minimum) attenuation (e.g., air holes). The relative x-ray intensities can be modulated by changing the aperture ratio between the blocking part and the opening part. One design uses a pair of fluence modulators with a uniform height but with a different aperture ratio (Figure 2.2, left bottom) and moves one of the modulators along the right–left direction to change their relative alignment, hence, the total aperture ratio, during the scan dynamically.[4] The amount of the horizontal shift required for the total aperture ratio modulation is very small. Another design stacks thin bowtie-shaped plates with opening slots between plates (Figure 2.2, right).[5] By rotating (or tilting) the device, the modulator blocks the peripheral x-rays more than central x-rays and changes the intensity of x-rays during the scan dynamically. The rotation angle required for the aperture ratio modulation is very small. It is our expectation that some of these fluence modulators will be implemented in CT systems in the near future; and they will be used for PCD-CT systems as well.

2.1.3 Calibration and Compensation Methods

There are two philosophically different approaches to dealing with distorted spectral data: corrections and compensation. Corrections attempt to undo the distortion process while compensation offsets the effect. Before discussing the two approaches, let us first define the terminology using Figure 2.2. Suppose that a forward imaging process to obtain an ideal x-ray spectrum y through an entire object x can be expressed as h: $y = h(x, a)$, where a is the initial x-ray intensity and spectrum exiting the bowtie filters. (Bold letters indicate tensors.) The spectrum y is then skewed to y' by PCD degradation factors g, i.e., $y' = g(y)$, which is then recorded as counts within N energy windows, i.e., $z = f(y')$. Note that the spectra y and y' can be described reasonably well by counts within narrow energy windows, e.g., 1 keV, or by using, e.g., 5–10 parameters.

One may be interested in *correcting* SRE and pileups, i.e., to estimate y from PCD data, z,[6] then reconstruct image x from y. The full energy spectrum y may be described by 5–10 parameters; however, estimating so many parameters from, e.g., 4-thresholded PCD data z, is an ill-posed problem. Different spectra y may produce the same set of counts, z. And complex crosstalk caused by SRE and pileups would make it even worse. We do not think it would work effectively and robustly. Nonetheless, a few approaches have been proposed.[7,8] They work well if, and only if, assumptions implicitly used as constraints, e.g., the object consists only of water, are correct.

We are interested in *compensating* SRE and pileups, i.e., to estimate x from z by iteratively solving the forward process $z = f(g(h(x)))$. In this chapter, we call this algorithm "PIECE," which stands for Physics-modeled Iterative reconstruction for Energy-sensitive photon Counting dEtector. PIECE incorporates a PCD model of SRE and pulse pileups and estimates either the imaged object or the sinogram using a maximum likelihood approach. The PCD degradation factors will be compensated for during the estimation process.[9-12] This can be formed as a well-posed problem; the method is depicted in Figure 2.3.

First, calibration is performed before the scan to obtain the x-ray intensity and the spectrum exiting the bowtie filters a for each sinogram pixel. Parameters for SRE and pulse pileups of PCDs will also be obtained. Thus, if we know the spectrum incident projected onto PCDs, y, we can calculate the recorded spectrum, y', by using the SRE and pulse pileup model. The expected counts of energy windows, z, can be calculated by integrating y' over the corresponding energy range: $z = f(y')$. Now, from a to z, the only missing link is how to obtain y from a, and we will use *material decomposition* to connect the link.

Let us explain the material decomposition[13] in this paragraph. The energy-dependent attenuation of the object at each pixel, $x(E)$, can be accurately modeled by a linear combination of two or three basis functions of energy, $f_k(E)$, and their coefficients, w_k, as $x(E) = \sum_k w_k f_k(E)$, where k is the index of the basis functions. Note that the attenuation model is exact if the number of basis functions is equal to or larger than the sum of the number of physics phenomena and the number of heavy elements inside the patient. It is exact regardless of the number of biological tissue types (e.g., muscle, fat, blood, skin, ligament, tendon, and bone). Two predominant physics phenomena, Compton (or

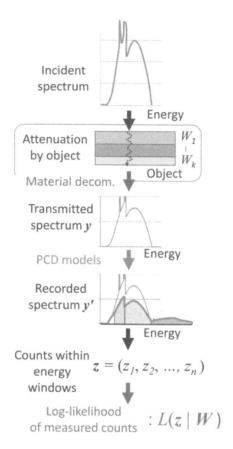

FIGURE 2.3
The model of the forward imaging process used in maximum likelihood methods to compensate for various spectral degradation factors. Figures are slightly modified from Ref. (1).

incoherent) scattering and photoelectric absorption, are sufficient to model the x-ray interactions with materials within the energy range of diagnostic x-ray. Rayleigh (or coherent) scattering occurs only in low energies and typically accounts for less than 5% of the diagnostic x-ray spectrum range. Pair production requires a photon energy of at least 1.02 MeV and plays no role in diagnostic imaging. Heavy elements include those used as contrast agents (e.g., iodine, gadolinium, or bismuth) and those in medical devices such as implants, stents, and bolts.

Now, let us return to the discussion of PIECE. Line integrals of basis function images w through the object, W, can be calculated for sinogram pixel: $W = \int w dr$. Then, using Beer's law, a and W, the transmitted x-ray spectrum y can be calculated. The entire forward imaging chain is now linked, and the only unknown information is the object we are imaging (or the thicknesses of basis functions), w_k or W_k. These will be estimated using a *maximum likelihood* approach that we will outline later.

We model the noisy PCD data $\hat{y}$ using a multivariate normal distribution, which takes into account the SRE and pulse pileups: $\tilde{z} \sim Normal(z, \Sigma)$, where z is the expected value and Σ is the covariance matrix. Both $\tilde{z}$ and Σ are joint functions of basis functions, W. And W are jointly estimated by maximizing the multivariate normal log-likelihood, i.e., $ln\,p(\tilde{z}\,|\,W)$. This process will work robustly and stably, as it is an overdetermined well-posed problem: the number of measurements (e.g., 4 with 4 energy windows) is larger than the number of unknowns (e.g., the thicknesses of two or three basis functions).

Multivariate normal distribution is more appropriate than Poisson distribution for modeling nonzero covariance of PCD data because the data are *not* Poisson distributed even without pulse pileups. Poisson assumes data are independent and not correlated. Thus, Poisson-noise model-based methods, which ignore the correlation of PCD data, would result in greater image noise.

We have recently developed a rudimentary version of PIECE, PIECE-1, which only models the intra-pixel, energetic cross-talk between energy windows.[14] We performed Monte Carlo simulations at high-count rates to evaluate the bias and noise standard deviation of the estimation. The results (Figure 2.4, left) show that PIECE-1 had very little bias and noise despite having very low detection efficiency (only 1–16%) when the water thickness was less than 10 cm. The method without the pileup model (green) had large biases when the water was thinner. The noise was significantly smaller when the covariance of multiple energy windows was used (circle), demonstrating the advantage of using covariance of data. Synthesized abdominal patient data were scanned by PCD-CT at 400 mAs (Figure 2.4, right). A significant

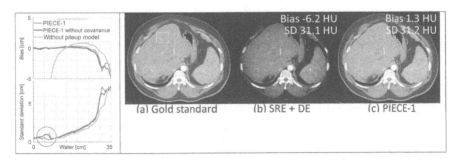

FIGURE 2.4

(Left) Both bias and noise of the estimated water thicknesses were improved by PIECE-1. (Right) Images of (a) gold standard, (b) reconstructed compensating for the SRE and detection efficiency (DE) (but ignoring pileups), and (c) reconstructed by PIECE-1. **Conditions:** A count rate exiting from the tube of 10^9 cps/mm^2; 1 cm water flat filter; detector dead time of 20 ns; pixel size of 0.5×0.5 mm^2; four energy thresholds set at 20, 50, 80, and 110 keV; and photopeak ratio of ~0.5 for SRE. The attenuators were 0–40 cm water and 0–5 cm bone.

cupping artifact toward the edge can be observed in the image reconstructed with the model that does not include pulse pileup (Figure 2.4b); this is attributable not to the beam-hardening effect but to pulse pileup effects. In contrast, the image reconstructed with PIECE-1(Figure 2.4(c)) shows no such artifacts.

2.1.4 PCD Models

The key to a successful PCD compensation is an accurate model of PCD degradation factors, g. It is logically possible, although it would not be practical, to perform PCD compensation successfully without any model. If the PCD is stable over a long period of time, one can acquire an extensive amount of calibration data to relate every possible x to PCD data $z = f(g(h(x)))$ with every possible combination of conditions (e.g., tube current, tube voltage, materials, and thicknesses of bowtie filters). This approach would not be practical, however, because the number of required calibration datasets is very large and PCD data may change by at least a few percentage points over time. It may be more reasonable to take an approach that is similar to the one implemented with EID-CT systems: an extensive calibration procedure performed less frequently (e.g., semiannually), and a quick calibration procedure employed every day, from which parameters necessary for PCD models are estimated and used to monitor the temporal change of PCD data for quality control. Both the model and the extensive calibration data acquired previously will be used to generate pseudo calibration data, which would be acquired if an extensive procedure was performed frequently.

We model SRE and pulse pileup separately. The integrated phenomena of the two factors are modeled by cascading models of attenuation, SRE, and pulse pileup.[15-17] Next, we discuss examples both of the pulse pileup and SRE models and the cascaded model.

The SRE can be integrated and described as a single spectral response function, which can be modeled based on measurements using radioisotopes or synchrotron radiation at a very low count rate.[18] Considering the stochastic nature of the SRE, spectral response function $SRF(E, E_0)$ models the probability density distribution of the recorded energy E, given the true photon energy E_0.[9,10] A small number of input energies E_0 can be used to measure SRF, and they will be interpolated to estimate SRF at desirable energies. When a polychromatic x-ray spectrum $S(E_0)$ is projected onto the PCD, the recorded spectrum can be calculated by the integration of the $SRF(E, E_0)$ weighted by $S(E_0)$ over E_0. Note that this process is usually *not* a convolution because $SRF(E, E_0)$ changes over E_0 (thus, the SRF is shift-variant). An example of the true and recorded spectra is shown in Figure 2.5. It can be seen that the SRE of the PCD blurs the spectrum and increases counts, especially at low energies.

The spectrum distortion caused by pulse pileup is most difficult to model because it is a very complex phenomenon. But it is necessary to model,

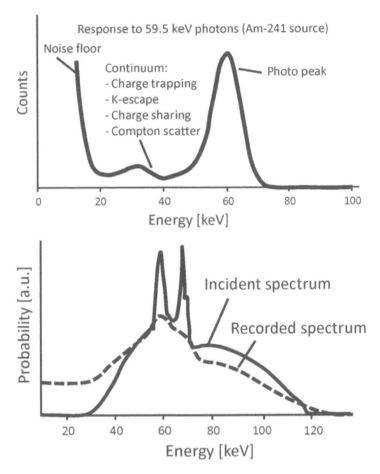

FIGURE 2.5
(Top) An illustration of a typical spectrum recorded by a PCD using Am-241. The spectrum is distorted even at a very low count rate (i.e., the pulse pileup effects are minimal). (Bottom) There is a significant discrepancy between the true and recorded polychromatic x-ray spectra. Figures are from Ref. (1).

because the output depends on the input count rates and spectra, and thus, depends on the object to be imaged. Simple models such as linear corrections or self-convolution[19] are not accurate for modeling complex mechanisms of distortion. Various pulse pileup models have been developed,[20-24] and we have developed a model[23,24] that satisfies the accuracy, efficiency, and ability to handle a large number of coincidence requirements for high input count rates. The pulse pileup model accounts for the (bipolar) shape of the pulse, the distribution function of time intervals between random events, and the transmitted spectrum as the probability density function. The model showed excellent agreement with Monte Carlo simulation[23] and with PCD data.[24] The coefficients of variation (i.e., the root mean square difference divided by the mean of measurements) were as small as 5.3–10.0% for dead-time loss up to 50% in a Monte Carlo simulation[23] and 7.2% with dead-time loss of 46% in a PCD experiment.[24]

The cascaded model of attenuation, SRE, and pulse pileups start with the spectrum incident onto the object. Using Beer's law to model the attenuation inside the object, the spectrum exiting from the object and incident onto PCD is calculated. The spectrum is then used as an input to SRF and the intermediate spectrum that results from SRE is calculated. Finally, the intermediate spectrum is used as an input to the pulse pileup model, which provides the expected recorded PCD data.

The cascaded model showed excellent agreement with the PCD data (Figure 2.6). The weighted coefficient of variation, or COVw (i.e., the root mean square difference weighted by the standard deviation of measurements,

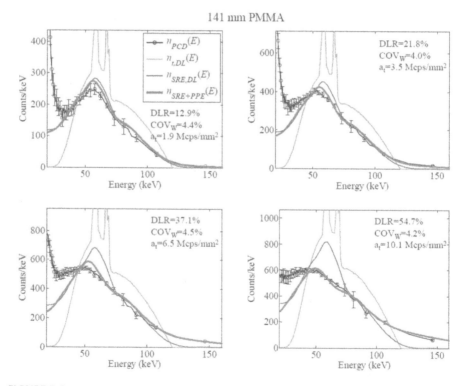

FIGURE 2.6

The spectrum recorded by a PCD, $n_{PCD}(E)$, was severely distorted by SRE and pulse pileups and there are significant discrepancies from the spectrum predicted by a linear model (i.e., the true spectrum linearly scaled by the dead-time loss ratio, DLR), $n_{t,DL}(E)$. The spectrum predicted by the SRE model and scaled by DLR, $n_{SRE,DL}(E)$, had better agreement with $n_{PCD}(E)$ than did $n_{t,DL}(E)$; however, the deviation increases with increasing DRL. The fully cascaded PCD model proposed in Ref. (17) accurately estimated the recorded spectrum over a wide range of DLRs (or count rates). a_t is the count rates incident onto the detector. Figures are from Ref. (17).

divided by the mean of measurements) averaged over all channels was as small as 1.5–6.7% for dead-time losses of (or detection efficiency reductions of) 1.1–55.2% with PMMA. In contrast, models which lack the pulse pileup model or both pulse pileup and SRE resulted in much larger coefficient of variation values: 1.7–36.3% without the pileup model, and 8.3–67.5% with neither pileup nor SRE models.[17]

Not many researchers have access to PCD-CT scanner and PCD data at present; thus, many wish to simulate data for their work. It is then critical to simulate realistic PCD data. In order to address the issue and help the community, we have developed a PCD cross-talk model[25-27] and made the software available to academic researchers in academic institutions, free of charge. The software is called "Photon Counting Toolkit (PcTK)"[28] and interested readers should visit pctk.jhu.edu for more information. Users can use PcTK to compute the expected recorded spectra with desirable PCD design specifications or simulate noisy PCD data (see Figure 2.7 for one of use cases).

2.1.5 Image Reconstruction

The fourth area for advancing and integrating imaging methods is to adapt advanced image reconstruction methods for photon counting CT data for the interior problem and spectral data.

Interior problem. Even with the earlier discussed PCD compensation schemes, photon counting data maybe inaccurate, especially for x-rays that go through the edge of the object or just outside the object when the object is off-center. Reconstructing images from such inaccurate data will result in undesirable artifacts. From the algorithmic point of view, this is a unique, softly-posed interior problem. The detector size defines the physical data truncation range. However, for acceptable data quality, only a subset of all detector channels may be used for reconstruction, e.g., because the count rates were high in the periphery. The usable range depends on the PCD compensation method and can be decided retrospectively for PCD-CT. Insight into this unique problem can be gained by studying the tradeoff between acceptable data quality and image fidelity using simulation and phantom studies.

There are two approaches to addressing the interior problem: (i) to estimate unmeasured data and "de-truncate" the projection data, and follow that with a standard image reconstruction method; or (ii) to reconstruct (quasi-)exact images only from the truncated measured data. Studying these methods for the softly-posed interior problem is certainly of interest.

For the first approach, various de-truncation methods have been proposed, which include empirical approaches aimed to decrease an abrupt change between the estimated and measured data,[29,30] or more mathematically rigorous approaches that use consistency conditions.[31,32] The use of prior conventional CT images for photon counting data has recently been proposed.[33]

Regarding the second approach, two important algorithms have recently been developed to solve the interior problem. First, when a small region

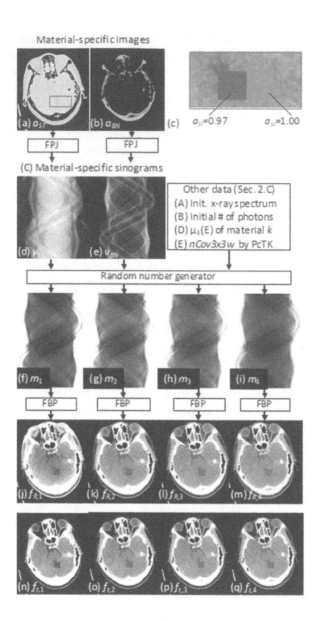

FIGURE 2.7
Materials prepared for and resulted from a simulated PCD-CT scan. (a–c) The density maps of soft tissue (a) and bone (b) with regions-of-interest with constant soft tissue density values (c). (d,e) Material-specific sinograms of soft tissue (h) and bone (i) generated from (a) and (b), respectively through forward projection (FPJ). (f–i) Noisy PCD projection data (counts) with cross-talk and spectral response effects of the four energy windows, generated using a random number generator and data (A–E) outlined in Sec. 2.C of Ref. (26). (j–m) CT images reconstructed from each of the PCD data (f–i) independently using filtered backprojection (FBP). (n–q) Monochromatic x-ray CT images synthesized from (a) and (b) at the center of the four energy windows used, 35 keV (n), 57 keV (o), 72 keV (p), and 100 keV (q). CT images are presented with the window width of 0.04 cm^{-1} and the window center at the value of $a_{ST} = 1.0$ region-of-interest (c). Figure is from Ref. (26).

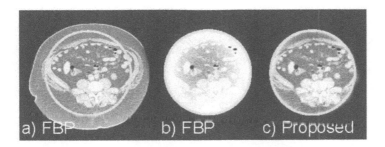

FIGURE 2.8

Reconstructed images (a) without or (b, c) with truncation outside the yellow circle, using (a, b) filtered backprojection (FBP) or (c) the proposed sequential method. The image reconstructed by the proposed method showed very little bias throughout the region-of-interest except near the edge of the region-of-interest, while the image appeared very similar to that reconstructed without truncation. Images are from Ref. (34).

located inside the region-of-interest is known, the region-of-interest image can be reconstructed exactly using a differentiated backprojection framework.[34-37] Second, if the region-of-interest is piece-wise constant, an exact image can be reconstructed using the total variation minimization algorithm without other *a priori* knowledge.[38,39] Clinical CT data satisfies neither of the requirements; however, it was demonstrated that quasi-exact region-of-interest images can be reconstructed even from noisy clinical CT projections by sequentially using filtered backprojection (FBP), total variation minimization, and differentiated backprojection (Figure 2.8).[40] Pixel values of a tiny flat region obtained by total variation minimization were used as *a priori* information during differentiated backprojection.

Spectral data. Spectral data that become available with PCDs provide room to investigate and develop new methods for improved contrast-to-noise ratio, material decomposition, and statistical reconstruction. A study[41] showed that weighting energy-window data by a factor of E^{-3}, where E is the effective energy of the window, improved the contrast-to-noise ratio of images (see Sec. V.A for various study results), and other weighting schemes have also been investigated.[42,43] The practical value of these methods in the presence of energetic cross-talk between energy windows, however, is not clear. A portion of signals obtained at lower energy windows come from higher energy photons, and photons that are supposed to be detected at lower energy windows may be counted by a higher energy window. Without appropriate handling of energetic cross-talk, energy-window-weighting approaches may enhance artifacts and biases. An application of local, highly constrained backprojection reconstruction (HYPR-LR) broke free from the tradeoff between the contrast and the noise of monoenergetic images,[44] although a challenge of this approach is how to handle the energetic cross-talk.

Recently, a new class of image reconstruction methods has been developed for PCD-CT.[45,46] The method, JE-MAP for joint estimation maximum *a posteriori*, jointly reconstructs basis function images and tissue maps of the object

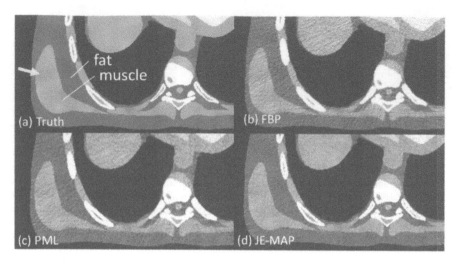

FIGURE 2.9
(a) The modified XCAT phantom with heterogeneous texture inside organs. (b–d) Images reconstructed by (b) filtered backprojection (FBP), (c) model-based iterative reconstruction, penalized maximum likelihood (PML), and (d) a joint estimation maximum *a posteriori* or JE-MAP. Figures are reprinted with modifications from Ref. (46).

directly from PCD data, making full use of the rich information that spectral PCDs acquire. JE-MAP is a new class of algorithm which integrates three steps—material decomposition, image reconstruction, and tissue characterization—into a single step. Using the knowledge if the pixels are at an organ boundary or inside an organ, JE-MAP decreased image noise effectively while maintaining the sharpness of organ boundaries and heterogeneous patterns inside organs (Figure 2.9).

In addition, there are several representational schemes for PCD images such as monoenergetic CT images, material-specific (e.g., iodine) density maps, effective atomic number maps, and electron density maps. Different types of images may be optimally obtained by using different algorithms. Integrating three steps—material decomposition, image reconstruction, and final output calculation—into a single step may improve the accuracy and precision of images.

2.2 Potential Benefits and Clinical Applications

Here, we outline the clinical merits and applications of PCD-CT from improved and evolutionary versions of what is currently available to innovative and revolutionary new ones. We have performed a simulation study to demonstrate some of the merits for coronary CT angiography that we

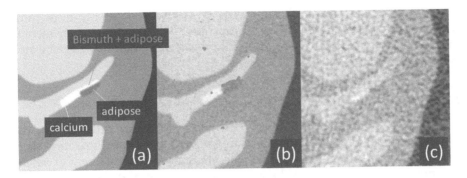

FIGURE 2.10
(a) A computer simulated XCAT phantom image with bismuth at the surface of fatty atherosclerosis in a coronary artery. (b, c) Reconstructed images of the phantom scanned at the equivalent dose using a PCD-CT (b) and an EID-CT (c). Densities of bismuth are shown in red in (b). The PCD image has a better contrast-to-noise ratio and appears sharper than the EID image. This is also an example of K-edge, molecular, and simultaneous multi-agent imaging. Images are from Ref. (47).

will discuss later (Figure 2.10). The scan conditions were as follows: 120 kVp; tube current modulation up to 667 mA for lateral direction, down to 200 mA for the AP direction; aluminum bowtie filter with thicknesses of 5–30 mm; focus-to-center, 600 mm; focus-to-detector, 1100 mm; 1892 channels for field-of-view of Ø500 mm for PCD-CT and 946 channels for EID-CT; 2560 projections per rotation. Images shown in Figure 2.7 were reconstructed while compensating for the spectral distortion due to SRE using a penalized maximum-likelihood approach[11] and FBP for PCD-CT, and FBP only for EID-CT.

2.2.1 Improved Contrast-to-Noise Ratio and Contrast of CT Images

The image quality metrics of CT will improve with PCDs and appropriate algorithms, and material decomposition will allow for reconstructing monoenergetic images at desirable energies. These improvements are significant for any applications, but are particularly important for molecular imaging, since weaker signals can be detected.

One point of caution is that, as shown in Ref. (48), results strongly depend on the conditions under which the studies are conducted. Close attention must be paid to factors such as the choice of objects and lesions, the degree of spectral distortion of PCDs, and the algorithms employed for compensation and image reconstruction. One simulation study showed that with optimal energy weighting, the contrast-to-noise ratios of PCD-CT images were better than those of EID-CT images by 15–57% depending on the materials.[42] When the spectral distortion caused by SRE was incorporated and compensated for, contrast-to-noise ratios of PCD-CT images were improved from EID-CT by 1.4–11.6% in one study[10] and by 40–63% in another study.[11] An experimental study using PCD-CT and clinical dual-energy CT, with contrast-to-noise ratio of oil and water, resulted in 57–96% improvement.[49] Another experimental study showed that the contrast-to-noise ratio of iodine solution against water increased by up to 20%.[50]

2.2.2 Dose Reductions of X-Ray Radiation and Contrast Agents

PCD-CT has the potential to improve the contrast-to-noise ratio of contrast-enhanced lesions at a given dose by as much as 30% or more. Expecting such an improvement, one could decrease the amount of contrast agent or radiation dose while maintaining the contrast-to-noise ratio of the lesion at the current level. The contrast dose reduction will be preferable for patients with renal function issues, while the radiation dose reduction will decrease a risk of cancers in general. Using the linear method shown in Appendix B of Ref. (1), the contrast dose might be reduced by 23% or the radiation dose by 41%. The amount of actual dose reduction achieved may be smaller than these values in practice, however, because the PCD-CT system and image reconstruction methods may be non-linear.

2.2.3 Improved Spatial Resolution

In order to handle the high-count rates required for clinical CT, the pixel size of PCDs will likely be smaller than that of EIDs: 0.2–0.5 mm for PCDs in contrast to 1.0–1.4 mm for EIDs. Each scintillator pixel of EIDs is surrounded by light reflectors that physically and optically separate pixels. The reflectors prevent light cross-talk between adjacent pixels, and direct scintillation lights to be collected by a photo diode underneath the scintillator pixel. The reflectors do not detect x-rays though, and thus, they decrease the geometrical efficiency of EIDs. The thickness of the reflectors is constant regardless of the pixel size; thus, the geometrical efficiency of EIDs decreases with a decrease in pixel size. This is the reason why the pixel size of EIDs cannot be as small as those of PCDs.

The intrinsic spatial resolution of PCD-CT images defined by the Nyquist frequency of the sampling condition will thus be superior to that of EID-CT images. Reconstructed images may become sharper and more accurate due to decreased partial volume effects from small structures such as calcium plaques, although it will come with increased noise.

2.2.4 Beam Hardening Artifacts

CT vendors have developed beam hardening correction methods for water and bone.[51] However, beam hardening artifacts with contrast agents remain a problem for cardiac images.[52,53] PCD-CT will address this problem and improve images where soft plaque, calcium/bone, and contrast-enhanced lumen are present.

2.2.5 Quantitative CT and X-Ray Imaging

Current CT pixel values are not as quantitative as one may think. They are measured in Hounsfield units, which are linearly related to the linear attenuation coefficients of x-rays at some energy. However, it is not clear which energy it is. The effective energies of the transmitted x-ray spectrum vary greatly during a scan, depending on factors such as fan/cone angles due to effects of bowtie filters, the attenuation of the object, projection angles, etc. Thus, the effective energy for an image pixel cannot be calculated. Pixel values of the same tissue vary as the effective energy varies. PCDs can make CT images quantitative using well defined energies. The physical properties of each image pixel can be accurately modeled using the concept of material decomposition and reconstruction from PCD data. The concentration of contrast agents at regions-of-interest can then be quantified, which will benefit applications such as cardiac perfusion CT. One problem of current perfusion CT is that it is necessary to subtract a baseline image from target images at different phases to calculate the enhancement due to the injection of the contrast agent. The subtraction will increase noise and misregistration due to motion results in inaccurate time-density-curves, and thus, perfusion measurements such as blood flow may be noisy and inaccurate. In addition, the calculated enhancement may change from scan to scan, because the pixel values of CT images are not quantitative. Measuring the concentrations of the contrast agent in target images without subtracting the baseline image, enabled by the quantitative PCD-based CT imaging methods, will improve the accuracy of perfusion CT and other applications.

2.2.6 Accurate K-Edge Imaging

Dual-energy CT provides only two measurements with different energies;[13] however, it is desirable to have three or more measurements for K-edge imaging for contrast-enhanced CT exams[54,55] and for corrections of various data quality degradation factors, as discussed earlier. A third basis function is necessary to model the attenuation curves of contrast agents with high atomic numbers (e.g., iodine, gadolinium, and barium) because the curves are discontinuous due to their material-specific K-shell binding energies (Figure 2.11). Using the material decomposition with a third basis function for an atom used in the contrast agent-of-interest will make it possible to quantify the spatial distribution of contrast agents on a pixel basis. This is called K-edge CT imaging, which will enable quantitative imaging of the contrast agent.

2.2.7 Simultaneous Multi-Agent Imaging

Simultaneous multi-agent imaging[9] for different functionalities may become possible. Large biological variations between animals and patients make it difficult to interpret measured quantities of agents. By injecting two agents

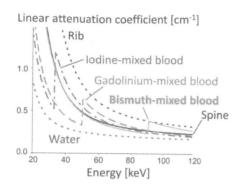
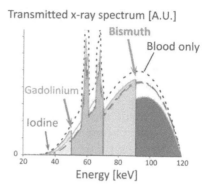

FIGURE 2.11

(Left) Energy-dependent linear attenuation coefficients of various materials. Contrasts between different materials are greater at lower energies in general. Four materials, spine, 0.49% w/w iodine-mixed blood, 0.26% w/w gadolinium-mixed blood, and 0.28% w/w bismuth-mixed blood, result in the same pixel value with the current EID-CT, although they have distinctly different attenuation curves. (Right) Transmitted spectra with 25 cm water and 5 cm blood without or with one of the three contrast agents. The K-edges of gadolinium and bismuth are clearly seen. Figures are from Ref. (1).

simultaneously, one with target receptors and labeled by one element and the other without receptors and labeled by another element, and imaging both simultaneously, the agent without receptors can be used as a control.[56] This will solve interpretation problems.

There are two blood supplies to the liver: hepatic artery and portal vein. Primary cancers receive 80% of their blood supply from the hepatic artery while the liver parenchyma receives 80% from the portal vein. Therefore, the liver is usually scanned at two different phases, one at the hepatic arterial phase and the other at the portal venous phase, which are separated by ~50 seconds. Patients are instructed to breath between the scans, which lead to misregistration between the two images. If two different contrast agents are administered, one early for the portal venous phase and the other later for the hepatic arterial phase, and the patient is scanned once, the single multi-agent image may present the distribution of two blood supplies.

2.2.8 Molecular CT with Nanoparticle Contrast Agents and Personalized Medicine

A new type of contrast agents may enable molecular CT imaging.[57-59] Nanoparticles of various size and function are labeled by atoms for CT imaging. So-called blood pool agents, which consist of large particles with a particle size of a few hundred nanometers (blood pool contrast agents),[60,61] stay in the system longer than 24 hours because they are not filtered out by the kidneys. Such large particles can carry more receptors and will increase chances of interaction with target sites, and thus, enhance target-specific therapy and imaging. For example, $\alpha_v\beta_3$-targeted nanoparticles[62-65] have been used to detect, characterize, and treat angiogenesis. Labeling particles for x-ray CT is achieved by attaching atoms with high atomic numbers (e.g., bismuth), which are preferable because signal-to-noise ratios are higher than iodine due to the following reasons: (1) they attenuate more photons with the same particle concentrations than those with lower numbers; and (2) there are more x-ray photons near the K-edges (see Figure 2.11).

There are many challenges to this development including toxicity, stability, and clearance for safety; uniformity of particle size for functionality; and particle concentration or uptake for functionality and signal detection. Significant investment from pharmaceutical companies is needed for manufacturing high quality agents, which will be challenging in an environment where there is both a limited market (due to its specific target) and rigorous regulatory hurdles to overcome. Nonetheless, nanomedicine research aligns well with NIH's goal of personalized medicine, and solutions may be, and should be, found. PCD-CT will be ideal for these biomedical applications and will play a vital role in advancing nanomedicine research.

References

1. Taguchi K, Iwanczyk JS. Vision 20/20: Single photon counting x-ray detectors in medical imaging. *Medical Physics*. 2013;40(10):100901.
2. Hsieh SS, Pelc NJ. The feasibility of a piecewise-linear dynamic bowtie filter. *Medical Physics*. 2013;40(3):031910.
3. Roessl E, Proksa R. Dynamic Beam-Shaper for High Flux Photon-Counting Computed Tomography. Workshop on Medical Applications of Spectroscopic X-ray Detectors; 2013; Geneva, Switzerland.
4. Stayman JW, Mathews A, Zbijewski W, et al. Fluence-field modulated x-ray CT using multiple aperture devices. *SPIE Medical Imaging*. 2016;9783:97830X.
5. Huck SM, Fung GSK, Parodi K, Stierstorfer K. The z-sbDBA, a new concept for a dynamic sheet-based fluence field modulator in x-ray CT. *Medical Physics*. 2020;47(10):4827–4837.
6. Miyajima S, Imagawa K, Matsumoto M. CdZnTe detector in diagnostic x-ray spectroscopy. *Medical Physics*. 2002;29(7):1421–1429.
7. Kappler S, Hoelzer S, Kraft E, Stierstorfer K, Flohr TG. Quantum-counting CT in the regime of count-rate paralysis: Introduction of the pile-up trigger method. Proc. SPIE 7661, Medical Imaging 2011: Physics of Medical Imaging; 2011; Orlando, FL.
8. Kraft E, Glasser F, Kappler S, Niederloehner D, Villard P. Experimental evaluation of the pile-up trigger method in a revised quantum-counting CT detector. Proc. SPIE 8313, Medical Imaging 2012: Physics of Medical Imaging; February 23, 2012, 2012; San Diego, CA.
9. Schlomka JP, Roessl E, Dorscheid R, et al. Experimental feasibility of multi-energy photon-counting K-edge imaging in pre-clinical computed tomography. *Physics in Medicine and Biology*. 2008;53(15):4031–4047.

10. Roessl E, Brendel B, Engel K, Schlomka J, Thran A, Proksa R. Sensitivity of photon-counting based K-edge imaging in x-ray computed tomography. *Medical Imaging, IEEE Transactions On.* 2011;30(9):1678–1690.
11. Srivastava S, Cammin J, Fung GSK, Tsui BMW, Taguchi K. Spectral response compensation for photon-counting clinical x-ray CT using sinogram restoration. SPIE Medical Imaging 2012: Physics of Medical Imaging; 2012; San Diego, CA.
12. Srivastava S, Taguchi K. Sinogram restoration algorithm for photon counting clinical x-ray CT with pulse pileup compensation. Proc. of The First International Meeting on Image Formation in X-Ray Computed Tomography; June 6–9, 2010, 2010; Salt Lake City, UT.
13. Alvarez RE, Macovski A. Energy-selective reconstructions in X-ray computerised tomography. *Physics in Medicine and Biology.* 1976;21(5):733–744.
14. Taguchi K, Nakada K, Amaya K. Compensation for spectral distortions due to spectral response and pulse pileup effects for photon counting CT. IEEE Nuclear Science Symposium and Medical Imaging Conference; 2013; Seoul, Korea.
15. Cammin J, Iwanczyk JS, Taguchi K. Spectral/Photon-counting Computed Tomography. In: Anastasio MA, Riviere PJL, eds. *Emerging Imaging Technologies in Medicine.* 1 ed.: Taylor & Francis Books; 2012:23–39. https://www.taylorfrancis.com/books/edit/10.1201/b13680/emerging-imaging-technologies-medicine-mark-anastasio-patrick-la-riviere
16. Cammin J, Xu J, Barber WC, Iwanczyk JS, Hartsough NE, Taguchi K. Modeling photon-counting detectors for X-ray CT: Spectral response and pulse pileup effects and evaluation using real data. Proc. SPIE 8668, Medical Imaging 2013: Physics of Medical Imaging; 2013; Lake Buena Vista (Orlando Area), FL.
17. Cammin J, Xu J, Barber WC, Iwanczyk JS, Hartsough NE, Taguchi K. A cascaded model of spectral distortions due to spectral response effects and pulse pileup effects in a photon-counting x-ray detector for CT. *Medical Physics.* 2014;41(4):041905.
18. Ding H, Molloi S. Image-based spectral distortion correction for photon-counting x-ray detectors. *Medical Physics.* 2012;39(4):1864–1876.
19. Guenzler R, Schuele V, Seeliger G, et al. A multisegment annular Si-detector system for RBS analysis. *Nuclear Instruments and Methods in Physics Research Section B: Beam Interactions with Materials and Atoms.* 1988;35(3-4):522–529.
20. Wielopolski L, Gardner RP. Prediction of the pulse-height spectral distribution caused by the peak pile-up effect. *Nuclear Instruments and Methods in Physics Research.* 1976;133:303–309.
21. Gardner RP, Wielopolski L. A generalized method for correcting pulse-height spectra for the peak pileup effect due to double sum pulses. *Nuclear Instruments and Methods in Physics Research Section A: Accelerators, Spectrometers, Detectors and Associated Equipment.* 1977;140:289–296.
22. Barradas NP, Reis MA. Accurate calculation of pileup effects in PIXE spectra from first principles. *X-ray Spectrometry.* 2006;35(4):232–237.
23. Taguchi K, Frey EC, Wang X, Iwanczyk JS, Barber WC. An analytical model of the effects of pulse pileup on the energy spectrum recorded by energy resolved photon counting x-ray detectors. *Medical Physics.* 2010;37(8):3957–3969.
24. Taguchi K, Zhang M, Frey EC, et al. Modeling the performance of a photon counting x-ray detector for CT: Energy response and pulse pileup effects. *Medical Physics.* 2011;38(2):1089–1102.

25. Taguchi K, Stierstorfer K, Polster C, Lee O, Kappler S. Spatio-energetic cross-talk in photon counting detectors: Numerical detector model (PcTK) and workflow for CT image quality assessment. *SPIE Medical Imaging 2018: Physics of Medical Imaging.* 2018;10573:1057310. https://www.spiedigitallibrary.org/conference-proceedings-of-spie/10573/1057310/Spatio-energetic-cross-talk-in-photon-counting-detectors--numerical/10.1117/12.2293881.full?SSO=1

26. Taguchi K, Stierstorfer K, Polster C, Lee O, Kappler S. Spatio-energetic cross-talk in photon counting detectors: Numerical detector model (PcTK) and workflow for CT image quality assessment. *Medical Physics.* 2018;45(5):1985–1998.

27. Taguchi K, Polster C, Lee O, Stierstorfer K, Kappler S. Spatio-energetic cross talk in photon counting detectors: Detector model and correlated Poisson data generator. *Medical Physics.* 2016;43(12):6386–6404.

28. Taguchi K. Photon Counting Toolkit (PcTK). pctk.jhu.edu Published 2018. Accessed.

29. Ohnesorge B, Flohr T, Schwarz K, Heiken JP, Bae KT. Efficient correction for CT image artifacts caused by objects extending outside the scan field of view. *Medical Physics.* 2000;27(1):39–46.

30. Zamyatin AA, Nakanishi S. Extension of the reconstruction field of view and truncation correction using sinogram decomposition. *Medical Physics.* 2007;34(5):1593–1604.

31. Hsieh J, Chao E, Thibault J, et al. A novel reconstruction algorithm to extend the CT scan field-of-view. *Medical Physics.* 2004;31(9):2385–2391.

32. Xu J, Taguchi K, Tsui BMW. Statistical projection completion in x-ray CT using consistency conditions. *Medical Imaging, IEEE Transactions on.* 2010;29(8):1528–1540.

33. Schmidt TG, Pektas F. Region-of-interest material decomposition from truncated energy-resolved CT. *Medical Physics.* 2011;38(10):5657–5666.

34. Courdurier M, Noo F, Defrise M, Kudo H. Solving the interior problem of computed tomography using a priori knowledge. *Inverse Problems.* 2008;24(6):065001.

35. Kudo H, Courdurier M, Noo F, Defrise M. Tiny a priori knowledge solves the interior problem in computed tomography. *Physics in Medicine and Biology.* 2008;53(9):2207–2231.

36. Yu H, Ye Y, Wang G. Interior reconstruction using the truncated Hilbert transform via singular value decomposition. *Journal of X-Ray Science and Technology.* 2008;16(4):243–251.

37. Wang G, Yu H, Ye Y. A scheme for multisource interior tomography. *Medical Physics.* 2009;36(8):3575–3581.

38. Yu H, Wang G. Compressed sensing based interior tomography. *Physics in Medicine and Biology.* 2009;54(9):2791–2805.

39. Yu H, Yang J, Jiang M, Wang G. Supplemental analysis on compressed sensing based interior tomography. *Physics in Medicine and Biology.* 2009;54(18): N425–N432.

40. Taguchi K, Xu J, Srivastava S, Tsui BMW, Cammin J, Tang Q. Interior region-of-interest reconstruction using a small, nearly piecewise constant subregion. *Medical Physics.* 2011;38(3):1307–1312.

41. Shikhaliev PM. Beam hardening artefacts in computed tomography with photon counting, charge integrating and energy weighting detectors: A simulation study. *Physics in Medicine and Biology.* 2005;50(24):5813–5827.

42. Schmidt TG. Optimal "image-based" weighting for energy-resolved CT. *Medical Physics.* 2009;36(7):3018–3027.

43. Schmidt TG. CT energy weighting in the presence of scatter and limited energy resolution. *Medical Physics.* 2010;37(3):1056–1067.
44. Leng S, Yu L, Wang J, Fletcher JG, Mistretta CA, McCollough CH. Noise reduction in spectral CT: Reducing dose and breaking the trade-off between image noise and energy bin selection. *Medical Physics.* 2011;38(9):4946–4957.
45. Nakada K, Taguchi K, Fung GSK, Amaya K. Maximum a posteriori reconstruction of CT images using pixel-based latent variable of tissue types. The Third International Conference on Image Formation in X-ray Computed Tomography; 2014; Salt Lake City, UT.
46. Nakada K, Taguchi K, Fung GSK, Amaya K. Joint estimation of tissue types and linear attenuation coefficients for photon counting CT. *Medical Physics.* 2015; 42(9):5329–5341.
47. Cammin J, Srivastava S, Fung GSK, Taguchi K. Spectral response compensation for photon counting clinical x-ray CT and application to coronary vulnerable plaque detection. Proc. of The Second International Meeting on Image Formation in X-Ray Computed Tomography; June 24-27, 2012, 2012; Salt Lake City, UT.
48. Polad MS. The upper limits of the SNR in radiography and CT with polyenergetic x-rays. *Physics in Medicine and Biology.* 2010;55(18):5317.
49. Cammin J, Srivastava S, Tang Q, et al. Compensation of nonlinear distortions in photon-counting spectral CT: Deadtime loss, spectral response, and beam hardening effects. SPIE Medical Imaging 2012: Physics of Medical Imaging; 2012; San Diego, CA.
50. Kappler S, Kraft E, Kreisler B, Schoeck F, Flohr TG. Imaging performance of a hybrid research prototype CT scanner with small-pixel counting detector. Workshop on Medical Applications of Spectroscopic X-ray Detectors; 2013; Geneva, Switzerland.
51. Hsieh J, Molthen RC, Dawson CA, Johnson RH. An iterative approach to the beam hardening correction in cone beam CT. *Medical Physics.* 2000;27(1):23–29.
52. So A, Hsieh J, Li J-Y, Lee T-Y. Beam hardening correction in CT myocardial perfusion measurement. *Physics in Medicine and Biology.* 2009;54(10):3031–3050.
53. Stenner P, Schmidt B, Allmendinger T, Flohr T, Kachelriess M. Dynamic iterative beam hardening correction (DIBHC) in myocardial perfusion imaging using contrast-enhanced computed tomography. *Investigative Radiology.* 2010;45(6):314–323.
54. Feuerlein S, Roessl E, Proksa R, et al. Multienergy photon-counting K-edge imaging: Potential for improved luminal depiction in vascular imaging. *Radiology.* 2008;249(3):1010–1016.
55. Roessl E, Proksa R. K-edge imaging in x-ray computed tomography using multi-bin photon counting detectors. *Physics in Medicine and Biology.* 2007;52(15): 4679–4696.
56. Li Y, Sheth VR, Liu G, Pagel MD. A self-calibrating PARACEST MRI contrast agent that detects esterase enzyme activity. *Contrast Media & Molecular Imaging.* 2011;6(4):219–228.
57. Jaffer FA, Weissleder R. Seeing within: Molecular imaging of the cardiovascular system. *Circulation Research.* 2004;94(4):433–445.
58. Pan D, Caruthers SD, Hu G, et al. Ligand-directed nanobialys as theranostic agent for drug delivery and manganese-based magnetic resonance imaging of vascular targets. *Journal of the American Chemical Society.* 2008;130(29):9186–9187.

59. Schmieder AH, Winter PM, Caruthers SD, et al. Molecular MR imaging of melanoma angiogenesis with 3-targeted paramagnetic nanoparticles. *Magnetic Resonance in Medicine.* 2005;53(3):621–627.
60. Hyafil F, Cornily J-C, Feig JE, et al. Noninvasive detection of macrophages using a nanoparticulate contrast agent for computed tomography. *Nature Medicine.* 2007;13(5):636–641.
61. Rabin O, Manuel Perez J, Grimm J, Wojtkiewicz G, Weissleder R. An x-ray computed tomography imaging agent based on long-circulating bismuth sulphide nanoparticles. *Nature Materials.* 2006;5(2):118–122.
62. Winter PM, Shukla HP, Caruthers SD, et al. Molecular imaging of human thrombus with computed tomography. *Academic Radiology.* 2005;12(5, Supplement 1):S9–S13.
63. Anderson SA, Rader RK, Westlin WF, et al. Magnetic resonance contrast enhancement of neovasculature with alpha-v beta-3-targeted nanoparticles. *Magnetic Resonance in Medicine.* 2000;44(3):433–439.
64. Flacke S, Fischer S, Scott MJ, et al. Novel MRI contrast agent for molecular imaging of fibrin: Implications for detecting vulnerable plaques. *Circulation.* 2001;104(11):1280–1285.
65. Winter PM, Caruthers SD, Kassner A, et al. Molecular imaging of angiogenesis in nascent Vx-2 rabbit tumors using a novel $\alpha_v\beta_3$-targeted nanoparticle and 1.5 tesla magnetic resonance imaging. *Cancer Research.* 2003; 63(18):5838–5843.

3

Optimized Energy Bins for K-Edge Imaging Using a Photon Counting Detector

Wang Zhe, Zhang Zhidu, Li Mohan, Wei Cunfeng, and Wei Long

CONTENTS

DOI: 10.1201/9781003218364-3

3.1 Introduction

K-edge imaging plays an important role in clinical examination. It uses the K-edge effect of the contrast agents to highlight the target area and solves the problem in distinguishing the target area and background tissues due to the small difference in attenuation [1, 2].

At present, the K-edge decomposition imaging is commonly realized using traditional dual-energy devices [3, 4]. There are four types of devices for dual-energy imaging: the sequence scan device, the dual-source device [5], the dual-layer detector device [6], and the fast kVp switching device [7]. The sequence scan device requires double exposures, which suffer from the motion artifacts [8]. The dual-source device has a much more complex system and the images obtained have different phases. The dual-layer detector device can obtain images at the same phase in one exposure but its energy resolution performance is relatively weak which goes against the quality of resulting images. The fast kVp switching device has a higher requirement of the imaging system and still has the problem of phase-matching. Besides, the energy mixing of the photons weakens the K-edge signal [9].

In recent years, the development of photon counting detector (PCD) provides a more advanced solution for dual- and multi-energy X-ray imaging [10]. Due to the ability to measure the energy of incident photons and compare it to multiple adjustable thresholds, a PCD can obtain images in several energy bins with a single exposure. Because of its good performance in energy resolution, PCD has a potential to solve the problems that beset the traditional devices and become an important topic in multi-energy imaging research [11, 12].

The energy bins used for the K-edge decomposition imaging based on the PCDs significantly affect the contrast of the processed images [13, 14]. The energy bins decided by the conventional theoretical attenuation method (TAM) are widely used for K-edge decomposition imaging that two energy bins are symmetrical on both sides of the theoretical K-edge position [15, 16]. However, limitation on energy resolution of the PCDs can lead to a distortion of the measured attenuation curve [17], which can also result in deviations in the K-edge detection. Several physical effects of the PCD are responsible for the degradation of energy resolution, including Compton scattering [18], charge sharing [19], pulse pileup [20], and fluorescence emission [21]. Moreover, the continuous X-ray spectrum also flattens the K-edge signal [22]. The weakened K-edge signal can further influence the quality of the decomposition image and increase the amount of the contrast agents.

Therefore, optimizing the energy bins is an effective way to improve the quality of the K-edge decomposition imaging with PCDs. Previous work mostly focused on optimizing the width of the energy bins used for K-edge imaging, He et al decided the energy bin width by signal difference to noise ratio (SDNR) next to the K-edge position [23]. Bo Meng et al used the redescribed signal to noise ratio (SNR) to obtain the energy bin width next to the

K-edge position [24]. Seung-Wan Lee et al carried out simulation work on optimizing the energy bins [25]. However, few studies considered the effects of the degraded energy resolution and the continuous X-ray spectrum on energy bin optimization which distort the attenuation curve and lead to a deviation of the energy bins. To take these negative factors into consideration, Silvia Pani et al selected the energy bins by mapping the spectrum passed through the contrast agents [25].

3.2 K-Edge Imaging

X-ray absorption edge imaging is a special enhanced imaging technique that is normally applied to the K-edge of sample material in the medical imaging domain. Because the attenuation characteristic changes notably in different materials, X-ray photons usually undergo different attenuation levels when they are penetrating different materials, even if they have the same energy. This phenomenon can be used to differentiate the marital type. Some chemical elements, including iodine, barium, gadolinium, etc., have their K-edges in the same energy range as the X-ray for typical medical use. The absorption edge forms a sudden change in the attenuation curve, which is continuous elsewhere. This means if two images are acquired under energy bins on both sides of the sample's absorption edge, the sample area will possess a large gray scale difference, while the background without the absorption edge will have similar gray scale values in these two images. By means of the decomposition algorithm, the background can be removed and the sample area will thus be enhanced.

3.2.1 Methods

Taking iodine and water as examples, the principle of K-edge imaging is shown in Figure 3.1, the theoretical mass attenuation coefficient data of water and iodine are obtained from the National Institute of Standards and Technology (NIST) [26]. The images on the left ($I(\mathbf{r}, E^L)$) and right ($I(\mathbf{r}, E^R)$) sides of the K-edge are represented as Equations 3.1 and 3.2.

$$I\left(\mathbf{r}, E^L\right) = I_0\left(E^L\right) \cdot \exp\left(-\left(\left(\frac{\mu}{\rho}\right)_K \left(E^L\right) \cdot (\rho d)_K(\mathbf{r}) + \left(\frac{\mu}{\rho}\right)_{bg}\left(E^L\right) \cdot (\rho d)_{bg}(\mathbf{r})\right)\right) \quad (3.1)$$

$$I\left(\mathbf{r}, E^R\right) = I_0\left(E^R\right) \cdot \exp\left(-\left(\left(\frac{\mu}{\rho}\right)_K \left(E^R\right) \cdot (\rho d)_K(\mathbf{r}) + \left(\frac{\mu}{\rho}\right)_{bg}\left(E^R\right) \cdot (\rho d)_{bg}(\mathbf{r})\right)\right) \quad (3.2)$$

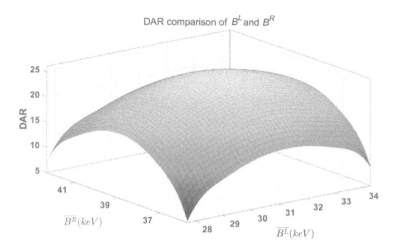

FIGURE 3.1
The mass attenuation coefficient of water and iodine for X-rays.

Where E^L and E^R are the energy bins on the left side and right side of the K-edge respectively, $\mathbf{r}$ represents the pixel position of the detector. $I_0(E^L)$ and $I_0(E^R)$ are the number of incident photons in the energy bins on both sides of the K-edge. $\left(\frac{\mu}{\rho}\right)_K (E^L)$ and $\left(\frac{\mu}{\rho}\right)_K (E^R)$ represent the mass attenuation coefficients on the left and right sides of material's K-edge. $\left(\frac{\mu}{\rho}\right)_{bg} (E^L)$ and $\left(\frac{\mu}{\rho}\right)_{bg} (E^R)$ represent the mass attenuation coefficients of background on the left and right sides of the K-edge. $(\rho d)_K (\mathbf{r})$ and $(\rho d)_{bg} (\mathbf{r})$ are the mass thicknesses of the iodine and the background.

The difference between $\left(\frac{\mu}{\rho}\right)_K (E^L)$ and $\left(\frac{\mu}{\rho}\right)_K (E^R)$ caused by the K-edge effect is much bigger than the difference between $\left(\frac{\mu}{\rho}\right)_{bg} (E^L)$ and $\left(\frac{\mu}{\rho}\right)_{bg} (E^R)$, thus we can realize K-edge imaging with different K-edge algorithms.

3.2.2 K-Edge Subtraction Algorithm (KSA)

The areas containing the contrast agents have a much larger attenuation difference than the background, thus the attenuation difference of the background is approximately zero and the KSA can highlight the area of the contrast agents [27].

The attenuation characteristics $T(\mathbf{r},E^L)$ and $T(\mathbf{r},E^R)$ for $I(\mathbf{r},E^L)$ and $I(\mathbf{r},E^R)$ can be expressed as Equations (3.3) and (3.4) according to Equations (3.1) and (3.2).

$$T\left(\mathbf{r},E^L\right)=\ln\left(I_0\left(E^L\right)\Big/I\left(\mathbf{r},E^L\right)\right)=\left(\frac{\mu}{\rho}\right)_K \left(E^L\right)\cdot(\rho d)_K (\mathbf{r})+\left(\frac{\mu}{\rho}\right)_{bg} \left(E^L\right)\cdot(\rho d)_{bg} (\mathbf{r}) \quad (3.3)$$

$$T\left(\mathbf{r},E^R\right)=\ln\left(I_0\left(E^R\right)\Big/I\left(\mathbf{r},E^R\right)\right)=\left(\frac{\mu}{\rho}\right)_K \left(E^R\right)\cdot(\rho d)_K (\mathbf{r})+\left(\frac{\mu}{\rho}\right)_{bg} \left(E^R\right)\cdot(\rho d)_{bg} (\mathbf{r}) \quad (3.4)$$

Since the difference of the attenuation is quite small for both sides of the K-edge in the background area, there is an approximate relationship $\left(\frac{\mu}{\rho}\right)_{bg}(E^L) \approx \left(\frac{\mu}{\rho}\right)_{bg}(E^R)$ in the KSA. The image of the iodine area can be expressed by:

$$(\rho d)_K^{KSA}(\mathbf{r}) \approx \frac{T(\mathbf{r}, E^R) - T(\mathbf{r}, E^L)}{\left(\frac{\mu}{\rho}\right)_K (E^R) - \left(\frac{\mu}{\rho}\right)_K (E^L)} \tag{3.5}$$

3.2.3 K-Edge Decomposition Algorithm (KDA)

The KSA is inaccurate because it ignores the width of energy bin. To better remove the background areas, a KDA is proposed [28, 29].

In the KDA, the difference of the attenuation of the background between the left and right energy bins is taken into consideration. By solving Equations (3.3) and (3.4) simultaneously, we can get $(\rho d)_K^{KDA}(\mathbf{r})$ and $(\rho d)_{bg}^{KDA}(\mathbf{r})$.

$$(\rho d)_K^{KDA}(\mathbf{r}) = \frac{\left(\frac{\mu}{\rho}\right)_{bg}(E^R) \cdot T(\mathbf{r}, E^L) - \left(\frac{\mu}{\rho}\right)_{bg}(E^L) \cdot T(\mathbf{r}, E^R)}{\left(\frac{\mu}{\rho}\right)_K (E^L) \cdot \left(\frac{\mu}{\rho}\right)_{bg}(E^R) - \left(\frac{\mu}{\rho}\right)_K (E^R) \cdot \left(\frac{\mu}{\rho}\right)_{bg}(E^L)} \tag{3.6}$$

$$(\rho d)_{bg}^{KDA}(\mathbf{r}) = \frac{\left(\frac{\mu}{\rho}\right)_K (E^R) \cdot T(\mathbf{r}, E^L) - \left(\frac{\mu}{\rho}\right)_K (E^L) \cdot T(\mathbf{r}, E^R)}{\left(\frac{\mu}{\rho}\right)_{bg}(E^L) \cdot \left(\frac{\mu}{\rho}\right)_K (E^R) - \left(\frac{\mu}{\rho}\right)_{bg}(E^R) \cdot \left(\frac{\mu}{\rho}\right)_K (E^L)} \tag{3.7}$$

3.3 Experimental Research of the Energy Bins for K-Edge Imaging

In this section, experimental research of the energy bins for the K-edge imaging based on PCDs is performed on phantom and mice with the KSA and KDA. In the phantom study, we use iodine as the material with the K-edge property and water as the background material. In the mice study, the material with the K-edge property is iodine and the background material is muscle. The results of the phantom and mice show that the optimized discrete energy bins determined by the threshold scan method (TSM) can get better K-edge imaging results than the conventional continuous energy bins used in the theoretical attenuation method (TAM).

3.3.1 Energy Bins Optimized

For K-edge imaging, the energy bins determined by the conventional method are shown in Figure 3.1, the left and right bins are symmetrical and continuous at the K-edge position, which is called the TAM. However, the energy bins used in this method do not maximize the difference of the attenuation for iodine. To optimize the energy bins for K-edge imaging, we perform the TSM to get the energy bins, which maximized the difference of the attenuation of the K-edge for the experiments.

The energy bins used in the TSM are selected from the spectrum obtained behind the area of the material which has the K-edge character. Our method to get the spectrum behind iodine is the threshold-scan using the PCD with a 1-keV step, as shown in Figure 3.2. The left and right bins are determined as the points near the K-edge position, where the derivative of the spectral intensity behind the detail of the material with K-edge approaches zero. For the energy bins with a certain width, the left energy bin which maximizes the intensity around the maximum point and the right energy bin which minimizes the intensity around the minimum point are used.

The energy bins used in the TAM are symmetrical to the K-edge and continuous, while the energy bins for TSM are figured out based on Figure 3.2. We use the same energy bin width in both TAM and TSM for comparison. The energy bins used in the TSM and the TAM are listed in Table 3.1.

3.3.2 Experimental Setup

In this section, the performance of TSM and TAM are compared experimentally with the KSA and KDA in phantom and mice.

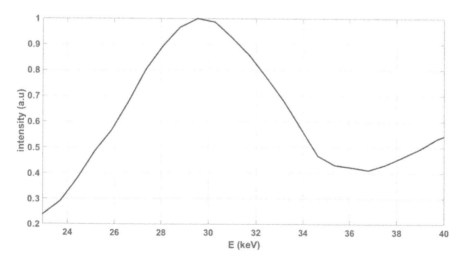

FIGURE 3.2
The spectrum used to identify the energy bins for the TSM.

TABLE 3.1

The Energy Bins for the TSM and the TAM

Method/Bins	Left Bin (keV)	Right Bin (keV)
TAM	28–33	33–38
TSM	27–32	34–39

3.3.2.1 Phantom and Mice

The phantom used in this study is shown in Figure 3.3, which is made of polymethyl methacrylate (PMMA) and filled up with water. There are five cylinders with the diameter of 0.5 cm inside the phantom which are filled up with the iohexol solution. The mass concentration of iodine for each cylinder is 15 mg I/mL, 10 mg I/mL, 5 mg I/mL, 3 mg I/mL, and 1 mg I/mL, respectively.

A 15.6-g weight -mouse is used for the experiment which is anesthetized and 0.2 mL Omnipaque (350 mg I/mL) is injected through the tail vein before the computed tomography (CT) scan.

The image quality is evaluated by the contrast to noise ratio (CNR) of the target area which is defined as Equation (3.8).

$$CNR = \frac{|m_1 - m_2|}{\sqrt{\frac{\left(\sigma_1^2 + \sigma_2^2\right)}{2}}} \tag{3.8}$$

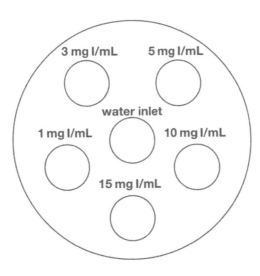

FIGURE 3.3
The phantom for the experiment.

Where m_1 and m_2 are the mean values of the target area and the background, σ_1 and σ_2 in the denominator represent the standard deviation (STD) of the target area and the background, respectively. A larger CNR indicates a better image quality.

3.3.2.2 Imaging System

The multi-energy device used in this study is a Small Animal Spectral CT which is independently developed by the Institute of High Energy Physics, Chinese Academy of Sciences. The PCD (XC-FLITE, XCounter, Sweden) uses a 0.75-mm thick CdTe photoconductor and has 1536×256 pixels with pixel size of 100×100 μm². The PCD has two adjustable energy thresholds. Due to the limitation of the threshold number of the PCD, the multi-energy images can be obtained by two or more scannings to simulate the multi-energy imaging system.

3.3.2.3 Image Acquisition and Reconstruction

The X-ray tube is operated at 80 kVp and the images are acquired at a dose level of 34 mAs for each scan. For the image reconstruction, we implement the KSA and KDA algorithms in the prereconstruction (projection) space. The results of Equations (3.5), (3.6) and (3.7) for each projection are calculated and then put into the Feldkamp-Davis-Kress (FDK) reconstruction algorithm [30] to get the reconstruction images of the phantom and mice. To compare the SNR of the different K-edge images, correction algorithms, such as ring artifacts removal, are not used in the stage of image processing.

3.3.3 Results

3.3.3.1 Phantom Results

The reconstruction images of the phantom are shown in Figure 3.4, and the CNR comparison of different imaging conditions is shown in Figure 3.5.

The phantom reconstruction results in Figure 3.4 show that the energy bins in the TSM can obtain a better image quality of iodine than that in the TAM. The KDA algorithm can better distinguish the iodine areas compared with the KSA algorithm indicating the KDA algorithm is more effective for the K-edge imaging. The water images of Figures 3.4(c) and 3.4(f) also support the results above.

Figure 3.5 shows that the CNRs of the reconstruction results with the energy bins of the TSM are improved compared to the TAM, the average increase of the CNR is ~39%. The CNRs of the reconstruction results with the KDA algorithm are improved compared with the KSA algorithm, the average increase of the CNR is ~43%.

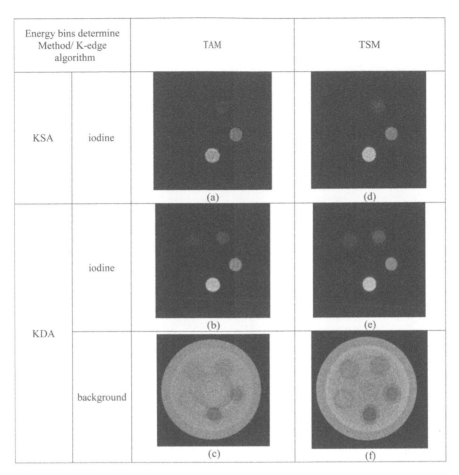

Energy bins determine Method/ K-edge algorithm		TAM	TSM
KSA	iodine	(a)	(d)
KDA	iodine	(b)	(e)
	background	(c)	(f)

FIGURE 3.4
Comparison of the reconstruction images of the phantom with different imaging conditions.

3.3.3.2 Mice Results

The reconstruction images of the mice are shown in Figure 3.6, we choose the kidney as the target area to demonstrate the K-edge imaging results.

The CNR comparison results for the kidney area of the mice are shown in Figure 3.7.

The comparison of Figures 3.6(a) and 3.6(b) as well as Figures 3.6(c) and 3.6(d) show that the KDA has a better performance in K-edge imaging than the KSA. In terms of the energy bins comparison, the results in the right column of Figure 3.6 are better than the left column which indicates that the energy bins determined by the TSM are more effective than the TAM.

The CNR comparison results in Figure 3.7 also provide a proof for the above conclusion. The TSM increases CNR by an average of ~31% compared

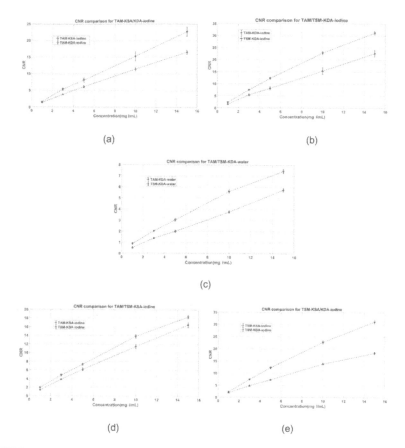

FIGURE 3.5
The CNR results for different imaging conditions. (a) The KSA and the KDA iodine comparison with the TAM energy bins; (b) The TAM and the TSM iodine comparison with the KDA; (c) The TAM and the TSM water comparison with the KDA; (d) The TAM and the TSM iodine comparison with the KSA; (e) The KSA and the KDA iodine comparison with the TSM energy bins.

with the TAM and the KDA increases CNR by an average of ~40.5% compared with the KSA.

3.4 An Optimized K-Edge Signal Extraction Method for K-Edge Imaging

In Section 3.3, the advantages of the TSM over the TAM for the K-edge imaging with the KSA and KDA in phantom and mice were experimentally studied. The results showed that the TSM can obtain better contrast agent images than the TAM in both the phantom and the mice reconstruction images, which proved that the discrete energy bins determined by the TSM are more effective than the continuous energy bins used in the TAM.

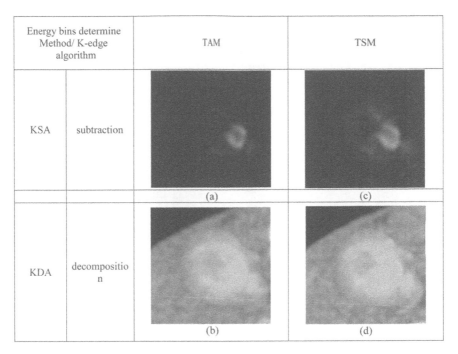

Energy bins determine Method/ K-edge algorithm		TAM	TSM
KSA	subtraction	(a)	(c)
KDA	decompositio n	(b)	(d)

FIGURE 3.6
The kidney reconstruction images of the mice. The upper row ((a) and (c)) is the subtraction images, the lower row ((b) and (d)) is the decomposition images overlaid with iodine (green) and background (gray).

But there are some limitations in the usage of TSM for K-edge imaging. Firstly, TSM selects the energy bins by mapping the spectrum of X-ray that has passed through the contrast agents, which is a complex operation for the practical application. Besides, the quality of the decomposition image is determined by not only the contrast but also the noise level, both widths and locations of energy bins used for K-edge imaging need to be optimized.

In this section, the Gaussian Spectrum Selection Method (GSSM) for multi-energy imaging to get higher-intensity K-edge signals is proposed. It takes the degraded energy resolution and the continuous X-ray spectrum into consideration by modeling the imaging system, and decides the optimal energy bins by energy region optimization constraints proposed in this research. GSSM can obtain both the widths and the locations of the optimized energy bins without the threshold scan process.

3.4.1 Imaging System Modeling

The theoretical mass attenuation curve of the material with K-edge changes to the shape shown in Figure 3.8 under the influence of R_E, which can be estimated using Gaussian convolution [17, 31]. The theoretical mass attenuation coefficient data is acquired from the NIST [27] and is recorded as $Att_0(E)$.

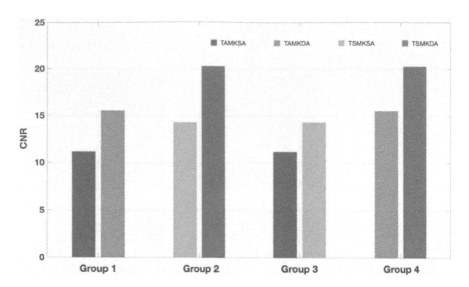

FIGURE 3.7
The CNR comparison results for the kidney area of the mice. The Group 1 is KSA and KDA comparison with TAM. The Group 2 is KSA and KDA comparison with TSM. The Group 3 is TAM and TSM comparison with KSA. The Group 4 is TAM and TSM comparison with KDA.

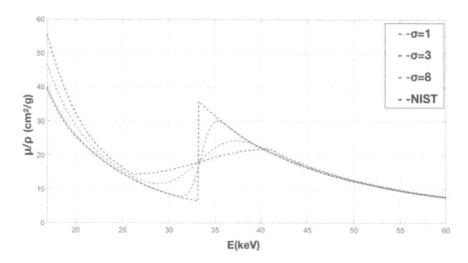

FIGURE 3.8
The theoretical mass attenuation curve and the mass attenuation curve with the influence of σ_E, the solid line is the theoretical mass attenuation curve and the dashed line is the result of the mass attenuation curve with the influence of σ_E.

The spectrum of the X-ray tube, $P_0(E)$, is estimated by the simulation software SpekCalc [32]. The process of Gaussian convolution is described in Equation (3.9):

$$Att_G(E) = Att_0(E) \otimes C(E),$$

$$P_G(E) = P_0(E) \otimes C(E). \tag{3.9}$$

Where

$$C(E) = \frac{1}{\sqrt{2\pi\sigma_E^2}} \exp\left(-\frac{E^2}{2\sigma_E^2}\right),$$

$$\sigma_E = \frac{R_E}{2.355} = \frac{R_0 \cdot E_K \cdot \sqrt{\dfrac{E_0}{E_K}}}{2.355} \tag{3.10}$$

$Att_G(E)$ is the mass attenuation curve after Gaussian convolution and $P_G(E)$ is the spectrum after Gaussian convolution. $C(E)$ is the Gaussian convolution kernel at the corresponding energy, which is determined by R_E. $C(E)$ has a complex form with energy on the overall mass attenuation curve, but at the K-edge position (E_K) it can be calculated as the Equation (3.10). The R_0 in the Equation (3.10) is the energy resolution at the known energy position E_0 and the E_K is the K-edge energy position of the material.

The influence of spectrum is expressed by the equivalent mass attenuation coefficient ($Att_{eq}(B)$) of the energy bin in Equation (3.11) [33], where B represents the corresponding energy bin:

$$Att_{eq}(B) = \frac{\displaystyle\int_B Att_G(E)P_G(E)dE}{\displaystyle\int_B P_G(E)dE} \tag{3.11}$$

We compare the equivalent attenuation coefficient curves of contrast agents derived from this model and TSM. In TSM, the measured attenuation characteristic curves of iodine and gadolinium are mapped by threshold scanning of the detector and converted to the equivalent attenuation coefficient. The comparisons with the results of GSSM are shown in Figure 3.9. The energy of the K-edge extreme points of iodine and gadolinium are shown in Table 3.2.

The results of GSSM are consistent on energy with the measured results, while the errors of the extreme points are less than 3%. There is a difference in Att_{eq} and this condition is more serious at lower energies. The possible reasons for this are Compton scattering and charge sharing, additional photons

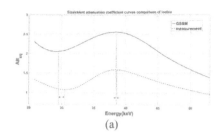
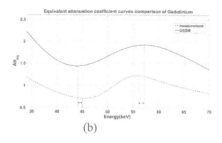

(a)   (b)

FIGURE 3.9

Comparison of the equivalent mass attenuation coefficient curves obtained by GSSM (solid line) and measurement (dashed line). The dash-dotted lines represent the energy positions of the extreme points and the arrows show the errors between GSSM and measurement. (a) Iodine results; (b) Gadolinium results.

at lower energies have been recorded which can result in an overestimation of photon number at low energy level.

3.4.2 Objective Function DAR

The quality of the decomposition image is determined by not only the contrast but also the noise level. This research proposes an objective function, difference of attenuation to relative standard deviation ratio (DAR) to maximize the K-edge signal; the formula of DAR is shown in Equation (3.12):

$$DAR = \frac{\Delta Att_{eq}}{N} \qquad (3.12)$$

Where ΔAtt_{eq} represents the difference of equivalent mass attenuation coefficients between the left and the right energy bins of the K-edge. N represents the total noise level of these two images. They are mutually restrictive: a wide bin can reduce the noise level while weakening ΔAtt_{eq} (shown in Figure 3.10). Conversely, a narrow bin can maintain the ΔAtt_{eq} while increasing the noise level of the image.

TABLE 3.2

The Energy Positions of the K-Edge Extreme Points of Iodine and Gadolinium

Material Method	Iodine			Gadolinium		
	GSSM	Measurement	Error (%)	GSSM	Measurement	Error (%)
Valley energy (keV)	29.51	30.42	2.99	44.16	45.02	1.91
Peak energy (keV)	38.54	38.61	0.18	57.1	56.15	1.69

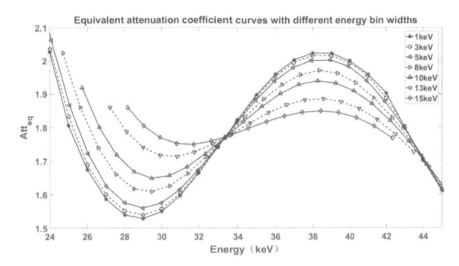

FIGURE 3.10
Equivalent mass attenuation coefficient curves of iodine with different energy bin widths.

The detailed derivation of DAR will not be discussed here. The final DAR expression is shown in Equation (3.13)

$$DAR(B^L, B^R) = \frac{(Att_{eq}(B^R) * (\rho d) - Att_{eq}(B^L) * (\rho d))^2}{\sqrt{\frac{1}{\left(\int_{B^L} P_G(E)dE\right) e^{-Att_{eq}(B^L)*(\rho d)}} + \frac{1}{\left(\int_{B^R} P_G(E)dE\right) e^{-Att_{eq}(B^R)*(\rho d)}}}}. \quad (3.13)$$

Where (ρd) is the mass thickness of the sample, B^L and B^R are the left and the right energy bins of K-edge. $Att_{eq}(B^L)$ and $Att_{eq}(B^R)$ are the equivalent mass attenuation coefficients of B^L and B^R respectively.

The optimal solution $\tilde{B}^L$ and $\tilde{B}^R$ is expressed in Equation (3.14). Since the difference of equivalent mass attenuation coefficients should be large enough in the contrast agents while should be small enough in the background between the two energy bins, the solution space of the optimal energy bins is limited to a very small area:

$$\max DAR\left(\tilde{B}^L, \tilde{B}^R\right),$$

$$\tilde{B}^L \in [E_K - n\sigma_E, E_K],$$

$$\tilde{B}^R \in [E_K, E_K + n\sigma_E]. \quad (3.14)$$

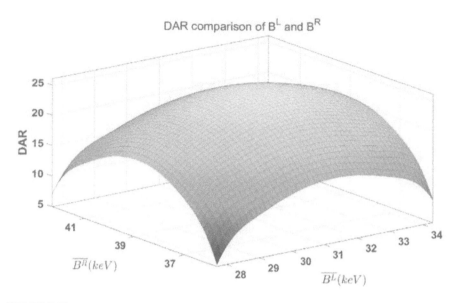

FIGURE 3.11

DAR comparison of B^L and B^R under 10 keV energy bin width, $\tilde{B}^L$ and $\tilde{B}^R$ represent the mid-values of B^L and B^R.

Where E_K is the energy at the K-edge position and n is an empirical constant, $\tilde{B}^L$ and $\tilde{B}^R$ represent the mid-values of B^L and B^R. In principle, the optimal solution of Equation (3.14) should be determined by analyzing the stationary points and the Hessian matrix of DAR. However, the analytic expressions of the derivatives for the DAR used in this research are difficult to get. Since the solution space is limited around the K-edge position, all energy bins in the solution space are calculated to determine the optimal solutions, which are corresponding to the maximum DAR.

For energy bin width of 10 keV in our experiments, corresponding DAR of B^L and B^R are shown in Figure 3.11.

DAR results of $\tilde{B}^L$ and $\tilde{B}^R$ for different energy bin widths shown in Figure 3.12 are computed. The optimal energy bin width is determined as 5 keV.

3.4.3 Experimental Setup

The imaging system used in this research is described in Section 3.3.2.2. The energy resolution of the detector in this system is 22.1% for 59.6 keV, which has been studied with several isotopes in our previous work [34]. The decomposition images of contrast agents obtained by GSSM and TAM with KDA were compared with phantoms. The X-ray tube is operated at 80 kVp and the images are acquired at a dose level of 9 mAs for each scan.

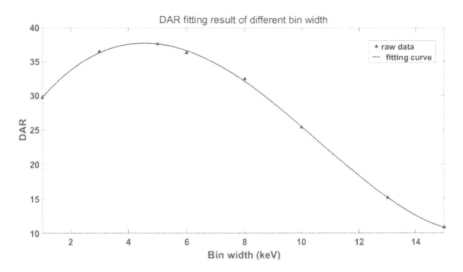

FIGURE 3.12
DAR results of different energy bin widths.

The phantom used in the experiment is made of PMMA as shown in Figure 3.13 with six holes filled with iodine contrast agents with different concentrations. The inner diameter of each hole is 1 mm. In Phantom A shown in Figure 3.13(a), concentrations of the iodine contrast agents are 0 mg I/mL, 20 mg I/mL, 25 mg I/mL, 50 mg I/mL, 75 mg I/mL, and 100 mg I/mL from right to left, respectively. Besides, a comparative experiment is designed by adding a complex background to the phantom, which is made of nylon and shown in Figure 3.13(b). The nylon strips, with a thickness of 1 mm, are used as a distraction of the iodine contrast agents in the phantom (the concentrations from right to left are 10 mg I/mL, 15 mg I/mL, 25 mg I/mL, 50 mg I/mL, 75 mg I/mL, 100 mg I/mL, respectively).

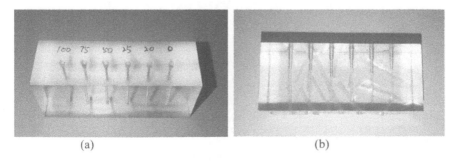

FIGURE 3.13
Phantoms used in the experiments: (a) phantom A, (b) phantom B with background of nylon strips.

3.4.4 Results

3.4.4.1 Imaging with Different Energy Bin Widths

The $\tilde{B}^L$ and $\tilde{B}^R$ for different energy bin widths are used for K-edge imaging to verify the optimal width of the energy bin selected by DAR, which is 5 keV shown in Section 3.4.2. The phantom A shown in Figure 3.13(a) is used in this study. The test widths of energy bin are set to 2 keV, 5 keV, 8 keV, and 13 keV for decomposition imaging. The corresponding experimental results are shown in Figure 3.14.

The experimental results are in good agreement with DAR results. Small energy bin width, such as 2 keV, provides good contrast of the iodine areas, but strong noise in the decomposition image (Figure 3.14(a)), while large energy bin width has a reverse effect (Figure 3.14(d)). The CNR result shows that the 5 keV energy bin width provides the best decomposition image quality which indicates the objective function DAR is effective.

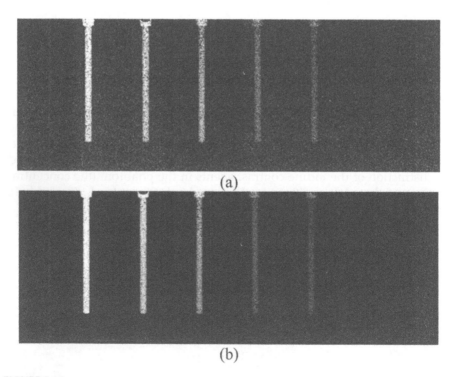

(a)

(b)

FIGURE 3.14

Comparison of GSSM decomposition images with different energy bin widths, (a–d) are decomposition images of 2 keV, 5 keV, 8 keV, and 13 keV, respectively; (e) comparison of CNR values of different concentrations (100 mg I/mL, 75 mg I/mL, 50 mg I/mL, 25 mg I/mL, and 20 mg I/mL from top to bottom) in the decomposition images.

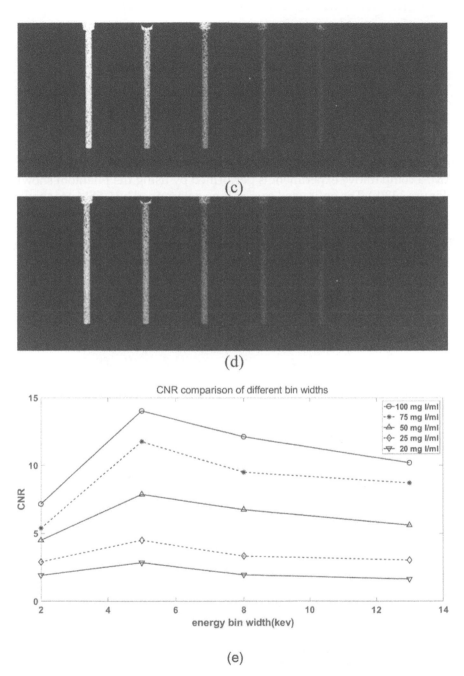

FIGURE 3.14
(*Continued*)

TABLE 3.3

The Energy Bins Used in Comparison Experiment

Method	Left Bin (keV)	Right Bin (keV)
TAM	28–33	33–38
GSSM	26–31	36–41

3.4.4.2 Comparative Experiment

To illustrate the effectiveness and the superiority of GSSM compared with TAM, a comparative experiment is performed by using the phantom shown in Figure 3.13(b). The optimized energy bins determined by GSSM are shown in Table 3.3. The same energy bin widths are used in TAM for comparison.

The X-ray and decomposition images are shown in Figure 3.15(a–c), the CNR quantitative comparison of the iodine areas in the decomposition images is shown in Figure 3.15(d).

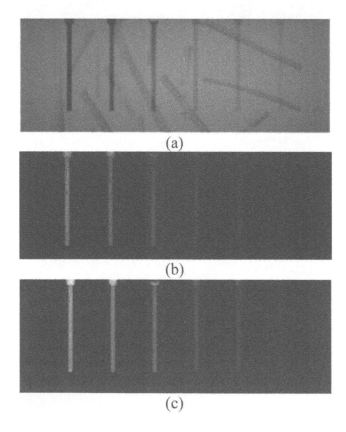

(a)

(b)

(c)

FIGURE 3.15

The results of comparison experiment, (a) X-ray image; (b) TAM decomposition image; (c) GSSM decomposition image; (d) CNR comparison for TAM and GSSM of the comparison experiment.

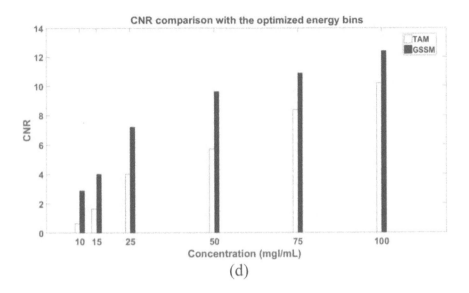

(d)

FIGURE 3.15
(*Continued*)

The normal X-ray image (Figure 3.15(a)) cannot distinguish the iodine areas while the images acquired by TAM and GSSM (Figures 3.15(b) and 3.15(c)) can exclude the interference of the background very well. It shows that the decomposition image acquired by GSSM extracts the K-edge signal of the iodine areas better than TAM. The CNR results are improved in GSSM compared with the TAM image. This improvement is more obvious in low concentration level because the K-edge signal drops in this condition, causing a greater deviation of energy bins and quality degradation of decomposition image.

3.5 Conclusion

The quality of the PCD-based K-edge imaging is largely determined by the strength of detected K-edge signal, which will be further influenced by the energy bins in practice. In this study, we explore the energy bin optimization strategy for the K-edge imaging using a PCD-based spectral CT. The energy bin optimization technique can either increase the image quality under the same K-edge imaging condition or decrease the requirement of contrast agent to get the same imaging result. Especially, by applying the GSSM to optimize the energy bins, we can remarkably improve the CNR of the area with low iodine concentration, meaning a good potential to reduce the usage of contrast agent.

References

1. Lewin JM, Isaacs PK, Vance V and Larke FJ, Dual-energy contrast-enhanced digital subtraction mammography: feasibility, Radiology, 229 (2003) 261–268.
2. Si-Mohamed S, Thivolet A, Bonnot PE et al. Improved peritoneal cavity and abdominal organ imaging using a biphasic contrast agent protocol and spectral photon counting computed tomography K-edge imaging. Invest Radiol., 53 (10) (2018) 629–639.
3. Zhang L, Zhou C, Schoepf U, Sheng H, Wu S, Krazinski A, Silverman J, Meinel F, Zhao Y, Zhang Z and Lu G, Dual-energy CT lung ventilation/perfusion imaging for diagnosing pulmonary embolism, Eur. Radiol., 23 (10) (2013) 2666–2675.
4. Badea C, Johnston S, Qi Y, Ghaghada K and Johnson G, Dual-energy micro-CT imaging for differentiation of iodine- and gold-based nanoparticles, Proc SPIE, 7961 (2011) 79611X.
5. Achenbach S, Anders K and Kalender WA, Dual-source cardiac computed tomography: image quality and dose considerations. Eur. Radiol., 18 (2008) 1188–1198.
6. Panetta D. Advances in x-ray detectors for clinical and preclinical computed tomography[J], Nucl. Instrum. Methods Phys. Res. A, 809 (2016) 2–12.
7. Maturen KE, Kleaveland PA, Kaza RK et al., Aortic endograft surveillance: use of fast-switch kVp dual-energy computed tomography with virtual noncontrast imaging, J. Comput. Assist. Tomogr., 35 (2011) 742–746.
8. Goo HW and Goo JM, Dual-energy CT: new horizon in medical imaging, Korean J. Radiol., 18 (2017) 555–569.
9. Wang X, Meier D, Taguchi K, Wagenaar DJ, Patt BE and Frey EC, Material separation in x-ray CT with energy resolved photon-counting detectors, Med. Phys., 38 (3) (2011) 1534–1546.
10. Brambilla A et al., Fast Cd-Te and Cd-Zn-Te semiconductor detector arrays for spectroscopic x-ray imaging, IEEE T. Nucl. Sci., 60 (2013) 408.
11. Taguchi K and Iwanczyk JS. Vision 20/20: single photon counting x-ray detectors in medical imaging, Med. Phys., 40 (10) (2013) 100901.
12. Schmitzberger FF, Fallenberg EM, Lawaczeck R, Hemmendorff M, Moa E,Danielsson M, Bick U, Diekmann S, Pollinger A, Engelken FJ and Diekmann F, Development of low-dose photon-counting contrast-enhanced tomosynthesis with spectral imaging, Radiology, 259 (2011) 558–564.
13. Shikhaliev PM, Photon counting spectral CT: improved material decomposition with K-edge-filtered x-rays[J]. Phys. Med. Biol., 57 (6) (2012) 1595.
14. Cunningham IA et al, A spatial-frequency dependent quantum accounting diagram and detective quantum efficiency model of signal and noise propagation in cascaded imaging systems, Med. Phys., 21 (1994) 417–427.
15. Anderson NG et al, Spectroscopic (multi-energy) CT distinguishes iodine and barium contrast material in MICE, Eur. Soc. Radiol., 20 (2010) 2126–2134.
16. Li L, Li R, Zhang S and Chen Z, Simultaneous x-ray fluorescence and K-edge CT imaging with photo-counting detectors, Proc. of SPIE, 9967 (2016) 99670F.
17. Ge Y, Zhang R, Li K and Chen GH, K-edge energy-based calibration method for photon counting detectors, Phys. Med. Biol., 63 (2018) 015022.
18. Sossin A, Tabary J, Rebuffel V et al. Influence of scattering on material quantification using multi-energy x-ray imaging, Proc IEEE Nucl. Sci. Symp. Med. Imaging Conf., 0 (2014) 1–5.

19. Kim JC, Anderson SE, Kaye W, Zhang F, Zhu Y, Kaye SJ and He Z, Charge sharing in common-grid pixelated CdZnTe detectors, Nucl. Instrum. Methods Phys. Res. A, 654 (1) (2010), 233–243.
20. Wang AS, Harrison D, Lobastov V and Tkaczyk JE, Pulse pileup statistics for energy discriminating photon counting x-ray detectors, Med. Phys., 38 (2011) 4265–4275.
21. Shikhaliev PM, Fritz SG and Chapman JW, Photon counting multienergy x-ray imaging: effect of the characteristic x-rays on detector performance, Med. Phys., 36 (2009) 5107.
22. Taguchi K, Frey EC and Wang X, An analytical model of the effects of pulse pileup on the energy spectrum recorded by energy resolved photon counting x-ray detectors, Med. Phys., 37 (2010) 3957–3969.
23. Peng He, Wei B, Cong W and Wang G, Optimization of K-edge imaging with spectral CT, Med. Phys., 39 (2012) 6572.
24. Meng B, Cong W, Xi Y and Wang G, Energy window optimization for x-ray K-edge tomographic imaging, IEEE Trans. Biomed. Eng., 63 (8) (2016) 1623–1630.
25. Lee S-W, Choi Y-N, Cho H-M, Lee Y-J, Ryu H-J and Kim H-J, A Monte Carlo simulation study of the effect of energy windows in computed tomography images based on an energy-resolved photon counting detector, Phys. Med. Biol., 57 (2012) 4931–4949.
26. Hubbell JH et al. Tables of x-ray mass attenuation coefficients and mass-energy absorption coefficients. In: Physical Reference Data. NIST Standard Reference Database 126, 1995. Available at http://physics.nist.gov/PhysRefData/XrayMassCoef/cover.html.
27. Pani S. and Saifuddin S, High energy resolution hyperspectral x-ray imaging for low-dose contrast-enhanced digital mammography, IEEE Trans. Med. Imag., 36 (2017) 1784–1795.
28. Roessl E and Proksa R, K-edge imaging in x-ray computed tomography using multi-bin photon counting detectors. Phys. Med. Biol., 52 (2007) 4679–4696.
29. Brambilla A et al., Basis material decomposition method for material discrimination with a new spectrometric x-ray imaging detector, J. Instrum., 12 (2017) P08014.
30. Feldkamp L et al, Practical cone-beam algorithm, J. Opt. Soc. Am. A, 1 (1984) 612–619.
31. Roessl E, Brendel B, Engel K-J, Schlomka J-P, Thran A, Proksa R, Sensitivity of photon-counting based K-edge imaging in x-ray computed tomography, IEEE Trans. Med. Imag., 30 (9) (2011) 1678–1690.
32. Poludniowski G, Landry G, DeBlois F et al. SpekCalc: a program to calculate photon spectra from tungsten anode x-ray tubes[J]. Phys. Med. Biol., 54 (19) (2009) N433–N438.
33. Cheng Z, Li M, Xu Q, Zhang Z, Hu J, Wei C, Wei L, Wang Z, Improved projection-based energy weighting for spectral CT, Radiat. Detect. Technol. Methods, 3 (2019) 28.
34. Li ZH, Yun MK, Jiang XP, Zhang ZD, Cheng ZW, Wei CF, Shi RJ, Wei L, Wang Z, Energy calibration method of the photon counting detector based on continuous x-ray spectrum[J], CT Theor. Appl., 27 (3) (2018) 363–372.

4

Spectral Photon-Counting CT System Based on Si-PM Coupled with Novel Ceramic Scintillators

H. Kiji, T. Toyoda, J. Kataoka, M. Arimoto,
S. Terazawa, S. Shiota, and H. Ikeda

CONTENTS

4.1 Introduction

X-ray computed tomography (CT) is widely used for visualizing internal organs and detecting tumors in the human body. X-ray CT systems irradiate the human body with X-rays propagating from different angles, and three-dimensional images of the human body can be obtained by applying numerical reconstruction algorithms, such as the two-dimensional Fourier transform. In particular, CT exhibits superior ability in the medical diagnosis of the chest (e.g., heart, aorta, bronchi, and lungs) and abdomen (e.g., liver and kidney).

A CT image reflects the distribution of the X-ray attenuation coefficients of the target materials, and the pixel values of the CT image are represented by the CT value. In this study, the CT value is a value relative to the linear attenuation coefficient of water. For example, the CT values of air, water, soft tissue, and bone are −1000, 0, 40–60, and 300–1000, respectively. Because the CT value of soft tissue is similar to that of water, which constitutes the greatest proportion in the human body, clear CT images with low noise are required to distinguish between water and soft tissues. In the diagnosis, contrast agents are injected into the human body to enhance the image contrast of blood vessels and organs. Iodine is used as a contrast agent in the CT scan to facilitate medical diagnosis.

X-ray CT imaging offers some advantages compared with other imaging modalities such as magnetic resonance imaging. First, small tumor tissues can be detected in the early stage owing to the superior spatial resolution (≤1 mm). Second, the short CT scan time (a few minutes) afforded not only reduces patient burden, but can also benefit emergency medical care.

Conventional CT detectors are composed of gadolinium oxysulfide (GOS) scintillators, coupled with photodiodes (PDs). Because the decay time of the GOS scintillator is on the order of a few microseconds, X-ray signals at an extremely high rate (e.g, ~100 MHz/mm^2) will overlap each other. The overlap of the signals results in insufficient information regarding the energy of incident X-ray photons. Consequently, the obtained CT images are monochromatic. Moreover, conventional CT systems require patients to be irradiated with high radiation doses (~10 mSv/scan) to distinguish between water and soft tissues. The radiation dose of X-ray CT is high compared with the annual radiation dose for healthy people (~2.4 mSv). Therefore, the radiation doses for patients who require regular CT scans for medical care must be reduced. In addition, when X-rays pass through a high-density material, the entire X-ray spectrum shifts to the high-energy side. Consequently, CT images contain beam hardening artifacts, which hinder accurate diagnosis. Hence, spectral photon counting CT (SPCCT) has been proposed as the next-generation X-ray CT [1–3].

SPCCT acquires the energy information of individual X-ray photons by obtaining the pulse height of the X-ray photon signal. In addition, the SPCCT system can suppress the effect of dark noise by setting the energy thresholds above the dark noise level; hence, the CT image quality is improved significantly. In SPCCT, clear CT images can be acquired with a low-radiation dose. Using the X-ray photons on the high-energy side, beam hardening artifacts can be suppressed. Furthermore, using the K-edge absorption feature of a material, multiple contrast agents can be identified and their absolute concentrations estimated. Therefore, multicolor CT images afforded in SPCCT provide better medical diagnoses.

The most typically studied systems utilize semiconductor devices, e.g., cadmium telluride (CdTe) or cadmium zinc telluride (CdZnTe). Because semiconductor detectors have fine energy resolutions, semiconductor-based CTs offer better energy discrimination and facilitate material identification easily.

For example, the whole-body SPCCT system (Siemens Healthcare, Forchheim, Germany) is based on a CdTe-based photon-counting detector. It was reported in 2017 that, compared with conventional energy-integrating CT, SPCCT demonstrated approximately 16% lower image noise and 21% higher contrast-to-noise ratio (CNR) of lung nodules *in vivo* [4]. In the same year, Symons et al. demonstrated the potential of photon-counting technology for improving the image quality of carotid and intracranial vessels of the human body [5]. It was reported in 2018 that whole-body SPCCT allowed the simultaneous material decomposition of multiple contrast agents (iodine, gadolinium, and bismuth) *in vivo* [6].

An animal SPCCT system (Philips Research, Haifa, Israel) based on 2-mm-thick CdZnTe sensors has been developed. In 2017, this SPCCT system demonstrated the discrimination of gold and iodine contrast agents in different organs *in vivo* [7]. In 2019, Mohamed demonstrated the possibility of performing simultaneous dual-contrast multiphase liver imaging using SPCCT *in vivo* [8].

Furthermore, the MARS SPCCT system (MARS Bioimaging Ltd., Christchurch, New Zealand) is based on a single CdZnTe assembled with a Medipix3RX chip. Marfo et al. demonstrated mouse imaging with fine energy resolution and high spatial resolution. In the experiment, the radiation dose was 70–90% lower than that of conventional micro CT [9].

However, the widespread use of semiconductor-based SPCCT is hindered by a few challenges. For example, the mobility of electrons and holes in CdTe or CdZnTe is extremely low. To tolerate extremely high X-ray fluxes in the 100 MHz/mm^2, range, the pixel size must be a few 100-μm or less. This easily causes charge sharing among neighboring pixels, thereby degrading CT image quality [1]. In addition, because semiconductor detectors do not have an internal amplification function, a charge-sensitive amplifier and a shaper are required for each channel. Hence, the readout circuit becomes overly complicated for processing fast signals. Furthermore, the development of a CdTe or CdZnTe detector with a large detection area is highly challenging and might hinder the clinical application of semiconductor-based SPCCT. In addition, the exorbitant cost of semiconductor detectors hinders the realization of semiconductor-based SPCCT. Hence, we herein propose a new SPCCT system that uses silicon photomultipliers (Si-PMs)/multipixel photon counters (MPPCs).

4.2 Concept of Si-PM-Based Spectral Photon-Counting CT

4.2.1 Principle of Si-PM/MPPC

An MPPC, which is a type of Si-PM, comprises multiple Geiger-mode avalanche photodiodes (APDs). An APD is a PD that offers an internal amplification function when a reverse voltage is applied. When a reverse voltage is applied to the PN junction, electrons and holes generated in the depletion layer drift to the N- and P-type sides because of the applied electric field, respectively. In addition, these carriers, which are accelerated by the electric field, produce new electron–hole pairs exponentially. This phenomenon is known as avalanche multiplication. Based on this principle, a weak light signal is multiplied and the effect of noise is reduced significantly. Because MPPCs have an extremely large internal gain ($M \sim 10^6$) and a fast temporal response (a few nanoseconds), it offers superior photon-counting capability.

MPPCs offer many advantages when used in CT applications. First, because an MPPC contains an internal signal amplification function, the readout circuit can be constructed simply without using a charge sensitive amplifier or shaper. Second, it can be easily replaced with conventional CT systems because the Si-PM-based system is based on an indirect-conversion-type detector using scintillators, similar to conventional CT. Third, the cost of the Si-PM-based SPCCT system is significantly lower than that of expensive semiconductor detectors.

However, we must consider the temperature characteristics of MPPCs. As the temperature of an MPPC increases, the lattice vibration of the crystal becomes active; subsequently, the probability of collision between carriers and phonons in a crystal increases before the carrier energy becomes sufficiently large. Consequently, the MPPC gain decreases as the temperature of the MPPC increases. This temperature characteristic is a barrier in utilizing MPPCs with SPCCT, in particular under extremely high X-ray intensities. The details and resolutions of this problem are provided in Section 4.3.2.

4.2.2 Comparison of CT Images with PD, APD and MPPC

To demonstrate the usefulness of MPPCs for X-ray CT, we first acquired CT images using a PD, an APD, and an MPPC in the "current" and "pulse" readout modes. The current mode corresponds to a mode in which the average photocurrent from the detector is recorded at a certain integration time, as in conventional CT systems. The X-ray projection data of objects were acquired by reading out the photocurrent values using a source measure unit (Keithley, Model 237) and a personal computer. The exposure time was 0.5 second per pixel. A GOS scintillator was used (Hitachi Metals, Ltd.), as in conventional CT systems. We used photosensor S8664-11 (Hamamatsu Photonics) both as a PD ($M = 1$) and an APD ($M = 50$) by changing the bias voltage. We used an

MPPC of the same size (S12571-010C, Hamamatsu Photonics) operating at $M = 1.35 \times 10^5$.

In the pulse mode, the number of X-ray pulses with different energy thresholds was counted, thereby enabling the acquisition of multicolor X-ray images. To withstand the maximum count rate, we used a single-crystal Ce:YAP scintillator, characterized by a fast decay time, $\tau \sim 25$ ns. The output pulses from the MPPC (S12571-050C) were first amplified using a current–voltage amplifier and then discriminated using multiple comparators with four energy thresholds of 20, 40, 60, and 80 keV. Subsequently, a counter card (Contec, CNT3204MT) was used to record the number of detected X-ray pulses every 0.5 second during CT scans, and the CT images were reconstructed using the projection data of the target phantoms.

We used a 6-cm-diameter cylindrical acrylic phantom filled with water (1.0 g/cm³), including a 2-cm-diameter cylinder filled with alcohol (0.78 g/cm³). The phantom was scanned with an X-ray tube voltage of 120 kV and a tube current of 0.2 mA. The resulting CT images obtained using the PD, APD, and MPPC in the current and pulse modes are shown in Figure 4.1. The images obtained using the APD (Figure 4.1(b)) and MPPC (Figure 4.1(c)) were of higher quality than those obtained using the PD (Figure 4.1(a)); as such, water and alcohol were clearly discriminated. In particular, using the MPPC in the pulse mode, the noise fluctuation for the acquired image reduced substantially compared with that acquired in the current mode, as shown in Figure 4.1(d).

We quantitatively evaluated the CT images acquired using the PD, APD, and MPPC by varying the tube current from 0.1 to 1.0 mA. In these experiments, the tube voltage was fixed at 120 kV. It is noteworthy that the radiation dose increased proportionally with the tube current, and a radiation dose of

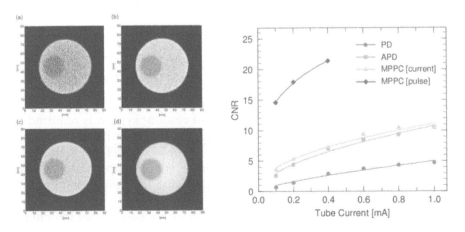

FIGURE 4.1
CT images (left) taken for the X-ray tube voltage of 120 kV with a tube current of 0.2 mA and the CNRs (right) as a function of the tube current for (a) PD, (b) APD, (c) MPPC (current) and (d) MPPC (pulse) (Figure is reconstructed from Figures 4.3 and 4.6 of H. Morita et al.) [10].

4.8 mGy/min was measured for a tube voltage of 120 kV and a tube current of 0.2 mA. We calculated the CNRs between water and alcohol. The CNR is defined as:

$$CNR = \frac{|\mu_M - \mu_B|}{\sigma_B} \tag{4.1}$$

where μ_M and μ_B are the linear attenuation coefficients of the target material and background, respectively, and σ_B is the standard deviation (SD) of the linear attenuation coefficient of the background. The regions of interest, ROI_M and ROI_B, were the regions of the target material (alcohol) and background (water), respectively. The CNR values of the CT images acquired using the APD, MPPC with the current mode, and MPPC with the pulse mode were 3.1, 3.7, and 12.6 times higher than that using the PD, respectively. We discovered that the contrast ratio enhanced when using the APD and MPPC; this implies that the MPPC-based SPCCT system enables low-dose CT. Figure 4.1 (right) shows the variations in the CNRs as a function of the tube current in the 0.1 − 1.0 mA range. By increasing the tube current, the CNR improved because the photon statistical noise reduced. Higher CNRs can be obtained using the APD and MPPC compared with using the PD, owing to internal amplification and suppression of dark currents. In particular, the CNR of the CT image acquired using the MPPC in the pulse mode was much higher than those of the others because the dark noise below the energy threshold was suppressed substantially by reading out the pulse signals only.

4.3 Development of Detector Array for Si-PM-Based SPCCT System

4.3.1 Dedicated Analog and Digital Large-Scale Integrated (LSI) Circuit

To realize a more practical CT system, multichannel CT systems must be developed. Furthermore, a compact electrical readout system is required to implement a high-density detector array. Hence, an LSI circuit dedicated to an SPCCT system is a key technology for enabling multichannel and high-speed analog and digital signal processing. Therefore, we developed a 16-channel MPPC-based SPCCT system (prototype) [11–13] and then expanded it into a 64-channel system (enhanced type) [14].

Figure 4.2 shows a brief schematic diagram of the 64-channel LSI circuit [15]. A pulse current I from the MPPC approaches the resistor R via a current conveyor and then converted to a voltage signal ($V = IR$). The output voltage signal is processed using normal comparators with six threshold levels (E_{th1}, E_{th2}, E_{th3}, E_{th4}, E_{th5}, and E_{th6}) and then converted to a digital output in

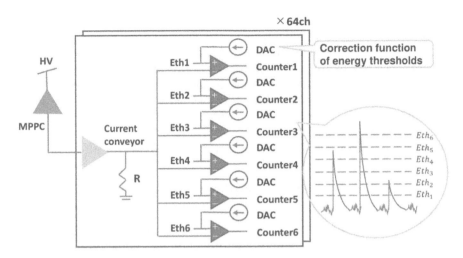

FIGURE 4.2
The schematic processing diagram of the developed 64-channel LSI [15].

six counters that provide the number of detected pulse counts. The pulse numbers with the six thresholds provide information regarding the X-ray photon energy, enabling multicolor X-ray CT to be performed. To correct the variations in the pedestal voltage level for each channel caused by impurities in the manufacturing process, we implemented a variable current source, whereby the current value for each channel was determined by the control registers in each channel. For analog processing, the LSI circuit must capture the pulse current from the MPPC promptly (i.e., within tens of nanoseconds) to preserve the original pulse shape. However, because MPPCs have a large detector capacitance (~100 pF), ordinary preamplifiers with extremely high input impedances cannot be used. To obtain a fast response from the input amplifier, we used a current conveyor. The current conveyor had a considerably low input impedance (~10Ω) and a significantly high output impedance (~10 kΩ). Its superior performance provided a high-speed response within tens of nanoseconds and a large voltage output. The output voltage was obtained via a resistor R, whose value can be set as 0.25, 0.5, or 1 kΩ. In this study, we used $R = 1 k\Omega$, which is suitable for the dynamic range determined by the light yield of the scintillator and the gain of the MPPC.

The 64-channel LSI circuit affords three major improvements compared with the previous 16-channel LSI circuit. First, as the number of channels increases from 16 to 64, we can obtain a wider image using a larger phantom (~60 mm). Second, the 64-channel LSI circuit has a gain correction capability to compensate for variations in the MPPC gain. Furthermore, it comprises digital-to-analog converters for setting precise threshold voltages to compensate for the intrinsic pedestal variations. Owing to the correction capability, the channel-by-channel deviation reduced from 12 mV (~3 keV) to

2 mV (~0.5 keV) [14]. Hence, the energy information acquisition for the 64-channel SPCCT improved substantially compared with that for the 16-channel SPCCT. Generally, a small energy uncertainty can affect the quality of CT images when setting the energy thresholds near the K-edge energy of contrast agents. The accuracy of estimating concentrations is improved when deviations in threshold energies are reduced [14]. Finally, when the number of energy thresholds increases from four to six and the available energy information increases, more reliable identification of phantom materials and more accurate estimation of contrast agent concentrations are afforded.

4.3.2 64-Channel MPPC Array Formed on Ceramic Substrate

Our SPCCT system encountered a problem under high-intensity X-ray beams, i.e., the gain of the MPPC decreased as its temperature increased owing to Joule heat generated by the signal current. This was primarily caused by the insufficient heat path of the MPPC formed on a glass–epoxy substrate. Hence, we developed a new 64-channel MPPC array formed on a ceramic substrate, as shown in Figure 4.3 (left). Typically, glass–epoxy has been used as the substrate of MPPCs and many electrical devices; however, it does not facilitate heat dissipation from the MPPC to the external environment. Ceramic offers better heat dissipation than glass–epoxy and is a suitable substrate material for MPPCs for counting X-ray photons in an extremely high X-ray intensity. In addition, the operation voltage of new MPPCs (pixel pitch: 15 μm) is typically 40 V, which is 10–20 V lower than that of conventional MPPCs (pixel pitch: 25 μm). Therefore, the Joule heat in each Geiger pixel in the MPPC arrays is lower.

First, we tested whether the ceramic MPPC array improved the gain variation owing to Joule heat. We irradiated the ceramic MPPC array and the glass–epoxy MPPC array with X-rays (150 kV; 0.1, 1, 2, and 3 mA) for 3 minutes

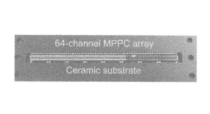 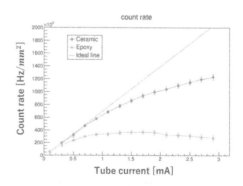

FIGURE 4.3
(Left) 64-channel MPPC array formed on a ceramic substrate, (Right) X-ray count rate over 80 keV in the glass-epoxy and ceramic MPPCs. Green and red points indicate the results of the conventional glass-epoxy and ceramic MPPCs, respectively. The blue line denotes an ideal line.

each and then compared their current changes. Immediately after the X-ray generator was turned on, the output currents decreased significantly owing to a substantial decrease in the MPPC gain due to the Joule heat generated in each MPPC pixel. This trend became more significant with increasing X-ray beam intensity. Specifically, the gain decrease rates of the glass–epoxy MPPC array were 16% and 23% at 1 and 3 mA, respectively. However, the gain decrease rates of the ceramic MPPC array were 3.4% and 7.1% at 1 and 3 mA, respectively. Therefore, we confirmed that the ceramic MPPC array inhibited the gain decrease under high-intensity X-rays.

Subsequently, we evaluated the rate tolerance of the ceramic and glass–epoxy MPPC arrays. We set a tube voltage of 150 kV and increased the tube current from 0.1 to 2.9 mA. We set the energy threshold at 80 keV and measured the X-ray photon counts above 80 keV. As shown in Figure 4.3 (right), the X-ray counts in the glass–epoxy MPPC saturated at ~1 mA and began declining from ~2 mA because of the gain decrease caused by Joule heat. However, the ceramic MPPC significantly improved the X-ray rate tolerance by suppressing the gain decrease. Specifically, the X-ray rate tolerance exceeding 80 keV strengthened by ~1 MHz/mm² at 2.9 mA (approximately six times higher than that of glass–epoxy MPPCs).

4.4 Imaging of Contrast Agents via SPCCT

4.4.1 Experimental Setup

The experimental setup of our SPCCT system for X-ray CT imaging is presented in Figure 4.4. The detector part comprised a 64-channel ceramic scintillator array coupled with a 64-channel MPPC array, in which yttrium–gadolinium–aluminum–gallium garnet scintillators (decay time ~70 ns, $1 \times 1 \times 1$ mm³, Hitachi Metals Ltd.) were used. A 1-mm-thick aluminum filter

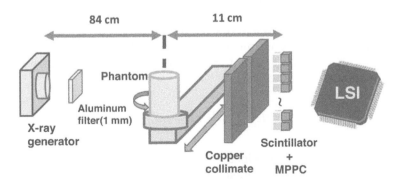

FIGURE 4.4
Experimental setup of the 64-channel MPPC-based SPCCT system [14].

was placed on the X-ray generator's output port to reduce low-energy X-ray contamination, and a 1-cm-thick copper collimator was placed in front of the detector array.

The settings of the imaging function using the 64-channel LSI were controlled using LabVIEW and LabVIEW-FPGA (National Instruments), where the setting parameters were the exposure times, threshold voltages, voltage corrections of the signal pedestal levels, etc. The SPCCT system acquired the projection images from 0° to 180° at a pitch angle of 4°. The projection images obtained were reconstructed using the filtered back projection with a rump filter.

4.4.2 K-Edge Imaging of Iodine and Gadolinium

Conventional CT is used to measure temporal changes in CT values of contrast agents as well as to detect tumor tissues in the human body. In this regard, multiple scanning of the liver is required, resulting in long radiation exposure time in conventional CTs. However, in regard to SPCCT, it has been suggested that the number of liver scans can be reduced by administering multiple contrast agents at different timings and then distinguishing them [7]. To achieve accurate medical diagnosis and reduce the radiation dose, we simultaneously performed the CT imaging of iodine and gadolinium using the MPPC-based SPCCT. The K-edge energies of iodine and gadolinium were 33.2 and 50.2 keV, respectively. We set the energy thresholds near the K-edge energies and attempted to identify two different contrast agents using the CT values in each energy band. We set pre-optimized energy thresholds (23, 31, 49, 70, 85, and 105 keV), a tube voltage of 120 kV, a tube current of 0.7 mA, and an exposure time of 0.5 second. We used a 10-cm-diameter cylinder phantom, including the iodine and gadolinium phantoms with different concentrations (2, 4, 6, 8, mg/mL) placed in a circular manner. The X-ray count rate correction was applied to reconstruct the CT images in the high-intensity X-ray beam.

Herein, we present a procedure for reconstructing material images and estimating the concentrations of contrast agents. As shown in Figure 4.5, when we set six energy thresholds in the LSI circuit, we were able to acquire six CT images with different energy windows. Subsequently, we estimated the absolute concentrations (mg/mL) of the contrast agents by applying K-edge imaging and the least-squares method. The equation of the least-squares method is expressed as:

$$J = \sum_{E} (CT_E - \rho \, Cal_E)^2 \tag{4.2}$$

where CT_E and ρ are the obtained CT value and concentration of the contrast agents, respectively. Cal_E denotes the calibration term acquired by the pretaken calibration phantom. We calculated the concentrations of the contrast agents in each pixel by combining the CT values with the acquired calibration

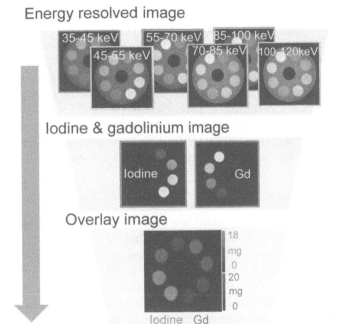

FIGURE 4.5
Procedure to make the concentration images of iodine and gadolinium [14].

term in the six energy windows. Hence, we were able to obtain concentration images of contrast agents such as iodine and gadolinium. Finally, we overlaid the iodine and gadolinium images using *ImageJ* software [16].

Low concentrations of iodine and gadolinium were identified and imaged in color, as shown in Figure 4.6. For the prepared concentrations of 2, 4, 6, and 8 mg/mL of iodine, the estimated concentrations were 1.5 ± 0.34, 4.8 ± 0.18,

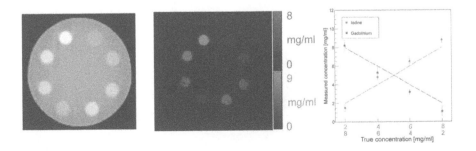

FIGURE 4.6
(Left) The mimicked conventional CT image using energies above 23 keV, (Middle) Overlaid image consisting of the iodine (red) and gadolinium (blue) phantoms, (Right) Estimated concentrations of iodine and gadolinium. Red and blue points denote iodine and gadolinium, respectively. The dotted lines denote the true values.

6.5 ± 0.31, 8.8 ± 0.29 mg/mL, respectively. For the prepared concentrations of 2, 4, 6, and 8 mg/mL of gadolinium, the estimated concentrations were 1.1 ± 0.52, 3.2 ± 0.23, 5.3 ± 0.76, 8.2 ± 0.55 mg/mL, respectively. This result shows that the MPPC-based SPCCT can identify multiple contrast agents and estimate their concentrations, although it yields a poorer energy resolution than those of semiconductor detectors, indicating the high potential of the MPPC-based SPCCT system to reduce the number of CT scans when administering multiple contrast agents for medical diagnosis. In the future, we will strive to improve the estimation accuracy on the low-concentration side by improving the image quality and the concentration estimation algorithm.

4.4.3 Mixed Phantom

When two different contrast agents are mixed in the human body, the absolute concentrations of each contrast agent must be measured for an accurate visualization. Hence, we performed the imaging of phantoms comprising a mixture of iodine and gadolinium at different concentrations. We used an 8-cm-diameter phantom filled with water, including two 2-cm-diameter phantoms (#1 and #2): Phantom #1 comprised 2.5 mg/mL of iodine and 5.0 mg/mL of gadolinium, whereas phantom #2 comprised 5.0 mg/mL iodine and 2.5 mg/mL of gadolinium. The measurement conditions were the same as those provided in Section 4.4.1.

First, we reconstructed a CT image that mimics a conventional CT image using X-ray photons in all energy bands above 23 keV (Figure 4.7). Despite the mixed proportions of the contrast agents, the two phantoms were similar to each other, as shown in Figure 4.7 (left top). The average CT values of phantoms #1 and #2 were 656 ± 17 and 601 ± 16 HU, respectively. The

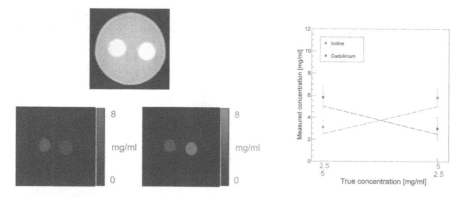

FIGURE 4.7
(Left) The mimicked conventional CT image using energies above 23 keV, iodine image (red) and gadolinium image (blue). The left phantom #1 consists of 2.5-mg/mL iodine and 5.0-mg/ mL gadolinium, and the right phantom #2 consists of 5.0-mg/mL iodine and 2.5-mg/mL gadolinium. (Right) True concentration (dotted lines) versus estimated concentration of the contrast agents of iodine and gadolinium.

difference between the two phantoms could not be identified based on only those values in the monochromatic case.

By assuming that the phantom included a mixture of iodine and gadolinium, the concentration images of iodine and gadolinium were obtained, as shown in Figure 4.7 (left bottom). For the prepared concentrations of 2.5 and 5.0 mg/mL of iodine, the estimated concentrations were 3.1 ± 0.67 and 5.8 ± 0.78 mg/mL, respectively. For the prepared concentrations with 2.5 and 5.0 mg/mL of gadolinium, the estimated concentrations were 3.0 ± 1.0 and 5.8 ± 1.1 mg/mL, respectively. The results show that the estimated concentrations were consistent with the true values within the 1-σ uncertainty, even when two different contrast agents were mixed. However, the uncertainties of the estimated concentrations were larger than those for the unmixed phantom. When calculating the concentrations of iodine and gadolinium, we derived two unknown parameters (i.e., concentrations of iodine and gadolinium). The least-squares formula for the mixed phantom had a smaller number of degrees of freedom by one compared with that of the unmixed phantom; hence, the uncertainty would be larger.

4.4.4 Comparison of Different X-Ray Intensities

In general, CT image quality improves as the X-ray photon statistics increase. Therefore, a clear image can be obtained by increasing the X-ray tube current. As mentioned in Section 4.3.2, our SPCCT system can suppress variations in the MPPC gain under an extremely high X-ray intensity when a ceramic MPPC array with excellent heat dissipation is used. In this section, we demonstrate the reduction in image noise by increasing the X-ray tube current.

To evaluate the image quality based on tube currents, we calculated the SD of the CT values and the CNR as quantitative indexes based on the CT image obtained. We set the ROIs for the background (water) and target object (iodine 10 mg/mL) in the CT image of the 35–45 keV band. It is noteworthy that SD represents the deviation of each pixel value in the background ROI.

We set the tube voltage to 120 kV, tube current to 0.1–1.0 mA, and exposure time to 0.4 second. Figure 4.8 shows the CNR and SD values for different tube currents in the 35–45 keV band. As shown in the plots, the CNR improved as the tube current was increased because the SD value (i.e., image noise) was reduced. In addition, the SD values obtained at 0.1 and 1.0 mA were 89.8 and 32.1, respectively (i.e., the SD became ~$10^{-1/2}$ times smaller when the number of X-ray photon was 10 times higher). This result is consistent with the Poisson theorem, i.e., the statistical uncertainty is inversely proportional to the square of the photon statistics.

By improvement of the CNR by increase of the tube current in this demonstration, we can distinguish between water and a low-contrast material (e.g., soft tissues or contrast agents with a low concentration). However, this

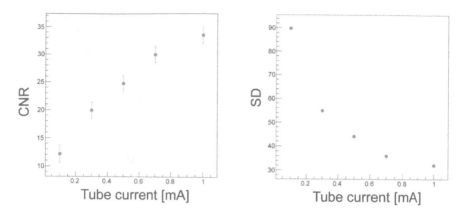

FIGURE 4.8
The obtained CNRs (left) and SDs (right) as a function of the tube current with the ceramic MPPC array. The tube voltage is 120 kV and the exposure time is 0.4 second.

result does not include any concern about radiation dose. Hence, we will investigate the tradeoff between radiation dose and CT image quality, which is required in clinical use of SPCCT.

4.5 New Application for Future Medicine

4.5.1 Visualization of Drug Delivery System

Two types of cancer treatment exist: surgery and chemotherapy. In chemotherapy, anticancer drugs are continuously administered to prevent cancer growth and/or reduce tumor size. Anticancer treatment typically impose a systemic effect to the entire human body. Additionally, the treatment is effective for extremely small cancers that cannot be confirmed via medical examination. Hence, the risk of cancer recurrence can be reduced via anticancer treatment. However, anticancer drugs damage normal cells and often cause serious side effects. Therefore, new cancer treatments using drug delivery systems (DDSs) have garnered significant attention [17].

DDSs are a technology that delivers the minimum amount of medicine to the target location. By encapsulating the anticancer drug in a liposome capsule and releasing the anticancer drug near the tumor, damage to normal cells caused by the anticancer drug can be reduced and side effects on the human body can be minimized. In a DDS, anticancer agents can be delivered to only tumor cells selectively owing to the enhanced-permeation-and-retention (EPR) effect. The EPR effect is based on the concept that because a tumor tissue contains gaps of ~200 nm among vascular endothelial cells whereas healthy vascular endothelial cells have no gaps, nanoparticles can

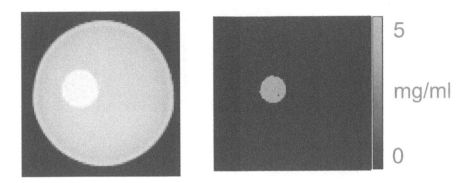

FIGURE 4.9
(Left) The mimicked conventional CT image using energies above 23 keV with a phantom of water and AuNPs, (Right) The obtained density map of 5-mg/ml AuNPs.

be selectively accumulated in the tumor tissue. Currently, DDS research is being actively conducted; however, there is no way to confirm whether the anticancer drug leaks out while it approaches the tumor and whether the anticancer drug is accumulated well in the tumor. Hence, liposomes as capsules are modified with contrast agents as a possible solution.

In this study, gold nanoparticles (AuNPs) were used for DDSs. AuNPs measuring tens of nanometers were modified for the targeted delivery of anticancer agents. Moreover, the K-edge energy of AuNPs is 80.7 keV, which is higher than those of iodine (33.2 keV) and gadolinium (50.2 keV). In SPCCT, differences in K-edge energies enable material decomposition even if they are simultaneously administered in the human body. In other words, by modifying the liposomes with iodine contrast agents and AuNPs with anticancer agents, as shown in Figure 4.10 (left), the *in vivo* kinetics of liposomes and anticancer agents can be tracked simultaneously using our SPCCT system.

First, we attempted to visualize only AuNPs using the MPPC-based SPCCT. We used an 8-cm-diameter phantom filled with water, including a 2-cm diameter phantom with 5 mg/mL of AuNPs. We set the tube voltage to 140 kV, tube current to 0.7 mA, and exposure time to 0.5 second. Consequently, we successfully identified AuNPs and water, as shown in Figure 4.9, and reconstructed the AuNP image by applying the material decomposition algorithm mentioned in Section 4.4.2. The estimated concentration of AuNPs was 4.8 ± 0.34 mg/mL with a high accuracy (the estimated relative error was 4%).

Subsequently, for DDS monitoring, we attempted the identification and simultaneous imaging of iodine and AuNPs using the MPPC-based SPCCT. We used an 8-cm-diameter phantom filled with water, including two 2-cm-diameter phantoms (5 mg/mL iodine and 5 mg/mL AuNPs). As shown in Figure 4.10 (right), iodine and AuNPs can be correctly identified and imaged in color, whereas iodine and AuNPs could not be distinguished

FIGURE 4.10
(Left) The schematic diagram of AuNPs encapsulated in liposomes, (Middle) The mimicked conventional CT image using energies above 23 keV, (Right) Overlaid image consisting of the iodine (red) and AuNP (green) phantoms.

in monochromatic images. The estimated concentrations were calculated by setting the ROIs in the center of each phantom to be 5.4 ± 0.1 mg/mL for iodine and 4.5 ± 0.2 mg/mL for AuNPs. The estimated relative errors of iodine and AuNPs were 8% and 10%, respectively, implying that we performed material decomposition imaging successfully and with high accuracy using our SPCCT system. In the future, we will attempt the CT imaging of low-concentration AuNPs to reduce burden on the human body, as well as the multicolor imaging of a mixed phantom comprising iodine and AuNPs, assuming a realistic condition for clinical applications.

4.5.2 Application of Machine Learning for Further Image Quality Improvement

The image quality of the AuNP phantom can be worse than that of iodine or gadolinium because AuNPs have the K-edge feature on the higher-energy side (80.7 keV) compared with iodine (33.2 keV) and gadolinium (50.2 keV), and the photon statistics of AuNPs for material decomposition are relatively low. In addition, the use of X-ray data in the narrow energy band ($\delta E = 10 - 20$keV) for multicolor imaging results in image deterioration. Therefore, to achieve accurate DDS monitoring using low-concentration materials, the image quality of SPCCT images must be improved.

Recently, machine-learning (ML) theory has been developed rapidly worldwide, particularly for medical image processing. For example, dictionary learning [18] is one of the ML techniques that has successfully improved the quality of CT images [19–21]. Furthermore, research pertaining to deep-learning-based models has progressed, and they have been applied to medical imaging in recent years [22]. U-Net is a typical deep-learning model that is widely used in medical imaging [23, 24]. Noise2Noise [25], which is another deep-learning model that can generate clean images by only viewing corrupted data, has recently garnered attention for improving image quality. In our study, we applied three types of ML models, i.e., dictionary learning, U-Net, and Noise2Noise to our SPCCT images [26].

We used an 8-cm-diameter cylinder phantom, including the iodine and AuNP phantoms with different concentrations (10 and 20 mg/mL). We confirmed the improvement in the estimation error by applying ML techniques. Regarding the deep-learning models, the uncertainties of concentration estimation were reduced successfully from 9.1 ± 4.2 to 9.7 ± 0.6 mg/mL for U-Net and to 9.4 ± 0.4 mg/mL for Noise2Noise using a 10-mg/mL AuNP phantom [26]. Currently, for clinical applications of DDS monitoring, our research group is targeting the imaging of mice administered with multiple contrast agents using our SPCCT system. In an actual DDS, the concentrations of iodine and AuNPs are typically ~0.1 mg/mL or less. Hence, we will extend the application of ML models to low-concentration agents and use these ML techniques to improve *in vivo* SPCCT images to obtain more precise material decomposition and concentration information.

4.6 Conclusion

SPCCT is a next-generation X-ray CT that can solve the problems of conventional CT, such as high radiation dose and lack of energy information. Whereas a semiconductor-based SPCCT system has yielded excellent results in *in vivo* imaging, its widespread use is hindered by the detector size and cost. Hence, we herein proposed an MPPC-based SPCCT system to realize an SPCCT system that is low in cost and simple in system configuration.

First, CT images acquired using a PD, an APD, and an MPPC were compared. The results demonstrated that the CNR of the CT image obtained using the MPPC was ~10 times better than obtained using the PD, indicating that the radiation dose for SPCCT can be reduced compared with that of conventional CT. Next, for more realistic CT imaging, we developed a 64-channel MPPC-based SPCCT system. The system reconstructed the color images of iodine and gadolinium, performed material decomposition, and estimated the concentrations. In addition, AuNPs, which are garnering significant attention as the next-generation DDS, were successfully visualized through our SPCCT system. Furthermore, the image quality of AuNPs has been improved by applying several machine-learning techniques.

The technical difficulty of SPCCT is evident from the fact that SPCCT has not been widely used worldwide, although more than 20 years have passed since SPCCT was first proposed. We believe that MPPC-based SPCCT is highly promising for practical applications and the widespread use of CT equipment. To realize SPCCT, both semiconductor- and MPPC-based SPCCT systems must be further improved. We will continue to develop MPPC-based SPCCT for clinical applications.

Acknowledgments

This work was supported by JSPS KAKENHI Grant Number JP15H05720, JP20H00669, JP19H04483, JST ERTO-FS Grant Number JPMJER 1905, the Naito Foundation, the Uehara Memorial Foundation, the Casio Science Promotion Foundation, the Key Researchers Development Program at Waseda University, and the JSPS Leading Initiative for Excellent Young Researchers program.

References

1. Martin J. Willemink, Photon counting CT: Technical principles and clinical prospects, Radiology. 2018; 289: 293–312.
2. Katsuyuki Taguchi, Vision 20/20: Single photon counting x-ray detectors in medical imaging, Med. Phys. October 2013; 40(10): 100901.
3. Salim Si-Mohamed, Review of an initial experience with an experimental spectral photon-counting computed tomography system, NIM-A. 21 November 2017; 873: 27–35.
4. Rolf Symons, Feasibility of dose-reduced chest CT with photon-counting detectors: Initial results in humans, Radiology. December 2017; 285(3): 980–989.
5. Rolf Symons et al, Photon-counting CT for vascular imaging of the head and neck: First in vivo human results, Invest. Radiol. March 2019; 53(3): 135–142.
6. Rolf Symons et al, Photon-counting CT for simultaneous imaging of multiple contrast agents in the abdomen: An in vivo study, Med. Phys. October 2017; 44(10): 5120–5127.
7. David P. Cormode et al, Multicolor spectral photon-counting computed tomography: In vivo dual contrast imaging with a high count rate scanner, Sci. Rep. 2017; 4784(1): 4784.
8. S. Mohamed et al, Spectral Photon-Counting Computed Tomography (SPCCT): In-vivo single-acquisition multiphase liver imaging with a dual contrast agent protocol, Scientific Reports, 2019; 8458.
9. Emmanuel Marfo et al, Assessment of material identification errors, image quality and radiation doses using small animal spectral photon-counting CT, IEEE. June 2020; 10.1109/TRPMS.2020.3003260
10. H Morita et al, Novel photon-counting low-dose computed tomography using a multi-pixel photon counter, Nucl. Instr. Meth. A. 2017; 857: 58–65.
11. M. Arimoto et al, Development of LSI for a new kind of photon-counting computed tomography using multipixel photon counters, Nucl. Instr. Meth. A. 2017; 912: 186–190.
12. T Maruhashi, et al, Evaluation of a novel photon-counting CT system using a 16-channel MPPC array for multicolor 3-D imaging, Nucl. Instr. Meth. A. 2018; 936: 5–9.
13. T Maruhashi et al, Demonstration of multiple contrast agent imaging for the next generation color X-ray CT, Nucl. Instr. Meth. A. 2020; 958: 162801.

14. Hiroaki Kiji et al, 64-channel photon-counting computed tomography using a new MPPC-CT, NIM-A. 21 December 2020; 984: 164610.
15. M Arimoto, et al, in prep.
16. M D Abramoff et al, Image processing with imageJ, Biophotonics Int. 2004; 11(7): 36–42.
17. S Thambiraj et al, An overview on applications of gold nanoparticle for early diagnosis and targeted drug delivery to prostate cancer, Recent Pat. Nanotechnol. 2018; 12(2): 110–131.
18. T Ivana and F Pascal, Dictionary learning IEEE sig, Proc. Maga. 2011; 28(2): 27–38.
19. D Karimi and R Ward, Reducing streak artifacts in computed tomography via sparse representation in coupled dictionaries, Med. Phys. 2016; 43: 1473–1486.
20. Z Zhiqin, C Yi, Y Hongpeng, L Yanxia and L Zhaodong, A novel dictionary learning approach for multi-modality medical image fusion, Neurocomputing. 2016; 214: 471–482.
21. A Oyama, S Kumagai, N Arai, T Takata, Y Saikawa, K Shiraishi, T Kobayashi and J Kotoku, Image quality improvement in cone-beam CT using the super-resolution technique, J. Radi. Res. 2018; 59(4): 501–510.
22. L Geert, K Thijs, E B Babak, A A S Arnaud, C Francesco, G Mohsen, Jeroen A W M van der Leek, van G Bram and I S Clara, A survey on deep learning in medical image analysis, Med. Image Anal. 2017; 42: 60–88.
23. H Dong, G Yang, F Liu, Y Mo and Y Guo, Automatic brain tumor detection and segmentation using U-net based fully convolutional networks, Med. Img. Unde. Anal. 2017; 506–517.
24. B Stephen et al, Deep segmentation networks predict survival of non-small cell lung cancer. Sci. Repo. 2019; 9: 17286.
25. J Lehtinen, J Munkberg, J Hasselgren, S Laine, T Karras, M Aittala and T Aila, Noise2Noise: Learning image restoration without clean data, ICML. 2018; 74: 4620–4631.
26. Takaya Toyoda et al, Application of machine-learning models to improve the image quality of photon-counting CT images. JINST. 2021; 16: 05021.

5

Spectral X-Ray Computed Micro Tomography: 3-Dimensional Chemical Imaging by Using a Pixelated Semiconductor Detector

Jonathan Sittner, Margarita Merkulova, Jose R. A. Godinho, Axel
D. Renno, Veerle Cnudde, Marijn Boone, Thomas De Schryver,
Denis Van Loo, Antti Roine, Jussi Liipo, and Bradley M. Guy

CONTENTS

5.1 Introduction

The accurate characterization of materials is indispensable in many fields such as geology, materials science and archeology (Cnudde et al., 2012; Freeth et al., 2006; Maire & Withers, 2014). Nevertheless, most characterization methods are limited to 2D, which implies the loss of information. Techniques such as scanning electron microscopy (SEM), electron probe microanalysis

DOI: 10.1201/9781003218364-5

(EPMA), or transmission electron microscopy (TEM) are commonly used in numerous application (Almgren et al., 2000; Bachmann et al., 2017, 2018; Findlay et al., 2009; Liipo et al., 2019; Mori & Yamada, 2007; Rinaldi & Llovet, 2015; Urban, 2008). However, usually, the investigated features of a sample are 3-dimensional (3D), thus a complete characterization would only be possible using a 3D technique.

The most common method for 3D investigations is X-ray computed tomography (micro-CT) that has been used for many years in medical science (Conroy & Vannier, 1987; Gawler et al., 1974), industrial inspection (Boone et al., 2014), paleontology (Boespflug et al., 1995), or geology (Cnudde & Boone, 2013; Godinho et al., 2019; Ketcham & Carlson, 2001). In CT, X-ray absorption images (radiographs) from different angles of the sample are recorded during the scan, which are then computationally reconstructed as a voxelized 3D image. The intensity of each voxel depends on the attenuation coefficient of the sample at a particular point. This intensity can be calculated using Lambert-Beer's law (Equation 5.1). Mathematically, it describes the attenuation of X-rays in the sample and therefore the transmitted intensity I (Attix, 1987; Beer, 1852; Lambert et al., 1760):

$$I = I_0 e^{-\int \mu(s)ds} \tag{5.1}$$

I_0 is the incident beam intensity and $\mu(s)$ is the linear attenuation coefficient along the X-ray path s (Attix, 1987; Beer, 1852; Lambert et al., 1760). Here, μ determines the intensity value for each point in a sample. This value depends on parameters such as material density ρ, the atomic number of elements Z and X-ray energy E (Attix, 1987; Cnudde & Boone, 2013; Knoll, 2000). Usually, the values of ρ and Z are unknown in a CT scan, making a material identification not possible. However, based on the attenuation contrast, i.e., using the grayscale values in the reconstructed CT image, materials can be distinguished (Cnudde et al., 2012; Ikeda et al., 2004; Meftah et al., 2019). In case of prior knowledge, some identification can even be performed.

To achieve 3D chemical information of materials, there have been different approaches. One example combines micro-CT with other analytical 2D methods such as SEM, automated mineralogy techniques, or 2D micro-X-ray fluorescence (micro-XRF). The focus of this combination is the calibration of the micro-CT data based on the high resolution 2D information (Boone et al., 2011; De Boever et al., 2015; Reyes et al., 2017). There are different approaches to obtain this high resolution 2D data: destructive by cutting the sample into several slices and analyzing them, non-destructive by measuring the surfaces of the sample, or a combination of both. This method using micro-CT and 2D methods is accurate as elements and minerals can be classified in high detail. However, it is time-consuming as it requires multiple scans with different analytical techniques and an image registration that is used to link both methods.

Another approach is X-ray absorption spectroscopy, which exploits changes in absorption edges (e.g., K-edges) of chemical elements, sharp discontinuities in the absorption spectra. These sharp edges appear as an abrupt increase from weak to strong absorption in the spectrum and occur at energies of an electronic transition of an atom meaning that this method is element-specific (Knoll, 2000). Based on the position of the absorption edge in the spectrum, one can determine different elements in the sample. One 3D method making use of this phenomenon is dual-energy CT (Alvarez & Rtacovski, 1976; Flohr et al., 2006; Graser et al., 2009; Pelgrim et al., 2017). Dual-energy CT uses two sequential CT scans of a sample with different incident X-ray energies below and above K-edges of chemical elements in investigated materials. This approach allows a decomposition of materials containing different chemical elements (Cann et al., 1982; Chandarana et al., 2011; Pelgrim et al., 2017). Dual-energy CT is a useful addition to conventional CT and it is commonly used in several research fields such as medical, biological and geological research (Al-Owihan et al., 2014; Graser et al., 2009; Iovea et al., 2009; Johnson et al., 2007). Nevertheless, dual-energy CT is time-consuming as it takes two scans instead of one and has its limitations with chemically complex samples that contain several elements with K-edges in the same measured energy range.

Besides conventional CT scanners with a polychromatic X-ray tube, there are different techniques for 3D chemical imaging based on monochromatic synchrotron radiation, such as absorption-edge tomography, which uses X-rays tuned to specific absorption edges (Bleuet et al., 2010; Stock, 2008). Additionally, 3D chemical information can also be complemented by X-ray absorption near-edge structure (XANES) and extended X-ray absorption fine structure (EXAFS), i.e., techniques sensitive to atomic environments of elements (Cnudde et al., 2009; Eisenberger & Kincaid, 1978; Merkulova et al., 2019; Penner-Hahn, 1998; Sayers et al., 1971; Yano & Yachandra, 2009). Alternately, there are synchrotron methods using X-ray fluorescence for 3D chemical imaging which offers higher sensitivity, at the cost of a slower scanning process (Cagno et al., 2011; Kanngießer et al., 2003). All of these methods provide high-resolution data and can even give additional information about oxidation states or the coordination environment in the sample (Merkulova et al., 2019; Yano & Yachandra, 2009). However, access to synchrotron facilities is much more limited than laboratory-based instruments.

Other 3D, but destructive techniques use focused ion beams (FIB) instead of X-rays such as time-of-flight secondary ion mass spectrometry (FIB TOF-SIMS) and FIB-SEM (Brison et al., 2013; Fletcher et al., 2011; Nygren et al., 2007; Robinson et al., 2012). FIB-based techniques provide high spatial and depth resolutions but are destructive due to the ablation of the material by the high intensity of the ion beam.Moreover, only small volumes can be scanned compared to other 3D techniques such as micro-CT. Another destructive high-resolution method for a 3D-material characterization is atom probe tomography, where the sample is evaporated from a small tip and single atoms are detected (Kelly & Miller, 2007).

As an alternative, the use of photon-counting X-ray detectors (PCD) is a promising approach in 3D imaging. They provide energy resolving or hyperspectral information without a monochromatic X-ray source as used in synchrotron facilities. Thus, they can be used in different applications, e.g., XRF or micro-CT (Campbell et al., 1998; Egan et al., 2015; Frey et al., 2007; Kumpova et al., 2018; Ren et al., 2018; Sittner et al., 2020; Taguchi et al., 2011; Van Assche et al., 2021; Wang et al., 2011). However, a major drawback of multipixel PCDs is long measurement times. In this study a 2D-photon-counting detector (PCD) is used in combination with a conventional CT scanner to decrease the acquisition time and perform faster scans. Based on the transmitted polychromatic X-ray spectrum, elements in a sample can be identified using the position of elemental K-edges. This approach is called spectral X-ray computed tomography (Sp-CT) and can be used as a 3D-material identification method.

In this work, we describe the general principle of Sp-CT, as well as the instrument and PCD. We present results of different samples with a focus on chemical characterization of geological samples and show the potential of this method in different fields of application. Parts of the results shown were already published in Sittner et al. (2020), but in this chapter we update the capabilities of the detector and complement the implications of this method with additional measurements on geological ore samples.

5.2 Principle of Spectral-CT

Various processes occur in the interaction between X-rays and matter, e.g., scattering, diffraction and absorption (Knoll, 2000). The linear attenuation (Equation 5.2) is the main phenomenon that gives contrast differences in phases. The attenuation is caused by four major processes: incoherent or Compton scattering, coherent scattering, photoelectric effect and pair production (Attix, 1987; Knoll, 2000). At low energies (50–100 keV) the photoelectric effect dominates the attenuation while at higher energies (5–10 MeV) Compton scattering is dominant. Pair production is only relevant for very high energies (Cnudde & Boone, 2013; Ketcham & Carlson, 2001; Knoll, 2000). For most laboratory micro-CT setups, the main contribution to X-ray attenuation is the photoelectric effect, which can roughly be expressed by (Equation 5.2) (Knoll, 2000).

$$\tau \cong constant \cdot \frac{Z^n}{E^{3.5}} \tag{5.2}$$

Here the exponent n varies between 4 and 5, meaning that the photoelectric effect increases steeply with the atomic number (Z). On the other hand, it decreases with increasing photon energy (E).

When the X-rays are attenuated by the sample, the spectrum of the primary X-rays emitted from the source is changed. Therefore, the spectrum of the X-rays transmitted through the sample contains information about the sample composition. Interpretation of the spectra resulted from an interaction of X-rays with a sample is the basis of Sp-CT. These spectra show different features such as abrupt changes in the transmission, which can be explained by absorption edges of certain elements. These sharp edges appear as transitions from weak to strong absorption, which is reversed in the transmission spectrum, i.e., at an absorption edge, the transmission decreases as a result of the increasing absorption. The absorption edges occur at energies of an electronic transition where the incoming X-ray photons correspond to the binding energy of the electron(s) in a particular electronic shell (K, L, M, etc.) of a specific atom. Thus, the technique is element-specific, i.e., one can determine different elements in the sample based on the energy position of the absorption edge (Knoll, 2000).

5.2.1 Photon-Counting Detector

The spectrum on which Sp-CT relies is detected by a PCD. To explain the difference between a conventional X-ray CT detector and a PCD, the following section summarizes the operation of both detectors. A conventional X-ray CT detector consists of a scintillator crystal (e.g., CsI), which converts the incoming X-rays into visible light. Optical sensors such as charge-coupled device (CCD) or complementary metal-oxide-semiconductor (CMOS) detect and transform the visible light into an electronic signal. The signal is proportional to the total energy of all photons deposited within the readout time, thus information about individual photons is lost (energy integrating detector) (Zhou et al., 2018). This method of detection is also called indirect because X-rays are transformed into light before the detection.

In contrast, a PCD uses a semiconductor crystal such as CdTe, CdZnTe, Selenium, or Silicon that directly converts the incoming X-ray photon into an electric charge. Therefore, it is also called direct detection (Prekas et al., 2012). The detector consists of a single semiconductor crystal, two electrodes and the readout electronics or application-specific integrated circuit (ASIC) (Figure 5.1). The cathode is a monolithic electrode and covers the whole semiconductor while the anode is pixelated and evenly distributed on the other side of the crystal (Ren et al., 2018). A bias voltage is applied between the two electrodes. The pixelated anode is connected to the ASIC by bump bonds which usually consist of Indium or lead/tin (Schulman, 2006). For a larger detector area, multiple of these modules with separate semiconductor crystals can be placed next to each other. If an X-ray photon hits the detector it generates electron-hole-pairs in the semiconductor and the applied voltage causes a separation of these charge carriers to the contacts (Figure 5.1). The readout electronics further process and amplify the signal. They usually consist of a charge-sensitive preamplifier, a pulse shaper and multiple

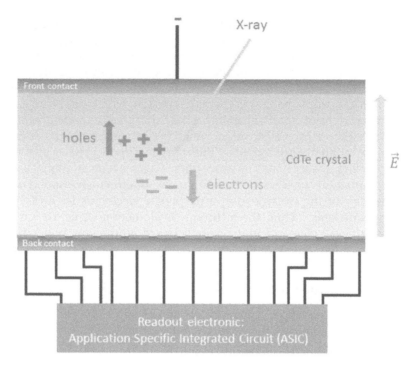

FIGURE 5.1
Schematic operation of a photon-counting detector with a CdTe crystal; the X-rays produce electron-hole-pairs, which are separated to the contacts due to the applied voltage in the CdTe crystal; the readout electronics further process and amplify the signal (Sittner et al., 2020).

pairs of voltage pulse height comparator and digital counter (Ren et al., 2018). The charge-sensitive preamplifier converts the current (electrical pulse) into a charge. The pulse shaper not only shapes the signal but also amplifies the signal and reduces the noise produced by the preamplifier. The voltage pulse height comparator sets different energy thresholds to determine which pulse is processed further. Thresholds can be set individually for each pixel to compensate for different variations in the electronics (Ren et al., 2018). In the end, the signal consists of a single voltage pulse, which is proportional to the energy of each individual detected X-ray photon.

5.2.2 Detector Artifacts

In an ideal detector, each X-ray photon would interact and produce a pulse with a voltage that represents its energy. However, in the case of real detectors, there are physical and technical limitations that affect the spectrum. Some photons do not interact at all with the detector and pass through. Some scatter from the detector itself or the shielding and produce secondary effects that lower the initial signal (Redus et al., 2009). The physical principles

TABLE 5.1

Characteristic X-Ray Energies for Cd and Te, in keV (Data from NIST X-Ray Transition Energies Database; Deslattes et al., 2003)

	K-α_1	K-α_2	K-β_1	K-β_2
Cd	23.17	22.98	26.09	26.06
Te	27.47	27.20	30.99	30.94

of interaction between photons and the detector substrate also lead to an incomplete collection of photon energy, e.g., charge sharing (Iniewski et al., 2007; Ren et al., 2018; Veale et al., 2011; Wang et al., 2011). Finally, the electronic readout can also be distorted by other factors, such as electronic noise (Knoll, 2000; Taguchi et al., 2011; Usman & Patil, 2018). All these artifacts are common in different types of detectors and must be considered when interpreting the spectrum.

In a CdTe detector, Cadmium (Cd) and Tellurium (Te) are excited by the incident X-rays and produce specific X-rays. This affects all electron transitions (K, L) but the effect of K radiation is the strongest. Both Cd and Te produce K-α and K-β peaks in the measured spectral range. Table 5.1 shows their characteristic X-ray energies. When the primary X-rays hit the detector, they can produce secondary X-rays with the specific energies of Cd or Te. When this specific X-ray radiation leaves the detector without further interaction, then this portion of energy can no longer be detected and an escape peak with a precisely defined lower energy is formed (Figure 5.2) (Dreier et al., 2018; Knoll, 2000; Tanaka et al., 2017).

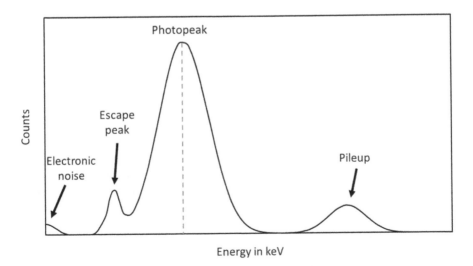

FIGURE 5.2

Schematic illustration of the spectral response of a CdTe detector to a photopeak with important detector artifacts such as electronic noise, escape peak and pileup (Sittner et al., 2020).

Pulse pileup is another known possible artifact, which describes a case of two X-ray photons hitting the detector within the readout time. This results in the detection of just one event and an artificial signal that consists of two summed X-ray photons (Knoll, 2000; Taguchi et al., 2011; Usman & Patil, 2018). Therefore, the counts are artificially increased at twice the energy, especially when the primary energy represents a peak value (Figure 5.2).

Moreover, low to intermediate energies are dominated by electronic noise. However, the low energy threshold of the detector blocks out the dark noise and eliminates most of these artifacts. Charge sharing is a phenomenon that occurs in every pixelated detector and is reported for CdTe detectors too (Maiorino et al., 2006; Pennicard et al., 2009). After an X-ray photon produced a charge cloud, it starts to drift toward the electrode and spreads laterally. Because the pixels are electrically separated by the bonds and not physically in the semiconductor crystal this charge cloud can spread over different neighboring pixels resulting in a sharing of induced current signals by multiple neighboring pixels. This reduces the measured signal in the specific pixel and contributes to a lower energy and lateral resolution (Iniewski et al., 2007; Ren et al., 2018; Veale et al., 2011; Wang et al., 2011).

5.2.3 Spectral-CT Workflow

Figure 5.3(a) shows the raw data from single radiographs, which were obtained with a photon-counting line detector (TESCAN PolyDet, see Section 5.3). The y-direction of the radiographs represents the energy separated into different bins and the x-direction gives spatial information. In the flat field radiograph with no sample, some dark vertical lines are caused by the structure and the readout electronics of the detector. Another visible feature is the bright horizontal band, which extends across the entire radiograph. This represents the tungsten (W) K-α fluorescence at 59.3 keV from the X-ray tube (NIST X-Ray Transition Energies Database; Deslattes et al., 2003). The W K-β fluorescence at 67.2 keV is also noticeable, but less intense (NIST X-Ray Transition Energies Database; Deslattes et al., 2003). The raw spectrum in Figure 5.3(b) shows the two peaks of W. The sample, in this case, a gold wire appears as a shadow in the image and after the normalization of both images, it is better visible (Figure 5.3(c)). The normalization eliminates all features that do not originate from the sample. To get a full scan with many radiographs from different angles, the sample needs to be rotated. The reconstruction of all radiographs measured with the line detector results in a 2D image of the sample (Figure 5.3(d)).

The reconstruction contains 128 images, one for each energy bin (see Section 5.3), showing not only the 2D spatial structure, but also the spectral information. In this case, the strong increase around 80 keV can be explained by the K-edge of Au (Figure 5.3(e)). To scan a 3D volume several 2D scans need to be stacked to get the 3D and spectral information.

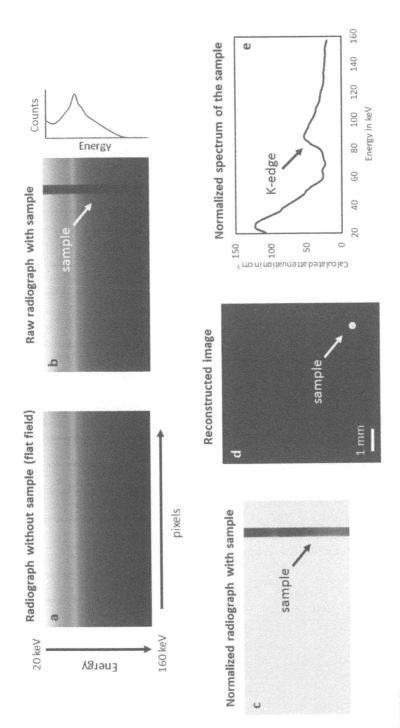

FIGURE 5.3

Radiographs and spectrum measured with a photon-counting detector (TESCAN PolyDet); (a) Radiograph without a sample (flat field). The vertical lines are caused by the readout electronics and the bright horizontal line represents the tungsten (W) fluorescence from the X-ray tube; (b) Radiograph of a sample and the resulting unnormalized spectrum. This spectrum indicates the peak from the W fluorescence; (c) Normalized radiograph with the sample. Normalized by subtracting the flat field from the radiograph with the sample; (d) Reconstructed 2D image with the sample; (e) Spectrum of the sample (gold) with the K-edge at 80 keV.

5.3 Methods

All scans presented in this study were performed with the TESCAN CoreTOM (Ghent, Belgium and Freiberg, Germany). Apart from the standard imaging detector, the CoreTOM is equipped with an additional photon-counting line sensor, the TESCAN PolyDet, which can automatically slide in front of the standard detector. The PolyDet is a CdTe X-ray detector with a sensor width of 307 mm. The direct detection sensor has an energy range from 20 up to 160 keV and can discriminate up to 128 energy bins (Table 5.2) (Sittner et al., 2020). In particular, the K-edge of elements from Cd up to U can be detected with this detector.

Samples with different complexities were investigated in this study. Several reference samples were measured to compare theoretical energy positions of K-edges with the measured positions. Pure elements, element oxides and minerals with different shapes were used as references. A detailed list of all materials is attached in Annexure A1. Furthermore, mixtures of pure element particles embedded in a low-absorbing matrix simulated a natural rock with various high attenuating minerals. For the mixture, <400 μm quartz grains (SiO_2) were mixed as a low-absorbing matrix mineral along with gold (Au), lead (Pb) and tungsten (W) particles. Epoxy resin was used to solidify the mixture in a cylindrical plastic container with a diameter of 1 cm. Finally, natural rock samples from the Au-U Witwatersrand Supergroup (South Africa) demonstrate potential applications of Sp-CT for geological samples. Different drill core samples with a diameter of 2.5 cm from the Kalkoenkrans Reef at the Welkom Gold field were measured with Sp-CT and micro-CT. Geologically the samples can be described as a pyritic quartz-pebble conglomerate. They contain high amounts of gold and uraninite (UO_2), which are usually associated with different carbon-bearing minerals, pyrite (FeS_2), galena (PbS), pyrrhotite ($Fe_{1-x}S$), gersdorffite (NiAsS), zircon ($ZrSiO_4$), or phyllosilicates such as muscovite ($KAl_2(Si_3Al)O_{10}(OH,F)_2$) or biotite ($K(Mg,Fe)_3(Al,Fe)Si_3O_{10}(OH,F)_2$) (Feather & Koen, 1975). Gold, uraninite and galena are of particular interest because they contain elements with K-edge energies in the spectral range of the detector (Au, U and Pb).

TABLE 5.2

General Specifications of the TESCAN PolyDet (Sittner et al., 2020)

Dimensions ($W \times H \times D$)	$398 \times 236 \times 76$ mm
Radiation sensor	Cadmium telluride (CdTe)
Number of pixels	384
Sensor width	307 mm
Energy range	20–160 keV
Energy resolution	9.1 keV FWHM at 96 keV

The measurement conditions were adapted to the respective samples and a detailed list is attached in Annexure A1. Most measurements use a tube voltage of 160 keV and a power of 10–25 W. All 360° scans were performed with 600 projection for each sample. Before each scan, a radiograph without sample (flat field) was measured and used to normalize the images later. The duration of each Sp-CT scan was about 22 minutes. The radiographs were reconstructed using the spectral reconstruction module in Acquila reconstruction by Tescan (version 1.14) with a fan beam reconstruction, a ring filter and a bad pixel correction (spot filter). In addition to the 2D scans, a 3D spectral scan was performed on the drill core samples. The scan consists of 45 single 2D scans stacked on top of each other. With the duration of 22 minutes per scan the total scanning time of the 3D volume was 16.5 hours.

5.4 Results

In the results section, we first describe the polychromatic spectrum of the X-ray tube detected by the PCD. Furthermore, we present the results of different reference sample measurements and the particle mixture embedded in a low-absorbing matrix. Finally, we show the results of mineral differentiation in the rock sample.

5.4.1 Polychromatic X-Ray Tube Spectrum

Figure 5.4 shows the unfiltered polychromatic spectra of the X-ray tube recorded with different maximum energies. The maximum energy of the tube in the individual scans is increasing starting at 40 keV and ending at 160 keV in steps of 20 keV with a constant power of 10 W. All spectra with maximum energy above 80 keV show the characteristic W K-α fluorescence at 59.3 keV caused by the tungsten target of the tube (NIST X-Ray Transition Energies Database; Deslattes et al., 2003). This normally distinct and sharp fluorescence peak is broadened by the PCDs energy response, which has an energy resolution of 9.1 keV FWHM at 96 keV. The less intense peak from W K-β fluorescence at 67.2 keV is visible in all spectra with maximum tube energy between 100 and 160 keV (NIST X-Ray Transition Energies Database; Deslattes et al., 2003). The W K-β peak appears on the right shoulder of the W K-α peak. Furthermore, the different spectra do not reach zero counts at their particular highest energy. There is a small shift toward higher energies in all spectra, which can be explained by an increase of the background due to pileup events of many signals.

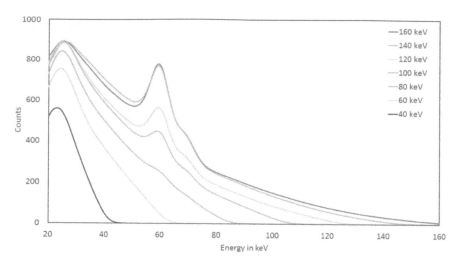

FIGURE 5.4
Polychromatic spectra of the X-ray tube measured with the CdTe detector with different maximum energies. The maximum energy of the tube starts at 60 keV and ends at 160 keV in steps of 20 keV with a power of 10 W and without a filter (Sittner et al., 2020).

5.4.2 Reference Samples

Different pure elements, element oxides and minerals were investigated in this study to compare the measured and theoretical spectra and the position of the K-edge energies. Figure 5.5 shows two measured and theoretical spectra and the first derivative of the measured spectra of lead (Pb) and Hafnium (Hf) (theoretical data from the NIST Standard Reference Database 66, Chantler, 2001). The first derivative is a good tool to compare the K-edge energy with its theoretical positions as the edge forms a distinct peak.

As it is seen from the spectra in Figure 5.5 the detector is capable of resolving the K-edges of particular elements in the spectral range. The measured spectra have differences compared to the theoretical spectra. One major difference is in the low energy range between 20 and 60 keV, in which the measured spectrum shows lower intensities. Furthermore, there is a small peak at 60 keV in both spectra, which is also visible in the first derivative. This is probably an artifact from the W K-α fluorescence. The K-edges of both elements, Pb and Hf, are visible and their energy position corresponds to the theoretical values. Widened-edge peaks on measured spectra are broader compared to theoretical spectra due to the energy resolution of the PCD as can be seen in Figure 5.4 for the W fluorescence lines. The Pb K-edge peak in the first derivative occurs at 90.0 keV, which is 2.0 keV higher than the theoretical position 88.0 keV (NIST X-Ray Transition Energies Database; Deslattes et al., 2003). The Hf K-edge peak in the first derivative occurs at 68.0 keV, which is 2.7 keV higher than a theoretical value at 65.3 keV (NIST X-Ray

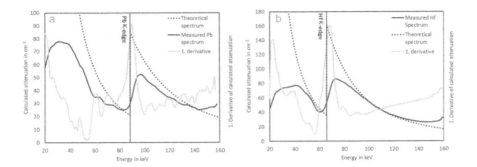

FIGURE 5.5
Measured spectra of lead (Pb) and Hafnium (Hf) with the first derivative of the measured spectra and theoretical spectra (data from the NIST Standard Reference Database 66, Chantler, 2001; and NIST X-Ray Transition Energies Database; Deslattes et al., 2003); (a) Theoretical and measured spectra of Pb with the first derivative of the measured spectrum and the theoretical position of the Pb K-edge (88.0 keV); (b) Theoretical and measured spectra of Hf with the first derivative of the measured spectrum and the theoretical position of the Hf K-edge (65.3 keV).

Transition Energies Database; Deslattes et al., 2003). A possible reason for this shift could be a detector artifact such as charge sharing.

This shift toward higher energies is noticeable in all measured reference samples. Figure 5.6 shows the comparison of the measured and the theoretical K-edge position for all sample materials. The shift in the range of 40–50 keV is slightly smaller compared to the rest of the spectral range with Samarium (Sm) showing the smallest shift. The rest of the elements follow

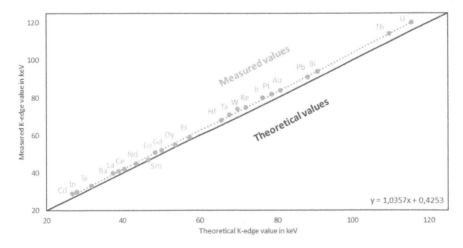

FIGURE 5.6
Measured and theoretical K-edge energy positions of all reference samples; the blue points represent the measured element K-edge values from the samples in Annexure A1, the black line shows theoretical K-edge positions (data from NIST X-Ray Transition Energies Database; Deslattes et al., 2003).

a trend that is parallel to the theoretical values. A linear regression of the measured elements was used to calculate new K-edge energies that were not measured. The coefficient of determination $R^2 = 0.9991$ is close to 1 indicating a small variation of the measured elements and a good prediction for new K-edge energies. The calculated K-edge energy of Tin (Sn) is at 30.6 keV using the formula in Figure 5.6 and the theoretical value is at 29.2 keV (NIST X-Ray Transition Energies Database; Deslattes et al., 2003).

Although the reference measurements show deviations from theoretical K-edge values, it is possible to determine elements based on their K-edge energy. Also elements with K-edges close to each other such as W and Rhenium (Re) are distinguishable. However, more detector and processing improvements are necessary to remove spectral artifacts and to achieve a better quality of the spectrum.

5.4.3 Particle Mixtures

A more complex sample is used for material identification in the presence of different elements. The sample consists of gold (Au), lead (Pb) and tungsten (W) particles which were mixed with quartz as a matrix material to simulate a chemically complex rock. This sample serves as a comparison to the single-particle measurements, whereby the position of the respective edges should be compared. Furthermore, the material identification of both a conventional micro-CT image and a Sp-CT image are compared.

Figure 5.7(a) and (b) shows the reconstructed micro-CT image of the mixture in comparison to the reconstructed and summed Sp-CT image (sum of all 128 energy bins). A major difference between the two images is spatial resolution. The voxel size of the scans differs from 11.5 μm of the micro-CT scan to 52.5 μm of the Sp-CT scan. The number of pixels of the PCD limits the resolution of the Sp-CT image and the individual particles appear blurry compared to the conventional micro-CT image.

The gray value histogram is typically used for material identification based on the reconstructed CT image (Figure 5.7(c)). Individual peaks in the histogram are assigned to the respective materials. In a perfect case, each material would have its peak in the histogram on which the reconstructed image can be segmented.

However, the histogram in Figure 5.7(c) shows only small peaks. The double peak at a gray value of around 5000 represents the resin in the sample and the matrix mineral, in this case, quartz. Even though the particles in the sample have a different elemental composition and density, the histogram shows no peaks on which the particles can be separated. This can be explained by the total number of voxels, whereby the individual particles represent only a small amount in contrast to the matrix and the air. Therefore, a material identification based on a conventional micro-CT scan is very difficult in that case.

With the Sp-CT scan, the spectra of the individual particles can be analyzed to identify the particle chemistry. Figure 5.7(d) shows three different spectra

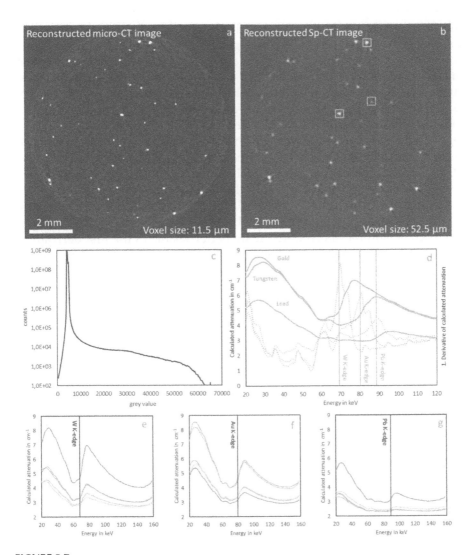

FIGURE 5.7
Reconstructed micro-CT and Sp-CT images, energy spectra and gray value histogram of the
particle mixture of quartz, gold (Au), lead (Pb) and tungsten (W); (a) Reconstructed micro-CT
image (160 keV, 15 W, no filter); (b) Reconstructed summed Sp-CT image (160 keV, 20 W, no
filter); (c) Gray value of the micro-CT image as a function of number of counts (gray value histogram), *y*-axis in logarithmic scale; (d) Energy spectra (solid lines) and first derivative (dotted
lines) of the different particles with the theoretical K-edge positions of Au, Pb and W; (e) Five
W spectra with the theoretical K-edge position; (f) Five Au spectra with the theoretical K-edge
positions; (g) Five Pb spectra with the theoretical K-edge positions (NIST X-Ray Transition
Energies Database; Deslattes et al., 2003).

and their derivatives from particles, which are marked in the reconstructed Sp-CT image (Figure 5.7(b)). Even though the gray values of the marked particles in the Sp-CT image are the same, the particle spectra show different characteristics and different positions of K-edges in the measured spectral range. The Pb-spectrum indicates the lowest signal with the K-edge energy as slightly higher than the theoretical value (NIST X-Ray Transition Energies Database; Deslattes et al., 2003). The small particle size results in an averaging with the surrounding pixels, which can explain the low signal of the Pb particle (partial volume effect; Godinho et al., 2019). The spectra of Au and W show higher intensities and well visible K-edges. Energy positions of both K-edges are higher than the theoretical values as discussed in Chapter 5.4.2 and it is easy to differentiate the Au and Pb particles based on their spectra. (NIST X-Ray Transition Energies Database; Deslattes et al., 2003). There are some artifacts in the spectra, which are better visible in the first derivatives of the spectra. Around 60 keV, there is a peak in all derivatives which can be explained by the W K-α fluorescence as seen in Figure 5.5. Based on this peak a possible escape peak is visible at around 45 keV. Furthermore, there are small peaks between 20 and 30 keV. These peaks are most likely results of the characteristic X-ray energies of Cd and Te (Table 5.1).

Figure 5.7(e)–(g) illustrates several spectra for each particle type in the sample. The intensities in all particle types vary but the K-edge positions of the specific elements are stable and comparable with the theory (NIST X-Ray Transition Energies Database; Deslattes et al., 2003). The energy range from 20 to 60 keV shows variations in the calculated attenuation. This could be explained by artifacts such as escape peaks and the effect of the W K-α fluorescence. As mentioned earlier, since the particles have different sizes an averaging of the particle with the matrix pixels can cause a variation in the spectral intensity. Furthermore, the position of the particle in the sample and surrounding particles can also influence the overall intensity, but not influence the K-edge energy position.

Based on the different K-edge positions all particles in the solid mixture can be identified because Sp-CT scan gives information about the elements present in the individual particles. In combination with the conventional micro-CT scan, additional high-resolution particle information is accessible.

5.4.4 Rock Samples

For a geological real case scenario, conglomeratic ore from the Au-U Witwatersrand Supergroup was measured using micro-CT and Sp-CT. First, a conventional CT scan was performed to identify areas of high attenuating minerals and second, Sp-CT was used to further chemically differentiate between certain minerals.

Figure 5.8 shows the reconstructed micro-CT and Sp-CT images with the spectra of different mineral phases. The drill core sample is from the

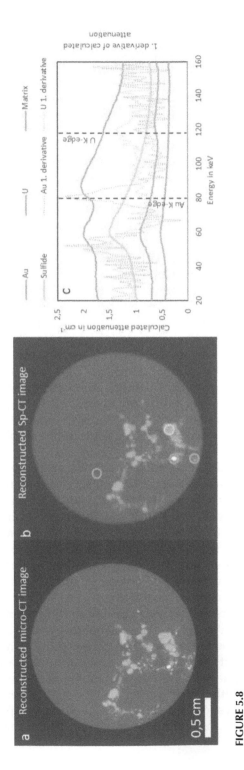

FIGURE 5.8

Reconstructed micro-CT image and Sp-CT image with spectra measured in the ore sample from Au-U Witwatersrand Supergroup; (a) Reconstructed micro-CT image (180 kV, 70 W, 2 mm Cu filter, 70 μm voxel size); (b) Summed reconstructed Sp-CT image (160 kV, 100 W, 2 mm Cu filter, 100 μm voxel size); (c) Spectra of different minerals indicated in the reconstructed Sp-CT image with the first derivative of gold (Au) and Uranium (U) and the theoretical K-edge energies of Au and U (dashed lines); NIST X-Ray Transition Energies Database; Deslattes et al., 2003).

Kalkoenkrans Reef (hanging wall) at the Welkom Gold field. The resolution of the two scans is different as described in Figure 5.7(a) and (b). This can be explained by the voxel size which differs from 70 μm in the CT scan to 100 μm in the Sp-CT scan. The minerals in the sample have different gray values indicating different attenuation coefficients. The matrix, which consists predominantly of quartz and lesser phyllosilicates, is low attenuating and there is no K-edge in the spectral range. Individual minerals of the matrix are not visible due to the 2 mm Cu filter, which filters the low energy range. This filter was used to gain more contrast for higher attenuating phases and reduce artifacts such as scattering. Sulfides, such as pyrite or pyrrhotite, are higher attenuating but neither shows a K-edge in the spectrum. However, all four spectra have a peak around 60 keV including the matrix spectrum. This can be explained by the artifact of the W fluorescence. The sulfides are commonly distributed between low attenuating matrix minerals, which can be seen on the right side of the micro-CT image (Figure 5.8). Several high attenuating phases with a size of ca. 10 μm are visible at the bottom of the images, indicated in the orange circle (Figure 5.8). The related spectrum taken in one of these small phases has a K-edge at around 119 keV and demonstrates the presence of U by comparing this spectrum with the reference material spectrum. Thus, the small high attenuating minerals contain U which in these samples is usually indicative of uraninite (UO_2), coffinite ($U(SiO_4)_{1-x}(OH)_{4x}$) or brannerite ($(U, Ca, Y, Ce)(Ti, Fe)_2O_6$). Another noticeable high attenuating phase indicated in blue with a size of 50 μm has a different spectrum. This spectrum has a K-edge positioned at around 80 keV, which corresponds to the K-edge of Au and thus suggests the presence of native gold in the sample typical for ores from Witwatersrand.

Figure 5.9 shows a colored 3D spectral CT scan of a drill core sample from the Kalkoenkrans Reef, which consists of several single 2D scans stacked on top of each other. This volume contains 3D spatial and spectral information. Figure 5.9 shows the energy of 90 keV as it provides a good contrast for the high attenuating phases in the sample. The scanned volume has a height of 36 mm and consists of 45 single 2D scans. In the different perspectives of the sample, the front view stands out. There are horizontal artifacts in the 3D volume, which are caused by the stack of several 2D scans. The top view only shows ring artifacts in the center of the volume.

The scanned area is characterized by many small high attenuating mineral phases associated with larger size low attenuating phases. As seen and discussed in Figure 5.8, the low attenuating phases are probably sulfides, such as pyrite or pyrrhotite. The spectra of three minerals are shown in Figure 5.9. The mineral phase in a red circle situated at the top of the volume has the spectrum with a low in intensity K-edge located at 119 keV. This K-edge position corresponds to the U K-edge, thus the spectrum represents uraninite, coffinite, or brannerite as shown in the measurements

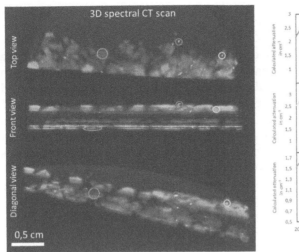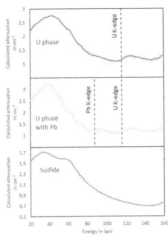

FIGURE 5.9

Three perspectives of the colored reconstructed 3D Sp-CT scan and the corresponding spectra of three minerals indicated in the 3D image measured in the ore sample from Au-U Witwatersrand Supergroup; the dashed lines indicate the position of K-edge energies of lead (Pb) and Uranium (U); (NIST X-Ray Transition Energies Database; Deslattes et al., 2003).

in Figure 5.8. Another high attenuating mineral, which is indicated by a yellow circle, shows an interesting spectrum with two K-edges: one at 119 keV corresponding to U K-edge and one at 90 keV. By comparing it with the reference spectra, the K-edge at 90 keV represents the Pb-Kedge. Thus this mineral phase contains both elements, U and Pb. In the Au-U Witwatersrand Supergroup uraninite often shows various degrees of alteration and replacements by different minerals such as quartz, phyllosilicates, or sulfides (Feather & Koen, 1975; Rantzsch et al., 2011). This mineral phase is most likely a U phase, which is altered and partially replaced by galena (PbS). Another possibility could be radiogenic Pb, which is common in U phases, especially in uraninite (Frimmel et al., 2009). Finally, a spectrum of a sulfide (probably pyrite) is shown in Figure 5.9 without any K-edge in the spectral range.

This example shows that Sp-CT can differentiate mineral phases within the spectral range in a drill core sample without any sample preparation based on a single 2D scan that takes 22 minutes. A 3D scan that combines both spatial and spectral information however requires long measurement times. This approach can be used as a first analytical step to define regions of interest, and subsequent high-resolution 2D methods can be used to determine more precise chemical composition of minerals.

5.5 Outlook

Sp-CT is a relatively new method for 3D material identification and offers a promising new potential. We have shown that it is possible to distinguish between K-edges in the spectral range and to identify materials in mixed samples using the K-edge energy without any sample preparation. Sp-CT opens exciting new possibilities for elemental and mineral analysis. With this technique, the 3D properties of particles can be measured and used for example in process mineralogy simulations. Currently, such simulations use 2D or bulk particle properties. With a combination of CT and Sp-CT, the 3D properties and chemical information of different particles can be extracted for an improved simulation. The analyzed volume compared to 2D techniques is also larger and therefore more representative. This is a major improvement to current simulations. Moreover, Sp-CT could potentially be used as an alternative technique for a regular characterization of ores and processed ores since more representative volumes can be analyzed. Sp-CT can also serve as a primary analytical method on which subsequent higher resolution methods like SEM or EPMA can be based. The next step in the development is the establishment of an automated way of material identification using the spectrum and the K-edge position. Furthermore, improvements on the detector and the spectrum are necessary to optimize the signal and to remove artifacts.

Acknowledgments

This research is part of the upscaling project "Resource Characterization: from 2D to 3D microscopy" and has received funding from the European Institute of Innovation and Technology (EIT), a body of the European Union, under the Horizon 2020, the EU Framework Program for Research and Innovation.

References

Al-Owihan, H., Al-Wadi, M., Thakur, S., Behbehani, S., Al-Jabari, N., Dernaika, M., & Koronfol, S. (2014). Advanced Rock Characterization by Dual-Energy CT Imaging: A Novel Method for Complex Reservoir Evaluation. *Conference Proceedings, IPTC 2014: International Petroleum Technology Conference, Jan 2014,* cp-395-00148 https://doi.org/10.3997/2214-4609-pdb.395.IPTC-17625-MS

Almgren, M., Edwards, K., & Karlsson, G. (2000). Cryo transmission electron microscopy of liposomes and related structures. *Colloids and Surfaces A: Physicochemical and Engineering Aspects, 174*(1–2), 3–21. https://doi.org/10.1016/S0927-7757(00)00516-1

Alvarez, R. E., & Rtacovski, A. (1976). Energy-selective reconstructions in x-ray computerized tomography. *Physics in Medicine and Biololgy, 21*(5), 733–744.

Attix, F. H. (1987). Introduction to radiological physics and radiation dosimetry. *International Journal of Radiation Applications and Instrumentation. Part A, 38*(2), 163. https://doi.org/10.1016/0883-2889(87)90017-7

Bachmann, K., Frenzel, M., Krause, J., & Gutzmer, J. (2017). Advanced identification and quantification of in-bearing minerals by scanning electron microscope-based image analysis. *Microscopy and Microanalysis, 23*(3), 527–537. https://doi.org/10.1017/S1431927617000460

Bachmann, K., Osbahr, I., Tolosana-Delgado, R., Chetty, D., & Gutzmer, J. (2018). Variation in platinum group mineral and base metal sulfide assemblages in the lower group chromitites of the Western Bushveld Complex, South Africa. *Canadian Mineralogist, 56*(5), 723–743. https://doi.org/10.3749/canmin.1700094

Beer. (1852). Bestimmung der Absorption des rothen Lichts in farbigen Flüssigkeiten. *Annalen Der Physik, 162*(5), 78–88. https://doi.org/10.1002/andp.18521620505

Bleuet, P., Gergaud, P., Lemelle, L., Bleuet, P., Tucoulou, R., Cloetens, P., Susini, J., Delette, G., & Simionovici, A. (2010). 3D chemical imaging based on a third-generation synchrotron source. *TrAC Trends in Analytical Chemistry, 29*(6), 518–527. https://doi.org/10.1016/j.trac.2010.02.011

Boespflug, X., Long, B. F. N., & Occhietti, S. (1995). CAT-scan in marine stratigraphy: A quantitative approach. *Marine Geology, 122*(4), 281–301. https://doi.org/10.1016/0025-3227(94)00129-9

Boone, M., Dewanckele, J., Boone, M., Cnudde, V., Silversmit, G., Van Ranst, E., Jacobs, P., Vincze, L., & Van Hoorebeke, L. (2011). Three-dimensional phase separation and identification in granite. *Geosphere, 7*(1), 79–86. https://doi.org/10.1130/GES00562.1

Boone, M. A., Nielsen, P., De Kock, T., Boone, M. N., Quaghebeur, M., & Cnudde, V. (2014). Monitoring of stainless-steel slag carbonation using X-ray computed microtomography. *Environmental Science and Technology, 48*(1), 674–680. https://doi.org/10.1021/es402767q

Brison, J., Robinson, M. A., Benoit, D. S. W., Muramoto, S., Stayton, P. S., & Castner, D. G. (2013). TOF-SIMS 3D imaging of native and non-native species within hela cells. *Analytical Chemistry, 85*(22), 10869–10877. https://doi.org/10.1021/ac402288d

Cagno, S., Nuyts, G., Bugani, S., De Vis, K., Schalm, O., Caen, J., Helfen, L., Cotte, M., Reischig, P., & Janssens, K. (2011). Evaluation of manganese-bodies removal in historical stained glass windows via SR-µ-XANES/XRF and SR-µ-CT. *Journal of Analytical Atomic Spectrometry, 26*(12), 2442–2451. https://doi.org/10.1039/c1ja10204d

Campbell, M., Heijne, E. H. M., Meddeler, G., Pernigotti, E., & Snoeys, W. (1998). *A Readout Chip for a 64 × 64 Pixel Matrix with 15-bit Single Photon Counting. CERN-ECP-97-010. CERN-ECP-97-10, 3 p.* https://doi.org/10.1109/NSSMIC.1997.672566

Cann, C. E., Gamsu, G., Birnberg, F. A., & Webb, W. R. (1982). Quantification of calcium in solitary pulmonary nodules using single- and dual-energy CT. *Radiology, 145*(2), 493–496. https://doi.org/10.1148/radiology.145.2.7134457

Chandarana, H., Megibow, A. J., Cohen, B. A., Srinivasan, R., Kim, D., Leidecker, C., & Macari, M. (2011). Iodine quantification with dual-energy CT: Phantom study and preliminary experience with renal masses. *American Journal of Roentgenology, 196*(6), W693–700. https://doi.org/10.2214/ajr.10.5541

Chantler, C. T. (2001). Detailed tabulation of atomic form factors, photoelectric absorption and scattering cross section, and mass attenuation coefficients in the vicinity of absorption edges in the soft X-ray (Z = 30-36, Z = 60-89, E = 0.1-10 keV)–addressing convergence issue. *Journal of Synchrotron Radiation, 8*(4), 1124. https://doi.org/10.1107/S0909049501008305

Cnudde, V., & Boone, M. N. (2013). High-resolution x-ray computed tomography in geosciences: A review of the current technology and applications. *Earth-Science Reviews, 123*, 1–17. https://doi.org/10.1016/j.earscirev.2013.04.003

Cnudde, V., Dewanckele, J., De Boever, W., Brabant, L., & De Kock, T. (2012). 3D characterization of grain size distributions in sandstone by means of x-ray computed tomography. In P. Sylvester (Ed.), *Quantitative Mineralogy and Microanalysis of Sediments and Sedimentary Rocks* (Vol. 42, pp. 99–113). Mineralogical Association of Canada (MAC).

Cnudde, V., Silversmit, G., Boone, M., Dewanckele, J., De Samber, B., Schoonjans, T., Van Loo, D., De Witte, Y., Elburg, M., Vincze, L., Van Hoorebeke, L., & Jacobs, P. (2009). Multi-disciplinary characterisation of a sandstone surface crust. *Science of the Total Environment, 407*(20), 5417–5427. https://doi.org/10.1016/j.scitotenv.2009.06.040

Conroy, G. C., & Vannier, M. W. (1987). Dental development of the Taung skull from computerized tomography. *Nature, 329*(6140), 625–627. https://doi.org/10.1038/329625a0

De Boever, W., Derluyn, H., Van Loo, D., Van Hoorebeke, L., & Cnudde, V. (2015). Data-fusion of high resolution X-ray CT, SEM and EDS for 3D and pseudo-3D chemical and structural characterization of sandstone. *Micron, 74*, 15–21. https://doi.org/10.1016/j.micron.2015.04.003

Deslattes, R. D., Kassler, E. G., Indelicato, P., De Billy, L., Lindroth, E., & Anton, J. (2003). X-ray transition energies: New approach to a comprehensive evaluation. *Reviews of Modern Physics, 75*(1), 35–99. https://doi.org/10.1103/RevModPhys.75.35

Dreier, E. S., Kehres, J., Khalil, M., Busi, M., Gu, Y., Feidenhans'l, R. K., & Olsen, U. L. (2018). Spectral correction algorithm for multispectral CdTe x-ray detectors. *Optical Engineering, 57*(5). https://doi.org/10.1117/1.OE.57.5.054117

Egan, C. K., Jacques, S. D. M., Wilson, M. D., Veale, M. C., Seller, P., Beale, A. M., Pattrick, R. A. D., Withers, P. J., & Cernik, R. J. (2015). 3D chemical imaging in the laboratory by hyperspectral x-ray computed tomography. *Scientific Reports, 5*. https://doi.org/10.1038/srep15979

Eisenberger, P., & Kincaid, B. M. (1978). EXAFS: New horizons in structure determinations. *Science, 200*(4349), 1441–1447. https://doi.org/10.1126/science.663627

Feather, C. E., & Koen, G. M. (1975). Mineralogy of the Witwatersrand reefs. *Minerals Science and Engineering, 7*, 189–224.

Findlay, S. D., Shibata, N., Sawada, H., Okunishi, E., Kondo, Y., Yamamoto, T., & Ikuhara, Y. (2009). Robust atomic resolution imaging of light elements using scanning transmission electron microscopy. *Applied Physics Letters, 95*(19), 10–13. https://doi.org/10.1063/1.3265946

Fletcher, J. S., Lockyer, N. P., & Vickerman, J. C. (2011). Developments in molecular SIMS depth profiling and 3D imaging of biological systems using polyatomic primary ions. *Mass Spectrometry Reviews, 30*(1), 142–174. https://doi.org/10.1002/mas.20275

Flohr, T. G., McCollough, C. H., Bruder, H., Petersilka, M., Gruber, K., Süß, C., Grasruck, M., Stierstorfer, K., Krauss, B., Raupach, R., Primak, A. N., Küttner, A., Achenbach, S., Becker, C., Kopp, A., & Ohnesorge, B. M. (2006). First performance evaluation of a dual-source CT (DSCT) system. *European Radiology, 16*(2), 256–268. https://doi.org/10.1007/s00330-005-2919-2

Freeth, T., Bitsakis, Y., Moussas, X., Seiradakis, J. H., Tselikas, A., Mangou, H., Zafeiropoulou, M., Hadland, R., Bate, D., Ramsey, A., Allen, M., Crawley, A., Hockley, P., Malzbender, T., Gelb, D., Ambrisco, W., & Edmunds, M. G. (2006). Decoding the ancient Greek astronomical calculator known as the Antikythera Mechanism. *Nature, 444*(7119), 587–591. https://doi.org/10.1038/nature05357

Frey, E. C., Taguchi, K., Kapusta, M., Xu, J., Orskaug, T., Ninive, I., Wagenaar, D., Patt, B., & Tsui, B. M. W. (2007). Microcomputed tomography with a photon-counting x-ray detector. *Medical Imaging 2007: Physics of Medical Imaging, 6510,* 65101R. https://doi.org/10.1117/12.711647

Frimmel, H., Emsbo, P., & Koenig, A. (2009). The source of Witwatersrand gold: Evidence from uraninite chemistry. *Proceedings of the Tenth Biennial SGA Meeting,* 353–355. http://www.geodynamik.geographie.uni-wuerzburg.de/fileadmin/09010000/_temp_/Frimmel_et_al._10th_SGA_Meeting_2009.pdf

Gawler, J., Sanders, M. D., Bull, J. W. D., du Boulay, G., & Marshall, J. (1974). Computer assisted tomography in orbital disease. *British Journal of Ophthalmology, 58*(6), 571–587. https://doi.org/10.1136/bjo.58.6.571

Godinho, J. R. A., Kern, M., Renno, A. D., & Gutzmer, J. (2019). Volume quantification in interphase voxels of ore minerals using 3D imaging. *Minerals Engineering, 144,* 106016. https://doi.org/10.1016/j.mineng.2019.106016

Graser, A., Johnson, T. R. C., Chandarana, H., & Macari, M. (2009). Dual energy CT: Preliminary observations and potential clinical applications in the abdomen. *European Radiology, 19*(1), 13–23. https://doi.org/10.1007/s00330-008-1122-7

Ikeda, S., Nakano, T., Tsuchiyama, A., Uesugi, K., Suzuki, Y., Nakamura, K., Nakashima, Y., & Yoshida, H. (2004). Nondestructive three-dimensional element-concentration mapping of a Cs-doped partially molten granite by x-ray computed tomography using synchrotron radiation. *American Mineralogist, 89,* 1304–1313.

Iniewski, K., Chen, H., Bindley, G., Kuvvetli, I., & Budtz-Jorgensen, C. (2007). Modeling charge-sharing effects in pixellated CZT detectors. *2007 IEEE Nuclear Science Symposium Conference Record, 6,* 4608–4611. https://doi.org/10.1109/NSSMIC.2007.4437135

Iovea, M., Oaie, G., Ricman, C., Mateiasi, G., Neagu, M., Szobotka, S., & Duliu, O. G. (2009). Dual-energy x-ray computer axial tomography and digital radiography investigation of cores and other objects of geological interest. *Engineering Geology, 103*(3), 119–126. https://doi.org/10.1016/j.enggeo.2008.06.018

Johnson, T. R. C., Krauß, B., Sedlmair, M., Grasruck, M., Bruder, H., Morhard, D., Fink, C., Weckbach, S., Lenhard, M., Schmidt, B., Flohr, T., Reiser, M. F., & Becker, C. R. (2007). Material differentiation by dual energy CT: Initial experience. *European Radiology, 17*(6), 1510–1517. https://doi.org/10.1007/s00330-006-0517-6

Kanngießer, B., Malzer, W., & Reiche, I. (2003). A new 3D micro x-ray fluorescence analysis set-up—First archaeometric applications. *Nuclear Instruments and Methods in Physics Research, Section B: Beam Interactions with Materials and Atoms, 211*(2), 259–264. https://doi.org/10.1016/S0168-583X(03)01321-1

Kelly, T. F., & Miller, M. K. (2007). Invited review article: Atom probe tomography. *Review of Scientific Instruments, 78*(3). https://doi.org/10.1063/1.2709758

Ketcham, R. A., & Carlson, W. D. (2001). Acquisition, optimization and interpretation of x-ray computed tomographic imagery: Applications to the geosciences. *Computers & Geosciences, 27*(4), 381–400.

Knoll, G. F. (2000). *Radiation Detection and Measurement*. New York, Wiley.

Kumpova, I., Vopalensky, M., Fila, T., Kytyr, D., Vavrik, D., Pichotka, M., Jakubek, J., Kersner, Z., Klon, J., Seitl, S., & Sobek, J. (2018). On-the-fly fast x-ray tomography using a CdTe pixelated detector—Application in mechanical testing. *IEEE Transactions on Nuclear Science, 65*(12), 2870–2876. https://doi.org/10.1109/TNS.2018.2873830

Lambert, J. H., Klett, M. J., & Detlefsen, C. P. (1760). *J. H. Lambert ... Photometria Sive De Mensura Et Gradibus Luminis, Colorum Et Umbrae*. Klett. http://slubdd.de/katalog?TN_libero_mab2)500288888

Liipo, J., Hicks, M., Takalo, V. P., Remes, A., Talikka, M., Khizanishvili, S., & Natsvlishvili, M. (2019). Geometallurgical characterization of South Georgian complex copper-gold ores. *Journal of the Southern African Institute of Mining and Metallurgy, 119*(4), 333–338. https://doi.org/10.17159/2411-9717/565/2019

Maiorino, M., Pellegrini, G., Blanchot, G., Chmeissani, M., Garcia, J., Martinez, R., Lozano, M., Puigdengoles, C., & Ullan, M. (2006). Charge-sharing observations with a CdTe pixel detector irradiated with a57Co source. *Nuclear Instruments and Methods in Physics Research, Section A, 563*(1), 177–181. https://doi.org/10.1016/j.nima.2006.01.090

Maire, E., & Withers, P. J. (2014). Quantitative x-ray tomography. *International Materials Reviews, 59*(1), 1–43. https://doi.org/10.1179/1743280413Y.0000000023

Meftah, R., Van Stappen, J., Berger, S., Jacqus, G., Laluet, J. Y., Guering, P. H., Van Hoorebeke, L., & Cnudde, V. (2019). X-ray computed tomography for characterization of expanded polystyrene (EPS) foam. *Materials, 12*(12). https://doi.org/10.3390/ma12121944

Merkulova, M., Murdzek, M., Mathon, O., Glatzel, P., Batanova, V., & Manceau, A. (2019). Evidence for syngenetic micro-inclusions of As 3+ - and As 5+ -containing Cu sulfides in hydrothermal pyrite. *American Mineralogist, 104*(2), 300–306. https://doi.org/10.2138/am-2019-6807

Mori, D., & Yamada, K. (2007). A review of recent applications of EPMA to evaluate the durability of concrete. *Journal of Advanced Concrete Technology, 5*, 285–298. https://doi.org/10.3151/jact.5.285

Nygren, H., Hagenhoff, B., Malmberg, P., Nilsson, M., & Richter, K. (2007). Bioimaging TOF-SIMS: High resolution 3D imaging of single cells. *Microscopy Research and Technique, 70*(11), 969–974. https://doi.org/10.1002/jemt.20502

Pelgrim, G. J., van Hamersvelt, R. W., Willemink, M. J., Schmidt, B. T., Flohr, T., Schilham, A., Milles, J., Oudkerk, M., Leiner, T., & Vliegenthart, R. (2017). Accuracy of iodine quantification using dual energy CT in latest generation dual source and dual layer CT. *European Radiology, 27*(9), 3904–3912. https://doi.org/10.1007/s00330-017-4752-9

Penner-Hahn, J. E. (1998). Structural characterization of the Mn site in the photosynthetic oxygen-evolving complex. In H. A. O. Hill, P. J. Sadler, & A. J. Thomson (Eds.), *Metal Sites in Proteins and Models Redox Centres* (pp. 1–36). Springer, Berlin, Heidelberg. https://doi.org/10.1007/3-540-62888-6_1

Pennicard, D., Fleta, C., Bates, R., O'Shea, V., Parkes, C., Pellegrini, G., Lozano, M., Marchal, J., & Tartoni, N. (2009). Charge sharing in double-sided 3D Medipix2 detectors. *Nuclear Instruments and Methods in Physics Research, Section A, 604*(1), 412–415. https://doi.org/10.1016/j.nima.2009.01.095

Prekas, G., Sabet, H., Bhandari, H., Derderian, G., Robertson, F., Kudrolli, H., Stapels, C., Christian, J., Kleinfelder, S., Cool, S., D'Aries, L., & Nagarkar, V. (2012). Direct and indirect detectors for X-ray photon counting systems. *IEEE Nuclear Science Symposium Conference Record.* https://doi.org/10.1109/NSSMIC.2011.6154354

Rantzsch, U., Gauert, C. D. K., van der Westhuizen, W. A., Duhamel, I., Cuney, M., & Beukes, G. J. (2011). Mineral chemical study of U-bearing minerals from the Dominion Reefs, South Africa. *Mineralium Deposita, 46*(2), 187–196. https://doi.org/10.1007/s00126-010-0317-4

Redus, R., Pantazis, J., Pantazis, T., Huber, A., & Cross, B. (2009). Characterization of CdTe detectors for quantitative x-ray spectroscopy. *Nuclear Science, IEEE Transactions On, 56*, 2524–2532. https://doi.org/10.1109/TNS.2009.2024149

Ren, L., Zheng, B., & Liu, H. (2018). Tutorial on x-ray photon counting detector characterization. *Journal of X-Ray Science and Technology, 26*(1), 1–28. IOS Press. https://doi.org/10.3233/XST-16210

Reyes, F., Lin, Q., Udoudo, O., Dodds, C., Lee, P. D., & Neethling, S. J. (2017). Calibrated x-ray micro-tomography for mineral ore quantification. *Minerals Engineering, 110*, 122–130. https://doi.org/10.1016/j.mineng.2017.04.015

Rinaldi, R., & Llovet, X. (2015). Electron probe microanalysis: A review of the past, present, and future. *Microscopy and Microanalysis, 21*(5), 1053–1069. https://doi.org/10.1017/S1431927615000409

Robinson, M. A., Graham, D. J., & Castner, D. G. (2012). ToF-SIMS depth profiling of cells: Z-correction, 3D imaging, and sputter rate of individual NIH/3T3 fibroblasts. *Analytical Chemistry, 84*(11), 4880–4885. https://doi.org/10.1021/ac300480g

Sayers, D. E., Stern, E. A., & Lytle, F. W. (1971). New technique for investigating noncrystalline structures: Fourier analysis of the extended x-ray-absorption fine structure. *Physical Review Letters, 27*(18), 1204–1207. https://doi.org/10.1103/PhysRevLett.27.1204

Schulman, T. (2006). *Si, CdTe and CdZnTe Radiation Detectors for Imaging Applications.* University of Helsinki, Finland.

Sittner, J., Godinho, J. R. A., Renno, A. D., Cnudde, V., Boone, M., De Schryver, T., Van Loo, D., Merkulova, M., Roine, A., & Liipo, J. (2020). Spectral x-ray computed micro tomography: 3-dimensional chemical imaging. *X-Ray Spectrometry, September*, 1–14. https://doi.org/10.1002/xrs.3200

Stock, S. R. (2008). Recent advances in x-ray microtomography applied to materials. *International Materials Reviews, 53*(3), 129–181. https://doi.org/10.1179/174328008X277803

Taguchi, K., Zhang, M., Frey, E. C., Wang, X., Iwanczyk, J. S., Nygard, E., Hartsough, N. E., Tsui, B. M. W., & Barber, W. C. (2011). Modeling the performance of a photon counting x-ray detector for CT: Energy response and pulse pileup effects. *Medical Physics, 38*(2), 1089–1102. https://doi.org/10.1118/1.3539602

Tanaka, R., Yuge, K., Kawai, J., & Alawadhi, H. (2017). Artificial peaks in energy dispersive x-ray spectra: Sum peaks, escape peaks, and diffraction peaks. *X-Ray Spectrometry, 46*(1), 5–11. John Wiley and Sons Ltd. https://doi.org/10.1002/xrs.2697

Urban, K. W. (2008). Studying atomic structures by aberration-corrected transmission electron microscopy. *Science, 321*(5888), 506–510. https://doi.org/10.1126/science.1152800

Usman, S., & Patil, A. (2018). Radiation detector dead time and pile up: A review of the status of science. *Nuclear Engineering and Technology, 50*(7), 1006–1016. Korean Nuclear Society. https://doi.org/10.1016/j.net.2018.06.014

Van Assche, F., Vanheule, S., Van Hoorebeke, L., & Boone, M. N. (2021). The spectral x-ray imaging data acquisition (Spexidaq) framework. *Sensors (Switzerland), 21*(2), 1–19. https://doi.org/10.3390/s21020563

Veale, M. C., Bell, S. J., Jones, L. L., Seller, P., Wilson, M. D., Allwork, C., Kitou, D., Sellin, P. J., Veeramani, P., & Cernik, R. C. (2011). An ASIC for the study of charge sharing effects in small pixel cdznte x-ray detectors. *IEEE Transactions on Nuclear Science, 58*(5 PART 2), 2357–2362. https://doi.org/10.1109/TNS.2011.2162746

Wang, X., Meier, D., Mikkelsen, S., Maehlum, G. E., Wagenaar, D. J., Tsui, B. M. W., Patt, B. E., & Frey, E. C. (2011). MicroCT with energy-resolved photon-counting detectors. *Physics in Medicine and Biology, 56*(9), 2791–2816. https://doi.org/10.1088/0031-9155/56/9/011

Yano, J., & Yachandra, V. K. (2009). X-ray absorption spectroscopy. *Photosynthesis Research, 102*(2), 241–254. https://doi.org/10.1007/s11120-009-9473-8

Zhou, W., Lane, J. I., Carlson, M. L., Bruesewitz, M. R., Witte, R. J., Koeller, K. K., Eckel, L. J., Carter, R. E., McCollough, C. H., & Leng, S. (2018). Comparison of a photon-counting-detector CT with an energy-integrating-detector CT for temporal bone imaging: A cadaveric study. *American Journal of Neuroradiology, 39*(9), 1733–1738. https://doi.org/10.3174/ajnr.A5768

Annexure A1

Summary of Investigated Sample Material with the Shape, Formula, Size, and Scan Settings

Sample	Shape	Chemical Formula	Size in mm	Tube Voltage in keV	Tube Power in W	Voxel Size in μm	Filter
Palladium	Wire	Pd	0.5	160	12	45.3	No
Silver	Sphere	Ag	0.1	160	24.5	24.5	No
Cadmium	Foil	Cd	0.125	160	10.0	45.5	No
Indium	Wire	In	0.5	160	10.0	45.5	No
Altaite	Particles	PbTe	0.5	160	15.0	15.0	No
Baryte	Particles	$BaSO_4$	2.0	160	20.0	30.1	No
Lanthanum	Foil	La	0.1	160	20.0	100.6	No

Sample	Shape	Chemical Formula	Size in mm	Tube Voltage in keV	Tube Power in W	Voxel Size in μm	Filter
Cerium oxide	Powder	CeO_2	0.05	160	27.2	27.2	No
Samarium oxide	Powder	Sm_2O_3	0.05	160	27.2	27.2	No
Neodymium oxide	Foil	Nd_2O_3	0.1	160	20.0	100.6	No
Europium oxide	Powder	Eu_2O_3	0.05	160	25.0	100.6	No
Gadolinium	Foil	Gd	0.1	160	20.0	100.6	No
Dysprosium	Foil	Dy	0.1	160	20.0	100.6	No
Erbium	Foil	Er	0.125	160	20.0	100.6	No
Hafnium	Wire	Hf	0.5	160	13.0	45.5	No
Tantalum	Foil	Ta	0.0256	160	10.0	45.5	No
Tungsten	Particle	W	0.25	160	27.2	27.2	No
Rhenium	Wire	Re	0.5	160	12.0	45.5	No
Iridium	Wire	Ir	0.25	160	12.0	45.5	No
Platinum	Wire	Pt	1.0	160	20.0	40.3	No
Gold	Particles	Au	0.1	160	23.0	44.3	No
Lead	Wire	Pb	0.1	160	20.0	40.3	No
Bismuth	Particles	Bi	1.0	160	25.0	45.5	No
Monazite	Particles	$(Ce, La, Nd, Th) PO_4$	1.0	160	25.0	40.3	No
Uraninite	Particle	UO_2	0.1	160	20.0	90.0	No
Au-W-Pb mixture CT	Mixture	SiO_2, Au, W, Pb	10	160	15.0	11.5	No
Au-W-Pb mixture Sp-CT	Mixture	SiO_2, Au, W, Pb	10	160	20.0	52.5	No
Witwatersrand sample 1 micro-CT scan	Drill core	–	25	180	70	70	2 mm Cu
Witwatersrand sample 1 Sp-CT-scan	Drill core	–	25	180	100	100	2 mm Cu
Witwatersrand sample 3D Sp-CT scan	Drill core	–	25	160	20	90	No

6

Range Verification by Means of Prompt-Gamma Detection in Particle Therapy[1]

Aleksandra Wrońska and Denis Dauvergne

CONTENTS

6.1 Introduction

Over 70 years ago Robert Wilson proposed the use ion beams for tumor treatment [2]. He combined the theoretical developments of Hans Bethe describing energy losses of protons and heavier ions traversing a medium, with the technical achievements of Ernest O. Lawrence, who built the first cyclotron. Currently, ion beam therapy (known also as particle therapy) is among most important methods of tumor treatment, next to surgery, chemotherapy, conventional radiotherapy and emerging immunotherapy (Nobel prize in medicine 2018).

DOI: 10.1201/9781003218364-6

Even though heavier ions exhibit larger biological effectiveness of the deposited dose and allow better dose conformation to the tumor volume, their use is strongly limited by much larger construction and maintenance costs as well as technological complexity of the treatment sites. Consequently, protons remain the main type of ions used for ion beam radiotherapy, as proposed in Wilson's original paper.

The field of particle therapy (PT) was rapidly developing over those seven decades: from the knowledge about how to fully exploit the potential of the Bragg peak, through the benefits of the fractionated delivery, to forming the plateau in the dose distribution via the spread-out Bragg-peak technique. In the 1990s there was a transfer of technology from research centers to hospitals. About the same time, commercial companies entered the game and around 2000 already the off-the-shelf solutions for ion beam therapy became available. The 21st century has brought further progress: multi-field irradiation, the use of gantries and scanning pencil beams, modern CT- and PET-assisted evaluation of treatment plans based on sophisticated computer simulations [3]. Those developments made the field flourish, the number of ion-beam therapy centers grew rapidly and is currently close to 100, with another 40 under construction [4].

PT takes advantage of the unique features of ion interaction with matter with its characteristic Bragg peak, allowing to almost arbitrarily shape the distribution of deposited dose. The well-defined ion range in the medium leads to a rapid falloff in the dose depth profile, allowing to spare deeper located tissues. However, the existence of the Bragg peak and its steep distal falloff in the dose depth profile which are the main advantages of charged heavy particles over X-rays make the proton and ion therapy more susceptible to errors. A detailed quantitative analysis of sources of proton beam range uncertainties, such as patient positioning, patient anatomical changes and translation of computed tomography (CT) images to water equivalent units, was performed by Paganetti [5]. In clinical practice, this leads to the necessity of using safety margins, i.e., enlarging the irradiated volume. Those margins are defined differently in each PT center, typical values vary from a few millimeters up to over a centimeter for deeply located malignancies. They are necessary to account for uncertainties of the beam range, but they also mean an extra dose delivered to patient's healthy tissues. The way to reduce beam range uncertainties and consequently safety margins would mean a lower dose delivered to a patient and thus lower risk of long-term side effects. This can be achieved by online monitoring of the deposited dose distribution during treatments [6–8]. The necessity to develop appropriate tools was pointed out by NuPECC and made one of the highlights in the 2014 report [9].

There are different approaches to ion beam therapy monitoring currently under development. Most of them are based on the idea of exploiting by-products of patient irradiation with a proton beam: prompt gamma radiation, which is the main focus of this chapter, but also acoustic wave [10, 11], secondary protons during carbon therapy [12], secondary electron bremsstrahlung [13], neutrons [14] and β^+-emitters [15]. For the latter, the range verification consists in imaging the decay of the β^+-emitters in two photons, by means of a positron emission tomography (PET) device. In this domain, spectacular progress was seen in the last years. Until recently, PET scans were considered useful for post-irradiation control rather than *in vivo* range verification [16] due to long acquisition times necessary to collect sufficient statistics resulting from limited acceptance of PET scanners and long lifetimes of β^+ emitters. This, in turn, opened a way to biological washout which blurred the image. Dendooven et al. proposed to eliminate this obstacle by focusing on short-lived β^+-emitters [17] and tested the experimental feasibility of such a solution [18]. A real breakthrough, however, came with the development of a high-acceptance, high-efficiency INSIDE setup currently operated at CNAO, Pavia [19]. In fact, this setup combines in-beam PET with secondary particle tracking. The first clinical test with a patient was reported in 2018 [20]. Range agreement within 1 mm was demonstrated, which proved the method and setup useful for verification of dose distributions in proton and carbon therapy. The issue of compliance of in-beam PET with the delivery mode remains, which necessitates either dual head (like for the former GSI prototype and INSIDE), or specific geometries like the dual ring Open-PET developed at NIRS [21]. In-beam PET records signal during beam pauses, which makes it suitable with low duty-cycle accelerators, otherwise in-room or offline PET imaging can be performed with commercial PET-CT, at the expense of the washout effect and long acquisition times. Note also that ongoing developments on ToF-PET would be beneficial toward real-time imaging to reduce background. Lastly, PET presents an additional asset in the case of carbon/oxygen therapy, since projectile-like β^+ emitters decay close to the Bragg peak location, giving rise to an enhanced signal in that region.

Of the listed types of secondary radiations, prompt gamma radiation (PG), typically of a few MeV energy (from 1–2 MeV up to about 7 MeV), has the advantage that it is produced instantaneously (ps time scale), and penetrates the tissue easily, mostly without any interaction on its way out. Thus, the information about the location of its origin is not distorted. The characteristics of the PG radiation allows to pinpoint the Bragg peak position [22–24]. Figure 6.1 illustrates the correlation between the deposited-dose depth-profile and the various secondary-particle vertex distributions caused by inelastic collisions in the case of proton beams. In the present case, the vertex distributions correspond to the particles with energy above 1 MeV emerging from a water cylinder of 15 cm diameter and 40 cm length, irradiated by 160 MeV protons (Geant4 simulations, adapted from ref. [25]). A clear correlation between PG vertices and dose is observed, although part of the PGs are emitted by secondary particles (mostly neutrons). Note, that the neutron vertex distribution is also well correlated to the projectile range, but, for the latter, the information on the vertex location will be blurred by scattering in the phantom before detection.

A variety of different approaches exploiting PGs are currently being developed, making use of different properties of PG radiation: its spatial, temporal

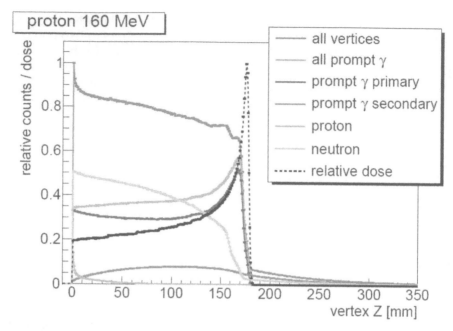

FIGURE 6.1
Relative vertex (emission point) distributions of secondary radiation induced by 160 MeV protons incident on a water phantom (15 cm diameter, 40 cm length). The vertices are scored when the particles emerge from the phantom surface with an energy greater than 1 MeV. The longitudinal dose distribution is also presented. Figure adapted from ref. [25].

or energy distributions. What they have in common are the conditions and the environment in which the developed setups must be operational. Those constraints and challenges are discussed in the Section 6.2. Then follows the description of different PG-based methods of range verification, where they have been divided into two groups: imaging and non-imaging techniques (Sections 6.3 and 6.4). The closing chapter offers a handful of concluding remarks and outlook.

6.2 Environment and Challenges

The nuclear collision yield, giving rise to secondary radiation, is typically of the order of a percent per cm path length for protons. Thus, this represents a fairly high rate at clinical beam intensities. In overall, all accelerators deliver pencil-beam spots ($10^6 - 10^8$ protons, $10^4 - 10^6$ carbon ions) in a fraction of a second (typically 10–100 ms), which, in turn, yields to PG emission rates of the order of 10^7 s^{-1} in 4π. A typical PG energy spectrum obtained in an

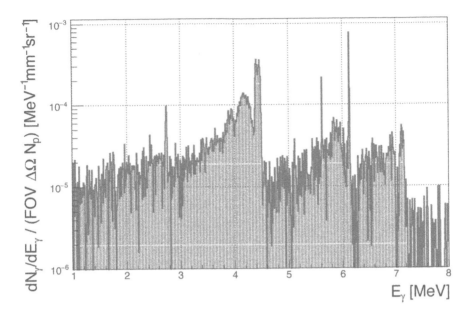

FIGURE 6.2

Energy spectrum of PG rays emitted from a 1-mm layer of a PMMA phantom located 2 mm before the Bragg peak, irradiated with a 70-MeV proton beam. The observed field of view was spatially separated from the proximal part of the phantom with a 20-cm air gap to suppress the radiation produced upstream of the depth range of interest. The data, recorded with an HPGe detector with an active Compton shield, is a part of the set presented in ref. [24]. The yield is normalized by the number of impinging protons, field of view and detector acceptance, corrected for the detection efficiency of HPGe, but not for the efficiency of the ACS (increase by 10–25%, depending on energy).

experiment with phantom irradiation is shown in Figure 6.2. As evoked by Pausch et al. [26], detectors need to cope with both high instantaneous rates, and also with strongly varying count rates within ms. In addition, instantaneous intensities vary strongly from one accelerator type to another. Indeed, the time structure of the delivered beam depends on the type of accelerator, and on the ion species used for treatment. Table 6.1 illustrates typical characteristics of proton and carbon ion beams at several accelerators (note that some other synchrotrons have higher intensities [27]; the numbers in Table 6.1 were taken from ref. [25]). So far, only cyclotrons, synchrotrons and synchrocyclotrons have been exploited for therapy. All machines have a microstructure with periods in the range of nanoseconds, which corresponds to the periodicity of the ion bunches inside the circular accelerator before extraction. On top of this microstructure, there is a macrostructure corresponding to the time separating two cycles of injection/acceleration/extraction, except for cyclotrons for which it is continuous. Therefore, the average PG yields given earlier may have much higher instantaneous values, depending on the accelerator duty cycle (~10% for cyclotrons and synchrotrons, ~10^{-3} for

TABLE 6.1

Typical Time Structures of Various Clinical Accelerator Types (Numbers from Ref. [25])

		Synchrotron (CNAO, HIT)		Cyclotron (IBA, Varian)	Synchro-Cyclotron (S2C2, IBA)
		C ions		Protons	
Typical intensity (ions/s)		10^7	10^9	10^{10}	$\sim 10^{10}$
Macrostructure	Period (s)	1–10			10^{-3}
Microstructure	Bunch width (ns)	20–50		0.5–2	8
	Period (ns)	100–200		10	16 (at extraction)
	Ions/bunch	2–5	200–500	200	4000

synchro-cyclotrons). This will have an impact for (i) time of flight measurement for background rejection and (ii) random coincidence rates when a signal corresponds to coincidences of several sub-detectors (Compton cameras).

For ions heavier than protons, on the one side fragmentation rates are higher, since the fragmentation of the projectile itself occurs, but, on the other side, the number of incident projectiles needed to provide a given physical dose is smaller than for protons, due to the $1/Z^2$ dependence of energy loss (Z being the atomic number), and the lower longitudinal and angular scattering of heavier ions. In addition, the Relative Biological Effectiveness (RBE) leads to a further reduction of the number of ions necessary to deposit the desired biological dose in the spread-out Bragg peak region with higher-Z ions. Therefore, the number of available PGs for a given beam spot decreases when Z increases: for carbon ions it is typically two orders of magnitude smaller than for protons.

6.3 Imaging Techniques

The approaches described in this section aim at range verification by means of the analysis of spatial distribution of PG vertices, i.e., are based on imaging. That can be performed in one, two or even three dimensions, depending on the complexity of the detection setup and analysis methods.

6.3.1 Knife-Edge Shaped Slit-Camera (KES)

A slit camera is probably the most natural approach, an analog of a pinhole camera, but providing one-dimensional (1D) imaging, see Figure 6.3. Feasibility studies were performed by several groups [28–30]. The worldwide

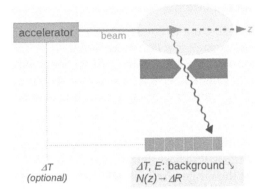

FIGURE 6.3
Principle of operation of a slit camera. PG rays are imaged using a single slit onto a position-sensitive detector. Spatial hit distribution allows to conclude about range shifts, while time and energy information may be used to suppress background. Figure adapted from ref. [1].

first test of a slit-camera setup in clinical conditions during patient treatment was reported in ref. [31] by the Dresden group in collaboration with OncoRay and IBA. Their camera was mounted on a movable trolley and consisted of a knife-edge shaped tungsten collimator and an array of 4 mm wide scintillating strips made of LYSO, readout by silicon photomultipliers (SiPMs). The treatment modality was passively scattered proton therapy and the authors focused on the monitoring of inter-fractional range variations rather than absolute range determination. Those variations were determined to be within ±2 mm.

The first clinical trial including prompt-gamma verification during a whole treatment was performed with a second knife-edge prototype from IBA at Philadelphia, using pencil-beam scanning conditions [32]. The analysis of the data sets acquired during six fractions of a brain treatment, performed by comparing measurements to simulations, revealed that the precision in range verification was better than the safety margins applied in the treatment plan.

A collimated system with many knife-edge shaped slits was studied by Ready et al. [33, 34]. In this design the slits formed a 2D pattern, thereby allowing 2D imaging. The gamma quanta were detected using a position-sensitive LFS detector. Experimental tests were performed with a 50-MeV proton beam of clinical beam current, showing the precision of relative Bragg peak localization of about 1 mm (2σ) at the delivery of only 1.8×10^8 protons. A serious limitation, though, was the distance of the collimator front face to the beam axis of only 8 cm. Moving the setup further away from the beam axis would require re-optimization of the collimator, but would also lead to a decrease of detector geometrical acceptance and thus to a smaller collected statistics. Measurements with larger proton beam energies are also needed to verify the performance in realistic proton therapy conditions.

6.3.2 Multi-Slit Camera (MPS)

A generalization of a slit-camera concept is a multi-slit camera, which presents the theoretical advantage of a non-restricted field of view (see Figure 6.4). Smeets et al. compared experimentally (using beam energies of 100, 160 and 230 MeV) two collimated setups: one with their optimized knife-edge shaped collimator (KES), and the other with a multi-slit collimator (MPS) of the same weight [35], the slits parallel to each other. Each collimator was combined with the same detector setup, described in the first paragraph of this section. Due to a smaller field of view, KES provided imaging of the Bragg peak region only, while MPS allowed to image also the beam entrance point. However, in terms of the range retrieval, KES was shown to require only half of the dose MPS needed to obtained the same statistical precision. Thus, the authors concluded KES to be the favorable option for further development. However, such conclusions are biased by the constraints put on the MPS camera, which was not independently optimized, as in ref. [36].

A similar collimated setup was built and tested experimentally by Park et al. [37]. The geometry of both the collimator (coarser pattern) and the detector (finer granulation) was somewhat different than in case of ref. [35]. The detector was made of CsI(Tl) and readout by photodiodes, as more radiation-hard than SiPMs. The authors presented reconstructed gamma depth profiles for different beam energies from the range 95–186 MeV, and for each beam energy for four different numbers of delivered protons, between 7.5×10^7 and 7.5×10^9. It turned out possible to locate the distal dose falloff (d90%), defined as the depth of distal 90% dose in a depth-dose curve, within about 2–3 mm of error for the spots which were irradiated with at least 3.8×10^8 protons, regardless of the beam energy. In a typical realistic treatment plan such a dose is delivered only to some of the spots.

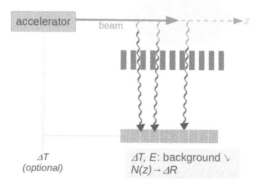

FIGURE 6.4
Principle of operation of a multi-parallel slit camera, similar to that of a single slit camera. However, due to a larger field of view, the setup allows to image not only the Bragg-peak region, but also the entrance point, providing full range control.

6.3.3 Gamma-Electron Vertex Imaging (GEVI)

In the gamma-electron vertex imaging (GEVI) method, PG rays are not detected directly. Instead, recoil electrons emitted from a beryllium converter as an effect of PG Compton scattering are detected and tracked. The concept was first proposed by a Korean group [38], that recently showed experimental results with test beams [39]. The electron tracking is done with the use of a three-stage hodoscope, shown schematically in Figure 6.5, consisting of two double-sided silicon strip detectors and a calorimeter made of plastic scintillator. The electron tracks are reconstructed based on the hit positions in the silicon detectors. An image is obtained from a back-projection of those tracks onto the imaging plane. Efficient background suppression is achieved by demanding a triple coincidence and imposing cuts on energy depositions in each detector module. The group performed test experiments with a high density polyethylene (HDPE) phantom and 6.24×10^9 delivered protons for seven beam energies between 90 and 180 MeV. For the purpose of beam range determination, the obtained two-dimensional (2D) images were projected onto the beam axis, forming 1D profiles. There, the observed distal falloff at each energy occurred at 94% of the proton range. According to the authors, the method allows to determine proton range with a precision of 2.7 mm without the necessity of using Monte Carlo simulations as reference. One should observe, though, that all results were obtained with the same phantom material. For phantoms with other elemental composition (or in case of a real patient treatment) the shape of the distal falloff may be different. Thus, it remains an open question whether the factor 0.94 is indeed

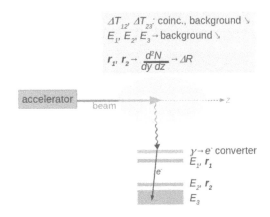

FIGURE 6.5
The GEVI method relies on detection if recoil electrons via Compton scattering of PG in a converter. Triggered by a triple coincidence, the system registers hit positions and energy depositions. Electron tracks are reconstructed and back-projected onto the imaging plane, forming a 2D image, which is then projected onto the beam axis, delivering information about beam range. Patterns of energy depositions are used to suppress background from other particle species. Figure adapted from ref. [1].

universal. What still needs to be explained is a much lower than expected imaging sensitivity of the system (1.6×10^{-6} vs. 1.2×10^{-5}), which is probably due to dead time. Nevertheless, the setup, mainly because of its small material budget and simplicity, is an appealing candidate for clinical applications.

6.3.4 Compton Camera (CC)

A direct reconstruction of a three-dimensional (3D) map of PG activity induced by an irradiation is possible with setups of the Compton camera (CC) type. In its classical form, such a setup consists of two modules: a scatterer and an absorber, as depicted in Figure 6.6. Its operation relies on detection of events in which an impinging gamma undergoes Compton scattering in the scatterer and subsequently is fully absorbed in the absorber, preferably via photoeffect. Registration of hit positions as well as associated energy depositions allows to reconstruct a cone of possible directions of the impinging gamma. Additional tracking of the Compton-scattered electron allows to constrain it further to a part of that cone. The image is formed by superposition of the reconstructed Compton cones and though in principle 3D, it is characterized by a rather poor resolution in the direction normal to the detector front face. This, however, can be cured by using a combination of two CCs observing the target from perpendicular directions. One additional difficulty in the case of PG detection is the broad spectrum of photon energies. This requires either the hypothesis of a full energy absorption, or the use of three-stage detection. The PG imaging by means of CCs of various designs was approached by several groups. In the following the main features of the studied CCs are described along with their obtained performance.

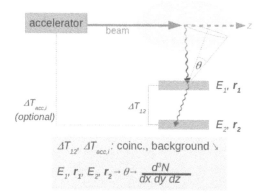

FIGURE 6.6
Principle of operation of a Compton camera. Hit positions and energy depositions in the two modules can be translated to a cone of possible directions of impinging PGs. Superposition of such cones delivers a 3D-map of PG activity. Timing information is used for background suppression. Figure adapted from ref. [1].

- Within the CLaRyS project, the Lyon group studied via simulations a CC with a scatterer made of double-sided silicon strip detectors and an absorber consisting of 100 streaked BGO crystals [40]. A semiconducting scatterer ensured very good position and energy resolution, and a scintillating absorber provided large detection efficiency. A similar setup was built in Munich by Thirolf et al., with a 0.5-mm thick Si scatterer and a monolithic LaBr$_3$ crystal readout by a multi-anode PMT as an absorber [41]. The image reconstruction procedure took into account also the electron track data. An image resolution of 3.7 mm was achieved when imaging point-like sources of 1.33 MeV energy. In order to increase the detection efficiency of the scatterer, it was considered to replace the scatterer with a pixelated GAGG detector developed in Japan (see here) [42].

- The Dresden group proposed and characterized a CC of the same type (semiconductor plus scintillator) [43]. For the scatterer, however, instead of the silicon detectors a set of two cross-strip CZT detectors was used, and the absorber was a single block of LSO, later replaced by a set of three segmented BGO detectors, each of them readout by a set of four classical photomultipliers. The tests performed with a 4.4-MeV gamma source proved the feasibility of imaging at such energies [44]. However, the determined setup efficiency and deduced number of expected registered PG events were by far insufficient for imaging at clinical dose rates. No extension to a clinical size was undertaken.

- The Baltimore group investigated a setup based solely on commercially available, semiconducting CZT detectors Polaris of the active area 2×2 cm^2 [45]. Those detectors have the advantage of excellent energy resolution, but have much worse time resolution and longer signals compared to fast scintillators. Polf et al. performed extensive performance tests of their multistage CC. Recently, results of measurements with a small-scale prototype under clinical conditions using HDPE phantoms were reported [46], proving the feasibility of 3D imaging. For that purpose the group developed sophisticated energy reconstruction and event selection methods. The paper reports sensitivity to detect beam range shifts of 2–3 mm, depending on the irradiation scheme. From the results of simulations benchmarked with the prototype measurements the authors conclude that a clinically relevant system of this architecture is feasible but requires a two-stage detector array as large as 4×12 cm^2.

- An orthogonal design using only scintillating detectors was proposed by the Valencia group. In the approach of Llosá et al., the MACACO Compton telescope consists of monolithic blocks of LaBr$_3$ scintillator readout by arrays of SiPMs [47]. In its first version, the setup consisted of two modules and its capability to image gamma sources of 2–7 MeV energy with 3–5 mm resolution was shown. Although the group demonstrated also the potential to observe range shifts of 10 mm for a 150-MeV beam, the available time resolution and detection efficiency still appeared as a problem. Further R&D comprised the use of different SiPM arrays and readout ASIC and allowed to improve the performance of a single detection plane. The imaging capability was demonstrated only with an array of ^{22}Na sources on a two-plane, not fully optimized setup featuring modified planes [48]. Results of beam tests are expected soon.

- Another scintillator-only design was proposed by the Japanese group. Their lightweight and compact (handheld) setup was built of modules consisting of small 'pixel' GAGG crystals of the dimensions $2 \times 2 \times 4$ mm^3 (2 layers) and of $2 \times 2 \times 2$ mm^3 (10 layers), readout by SiPMs [49]. The initial version of the setup did not have sufficient resolution to deliver conclusive information about possible range shifts in proton therapy. The group introduced modifications, among others reducing all pixels to be cubes of 2 mm and substituting SiPMs with multi-anode PMTs. In the following tests an attempt to image 4.4 MeV gammas emitted from PMMA phantoms irradiated with a 70-MeV proton beam was undertaken [50]. However, the measurements were performed at the conditions far from clinical-beam current of 3 pA and measurement time of 5 h allowed to collect vast statistics with a suppressed contribution of random coincidences. Although the capability to use 4.4 MeV PG rays for imaging was demonstrated, no quantitative analysis of range retrieval or feasibility of use in clinical conditions was presented.

- The SiFi-CC project pursued by the Cracow-Aachen collaboration aims at building a CC composed of fibers made of heavy scintillator readout by SiPMs. Laboratory tests and Monte Carlo simulations indicated that the optimal building blocks for such a setup would be $1 \times 1 \times 100$ mm^3 LYSO fibers [51]. High granularity of the detectors will allow to reduce the problem of pile-ups occurring at high-count rates. The core of the data acquisition system will be custom FPGA-based boards developed for the J-PET project [52], adapted to feature the ADC functionality. Based on the simulations, the expected imaging sensitivity of 0.001% and the contribution of random coincidences at clinical conditions make imaging of a single beam spot an achievable goal [53]. Currently the group is constructing a small-scale prototype, which will allow one to optimize detector design and performance.

- An alternative approach to Compton imaging was proposed by the Dresden group in collaboration with OncoRay [54]. The setup comprising Directional Gamma Radiation Detector (a segmented scintillator) is used in the modality known as Single Plane Compton

Imaging, taking advantage of the angular distribution of Compton scattering. However, the initial tests with two $2'' \times 2''$ CeBr$_3$ detectors showed that conclusive results useful for range verification require $10^3 - 10^4$ times the statistics obtained for a single spot. Although this seems hardly feasible to achieve, this modality of detection and data analysis can also be applied to conventional, multistage setups, and combined with cone-reconstruction analysis.

In parallel to hardware developments, a lot of effort is put into improving event selection and image reconstruction algorithms. The simplest reconstruction method is the line-cone reconstruction (intersection of the Compton cone with the incident ion trajectory). Livingstone et al. showed recently that this method is optimized when high timing resolution is used to select the right intersection point, corresponding to full Compton absorption [55]. Advanced reconstruction algorithms can handle more complex event topologies, e.g., three-interaction events which allow to determine the energy of the primary gamma [46]. Another approach, reported recently in [56], takes into account the probability of incomplete energy deposition of a primary gamma in the detector. The aspect of computation speed was addressed in ref. [57], where it was shown that a fast line-cone reconstruction technique yields significantly lower precision than an iterative maximum likelihood expectation maximization algorithm. An issue related to the yield of random coincidences between the several detection stages was addressed by several authors [57–59]. A solution for Compton imaging of PG should feature a compromise between the requirement of maximized detection efficiency and reduced background. The tests performed so far indicated that good time resolution allowing to reduce coincidence time and dead time to a minimum, high-throughput electronics and data acquisition system, and a large degree of granularity seem to be key features to build a setup operational at clinical conditions.

6.3.5 3D Imaging with Dual PG Coincidence

Panaino et al. make use of coincidence between two consecutive gamma-rays from the de-excitation of ^{16}O (2.75 MeV followed by 6.13 MeV with 25 ps decay time) to reconstruct in 3D the emission vertex position, by means of a nearly-4π multi-detector of high temporal resolution [60]. This is made possible since the two gamma emissions occur at the same location. In addition, the corresponding excitation cross section by proton impact is maximum at energies close to 13–14 MeV, thus the vertex distribution is maximum very close to the Bragg peak. Although it seems hardly conceivable to install a 4π multi-detector with spectroscopic and time resolution capabilities to perform such an imaging in clinical environment, this example, making use of time, spectroscopic and position information, presents an interesting approach of making full use of the setup potential.

6.4 Non-Imaging Techniques

The second group of methods of beam range verification comprises non-imaging techniques, which take advantage of the correlation between the beam range and the features of temporal or spectral characteristics of PG emission.

6.4.1 Prompt-Gamma Timing (PGT)

Instead of recording the spatial distributions of PG radiation, Golnik et al. proposed a novel concept of range assessment by prompt-gamma timing (PGT) [61]. The concept is based on the fact, that the transient time of ions in the patient tissue before stopping depends on the proton range. Only during this time PG photons can be produced. Since excited nuclei lifetimes associated to PG emission are shorter than this transient time, the ToF distribution of PG detection relative to the impact of protons is also correlated to the ion range. The concept was somehow approached by Testa et al. by considering ToF distributions of PG emitted from a collimator-restricted field of view, containing the Bragg peak position [62]. In practice, the easiest way to implement PGT is to use a fast detector for PG detection (preferably in backward direction so that flight times of the transient ion and PG are adding), and to synchronize the detection with the accelerator HF signal to provide a stop for ToF measurement, as depicted in Figure 6.7. However, this necessitates to adapt the time calibration whenever the beam energy changes. Moreover, the beam pulse width is an intrinsic limitation of the method, since the time distribution results from the convolution of individual transient times with

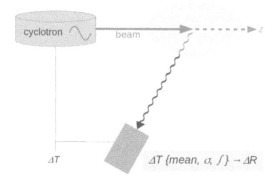

FIGURE 6.7
Principle of the PGT and PGPI methods. Both of them rely on recorded distributions of time between beam bunch and registered gamma rays. Analysis of the mean, width and integral of those distributions yields information about beam range. Note that ΔT can be obtained by means of a beam monitor triggering beam pulses, instead of the cyclotron HF signal. Figure adapted from ref. [1].

the pulse width. In particular this is not possible with long lasting beam pulses as delivered by synchrotrons. Alternatively to the cyclotron HF, a signal from a beam pulse monitor can be used for direct ToF measurement. The distribution of time elapsed between the two signals, the so-called PGT spectrum, allows to conclude about the proton range by comparing it with the PGT spectrum modelled in the simulations. The simplicity of the setup is an undeniable advantage of this approach.

After initial tests constituting a proof of principle [63], developments toward translation of the method to clinics have been undertaken and reported recently [64]. For this purpose, a series of measurements was performed using a phantom with air cavities and a close-to-realistic treatment plan. The detection unit consisted of a $CeBr_3$ crystal coupled to a classical photomultiplier tube (PMT). The scintillator material was chosen for its short decay time and excellent energy resolution. The signals were fed to and processed in a custom FEE unit U100 attached to the PMT, analyzed for a single spot or summed up over the whole iso-energy layer. In the first case there was clearly insufficient statistics, calling for additional detection units. Finally, the PGT spectra for a layer of about 10^2 iso-energy spots were conclusive and provided a precision of 2–3 mm for eight detection units. Further developments by the group are in progress, including the setup of extra detector units, and the increase of their rate capacity, as well as more tests in the clinical conditions.

Marcatili et al. studied the possibility to perform high time-resolution PGT at reduced beam intensity, using a diamond beam trigger working in single-proton counting mode, and fast gamma-ray scintillation detectors (BaF_2 and $LaBr_3$) [65]. An experimental resolution of 100 ps rms was obtained on the PG-ToF with 65 MeV proton beams using such a system. The simulations undertaken, accounting for such a resolution with the same detection setup, led to a sensitivity to detect an air cavity of 3 mm located at about 1 cm from the end of the range with 95% confidence level with a single beam spot of 10^8 protons. The improvement with respect to the spot-based PGT technique comes from (i) the independence on the beam spot duration and (ii) the better ToF to proton-position correlation due to the excellent timing resolution. A reconstruction method is proposed in ref. [66].

6.4.2 Prompt-Gamma Peak Integrals (PGPI)

In its original form, the PGT method discussed earlier uses the statistical momenta of the timing distributions, i.e., its mean and width. Krimmer et al. proposed instead of those features to measure peak integrals of the PGT spectra, while relaxing the constraints on time resolution to a level of 1–2 ns, so as to select only PG emitted inside the patient tissues, and not from the beam nozzle, and to eliminate neutron-induced background. This variation is called Prompt Gamma Peak Integral (PGPI). In an experiment with $LaBr_3$ and BaF_2 detectors, a PMMA phantom and a beam of about 65 MeV energy passing through a modulator wheel, the sensitivity of peak integrals to the proton range in the phantom was demonstrated. The used energy range is a lower limit of that used clinically, a favorable one for PG measurements due to lower count rate and neutron background. Thus the authors used Monte Carlo methods, benchmarked at that energy, to simulate the setup response at higher energies. A precision of about 3 mm in range verification was obtained for a 10^8-proton pencil beam and a single detector of 25 msr solid angle. In the multiple-detector setup it is possible to additionally detect transverse target misplacements by comparing signals from different detector units.

6.4.3 Prompt-Gamma Spectroscopy (PGS)

PGS is an approach proposed by Verburg et al. [67, 68] from MGH Boston, exploiting spectral characteristics of the PG emission. It requires the registration of PG spectrum with a spectroscopic detector, collimated and focused on a part of the beam path a few millimeters upstream of the Bragg peak. Figure 6.8 shows a schematic of such a setup. Cross sections of inelastic proton interactions with the traversed medium are energy dependent. Moreover, depth profiles of gamma emission for different discrete transitions, e.g., from the reactions such as $^{12}C(p, p\gamma_{4.44\text{MeV}})^{12}C$ and $^{16}O(p, p\gamma_{6.13\text{MeV}})^{16}O$, are different. Therefore, the ratios of yields of gammas from various spectral lines can be recalculated to the mean proton energy in the observed slice and, consequently, connected to the residual range. Last but not least, the mentioned ratios allow one to conclude about the elemental composition of the tissues in the observed region. The proof-of-principle experiments were performed using an $LaBr_3$ detector with an active anti-Compton

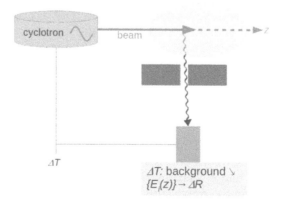

FIGURE 6.8
Principle of the PGS method: PG energy is registered by a spectroscopic detector. From yields ratios of PGs originating from different discrete transitions one can deduce the residual range. Physical background is reduced by the ACS and time filtering. Figure adapted from ref. [1].

shield (ACS) made of BGO, both with classical PMTs readout. As a collimator a 15-cm thick lead brick was used. The ACS response as well as the hit time information relative to the cyclotron timing were used to suppress Compton continuum and neutron-induced components in the spectra, respectively. However, in the first experiments also the small statistics issue appeared and a summation over a whole iso-energy layer had to be performed. To overcome this, a new setup with eight $LaBr_3$ detectors, a tungsten collimator and a dedicated custom high-throughput electronics ($10^7 s^{-1}$) was constructed as a clinical prototype [69]. A support frame made it possible to adjust the apparatus to the irradiation direction. Using sophisticated algorithms of background suppression the group achieved a 1.1-mm precision of beam range verification in measurements with a water phantom and a proton beam of clinically relevant current and dose. As the next steps, commissioning with an anthropomorphic phantom and clinical study with patients are planned.

The application of the PGS method for heavier ion beams was recently presented by Dal Bello et al. [70]. The measurements were performed in HIT Heidelberg using proton, ^{4}He, ^{12}C and ^{16}O beams accelerated in a synchrotron, with most effort devoted to study the ^{12}C beams. The setup comprised a $CeBr_3$ detector with a BGO ACS, both equipped with classical PMTs and readout by Flash ADC modules, and a 12.5-cm thick tungsten collimator. Time reference information was provided by a beam trigger built of scintillating fibers located in front of the target. The group showed that the developed small-scale prototype setup along with the data analysis routines allowed for absolute measurement of the Bragg peak position for synchrotron-accelerated ^{12}C beams at clinically relevant energies and intensities. However, the reported submillimetric precision of range retrieval required delivery of $8 \cdot 10^9$ primary ions, which is much more than the total in a typical treatment plan. Realistically, range verification at 2 Gy dose with 2-mm precision on a single-spot basis seems unfeasible, and on a plane basis would require an increase of the setup efficiency of 15–40 times, depending on the beam ion species.

6.5 Conclusion and Outlook

Various approaches to online monitoring of PT have been presented. Some of them have already been successfully tested clinically with patients. The only commissioned setup to date providing a 3D-verification of the dose is the PET scanner INSIDE. Up to now, online control with PG has been restricted to 1D range verification and proton beams. It remains a challenge to construct a 3D-imaging device exploiting PG radiation which would offer similar performance with a smaller material budget and dimensions.

Some authors suggest that the prospective future developments should be based on a more generalized approach and make full use of information delivered by a cutting-edge, optimized setup [26]. The 3D imaging with PG, e.g., like that presented in Section 6.3.5, combined with PGPI and PGT with excellent timing resolution could serve as an example. The use of PG with heavier ions (from helium to oxygen) raises the issue of lower statistics and higher neutron background. Therefore, different strategies may be adopted for beams of different ion species. For carbon ions, integral counting methods combined with the use of ToF are certainly among the most promising options.

Robust identification of range errors by PG detection techniques requires appropriate algorithms. Powerful tools of machine learning have been employed to interpret the data obtained from simulations and experiment. For instance, using the existing IBA-camera, Khamfongkhruea et al. used training data sets to identify the possible sources of errors (e.g., CT-based range prediction, patient setup, anatomical changes between fractions) [71]. Then they built a decision tree in order to classify the error sources. This is a first step using learning approaches, and further developments are needed before clinical implementation, in particular the use of realistic PG acquisition (including noise and reduced statistics). Another attempt of using machine learning methods to convert simulated PG vertex distribution into the deposited dose distribution was presented in ref. [72]. The natural next step should be to apply it to experimental data. Note that, more generally, Artificial Intelligence is rapidly taking part in the improvement of PG range verification (see e.g., refs [73, 74] for filtering and reconstruction in Compton imaging).

The influence of counting rates has been discussed for various detection modalities. In particular, range verification by means of PG detection at clinical beam intensities needs to cope, on the one side, with high instantaneous counting rates, rapidly varying (beam on and off at every spot and even at each accelerator pulse). On the other side, all verification methods are facing the issue of counting statistics in order to reach the best precision with the shortest irradiation period, possibly a single irradiation spot. This was discussed in the paper by Pausch et al. [26]. Verification methods such as Compton imaging seem to require reduced intensities, at least for a few spots. Dauvergne et al. reviewed the advantages of beam intensity reduction for a few spots at the beginning of a treatment fraction, so that irradiation can be controlled in single ion counting conditions, by means of a fast beam monitor and secondary radiation detection. Obvious advantages for Compton imaging are presented (reduction of random coincidences, real time imaging with reduced background using line-cone reconstruction enabling ToF-based filtering with centimeter precision, and improving the precision of this reconstruction method to the level of MLEM), but the technique would also enable PGT at 100 ps scale and bring benefits to proton radiography [75]. The next decade will surely bring new, exciting developments that will hopefully result in clinical systems for online verification of deposited dose distribution, making PT safer for patients.

Acknowledgments

AW is grateful to the Polish National Science Centre for supporting her work on prompt-gamma imaging with the SONATA BIS grant no. 2017/26/E/ST2/00618. DD is supported by the Laboratory of Excellence LabEx PRIMES (ANR-11-LABX-0063) and the ITMO-Cancer. Both authors would like to thank warmly Jean Michel Létang, Katarzyna Rusiecka and Jonas Kasper for their help by reading this manuscript.

Note

1. The chapter has been written based on ref. [1].

References

1. Wrońska A 2020 *Journal of Physics: Conference Series* **1561** 012021
2. Wilson R R 1946 *Radiology* **47** 487–91.
3. Linz U (ed) 2012 *Ion Beam Therapy* (Heidelberg Dordrecht London, New York: Springer).
4. PTCOG web page https://www.ptcog.ch/index.php/facilities-in-operation
5. Paganetti H 2012 *Physics in Medicine and Biology* **57** R99–R117.
6. Knopf A C and Lomax A 2013 *Physics in Medicine and Biology* **58** R131–60.
7. Kraan A C 2015 *Frontiers in Oncology* **5** 150.
8. Parodi K 2019 *The British Journal of Radiology* **93** 20190787.
9. NuPECC report 2014: Nuclear Physics for Medicine http://www.nupecc.org/pub/npmed2014.pdf
10. Parodi K and Assmann W 2015 *Modern Physics Letters A* **30** 1540025.
11. Lehrack S, Assmann W, Bender M, Severin D, Trautmann C, Schreiber J and Parodi K 2020 *Nuclear Instruments and Methods in Physics Research Section A* **950** 162935.
12. Traini G et al. 2019 *Physica Medica* **65** 84–93.
13. Yamaguchi M, Nagao Y, Ando K, Yamamoto S, Toshito T, Kataoka J and Kawachi N 2016 *Nuclear Instruments and Methods in Physics Research Section A* **833** 199–207.
14. Marafini M, Gasparini L, Mirabelli R, Pinci D, Patera V, Sciubba A, Spiriti E, Stoppa D, Traini G and Sarti A 2017 *Physics in Medicine and Biology* **62** 3299–312.
15. Enghardt W, Crespo P, Fiedler F, Hinz R, Parodi K, Pawelke J and Pönisch F 2004 *Nuclear Instruments and Methods in Physics Research Section A* **525** 284–8.
16. Parodi K 2015 *Medical Physics* **42** 7153–68.

17. Dendooven P, Buitenhuis H J, Diblen F, Heeres P N, Biegun A K, Fiedler F, Van Goethem M J, Van Der Graaf E R and Brandenburg S 2015 *Physics in Medicine and Biology* **60** 8923–47.
18. Ozoemelam I S, van der Graaf E R, van Goethem M J, Kapusta M, Zhang N, Brandenburg S and Dendooven P 2020 *Physics in Medicine and Biology* **65** 245013.
19. Bisogni M G et al. 2016 *Journal of Medical Imaging* **4** 011005.
20. Ferrero V et al. 2018 *Scientific Reports* **8** 4100.
21. Yamaya T 2017 *Journal of Physics: Conference Series* **777** 012023.
22. Min C H, Kim C H, Youn M Y and Kim J W 2006 *Applied Physics Letters* **89** 183517.
23. Pinto M et al. 2015 *Physics in Medicine and Biology* **60** 565–94.
24. Kelleter L et al. 2017 *Physica Medica* **34** 7–17.
25. Krimmer J, Dauvergne D, Létang J M and Testa E 2018 *Nuclear Instruments and Methods in Physics Research Section A* **878** 58–73.
26. Pausch G, Berthold J, Enghardt W, Römer K, Straessner A, Wagner A, Werner T and Kögler T 2020 *Nuclear Instruments and Methods in Physics Research Section A* **954** 161227.
27. Kanazawa M, Endo M, Himukai T, Kitamura M, Mizota M, Nakagawara A, Sato H, Shioyama Y, Totoki T and Tsunashima Y 2017 Scanning Irradiation System at SAGA-HIMAT. *Proc. of International Particle Accelerator Conference (IPAC'17), Copenhagen, Denmark, 14–19 May, 2017*(JACOW, Geneva, Switzerland) pp 4698–700. ISBN 978-3-95450-182-3 URL http://accelconf.web.cern.ch/ipac2017/doi/JACoW-IPAC2017-THPVA101.html
28. Smeets J et al. 2012 *Physics in Medicine and Biology* **57** 3371–405.
29. Perali I et al. 2014 *Physics in Medicine and Biology* **59** 5849–71.
30. Cambraia Lopes P et al. 2015 *Physics in Medicine and Biology* **60** 6063–85.
31. Richter C et al. 2016 *Radiotherapy and Oncology* **118** 232–7.
32. Xie Y, Bentefour H, Janssens G, Smeets J, Stappen F V, Hotoiu L, Yin L, Dolney D, Avery S, O'Grady F, Prieels D, McDonough J, Solberg T D, Lustig R, Lin A and Teo B K K 2017 *International Journal of Radiation Oncology • Biology • Physics* **99** P210–18.
33. Ready J, Negut V, Mihailescu L and Vetter K 2016 *Medical Physics* **43** 3717.
34. Ready J F 2016 *Development of a Multi-knife-edge Slit Collimator for Prompt Gamma Ray Imaging during Proton Beam Cancer Therapy.* PhD University of California, Berkeley. URL http://adsabs.harvard.edu/abs/2016PhDT.......151R
35. Smeets J, Roellinghoff F, Janssens G, Perali I, Celani A, Fiorini C, Freud N, Testa E and Prieels D 2016 *Frontiers in Oncology* **6** 156.
36. Pinto M, Dauvergne D, Freud N, Krimmer J, Létang J M, Ray C, Roellinghoff F and Testa E 2014 *Physics in Medicine and Biology* **59** 7653–74.
37. Park J H, Kim S H, Ku Y, Kim C H, Lee H R, Jeong J H, Lee S B and Shin D H 2019 *Nuclear Engineering and Technology* **51** 1406–16.
38. Kim C H, Park J H, Seo H and Lee H R 2012 *Medical Physics* **39** 1001–5.
39. Kim C H, Lee H R, Kim S H, Park J H, Cho S and Jung W G 2018 *Applied Physics Letters* **113** 114101–5.
40. Krimmer J et al. 2015 *Nuclear Instruments and Methods in Physics Research Section A* **787** 98–101.
41. Aldawood S et al. 2017 *Radiation Physics and Chemistry* **140** 190–7.

42. Liprandi S et al. 2017 Characterization of a Compton camera setup with monolithic LaBr3(Ce) absorber and segmented GAGG scatter detectors. *Proceedings of 2017 IEEE Nuclear Science Symposium and Medical Imaging Conference (NSS/MIC)* (IEEE) ISBN 978-1-5386-2282-7.
43. Hueso-González F et al. 2014 *Journal of Instrumentation* **9** P05002.
44. Golnik C et al. 2016 *Journal of Instrumentation* **11** P06009.
45. Polf J C, Avery S, Mackin D S and Beddar S 2015 *Physics in Medicine and Biology* **60** 7085–99.
46. Draeger E, Mackin D, Peterson S, Chen H, Avery S, Beddar S and Polf J C 2018 *Physics in Medicine and Biology* **63** 035019.
47. Llosá G et al. 2013 *Nuclear Instruments and Methods in Physics Research, Section A* **718** 130–3.
48. Barrio J, Etxebeste A, Granado L, Muñoz E, Oliver J F, Ros A, Roser J, Solaz C and Llosá G 2018 *Nuclear Instruments and Methods in Physics Research Section A* **912** 48–52.
49. Taya T, Kataoka J, Kishimoto A, Iwamoto Y, Koide A, Nishio T, Kabuki S and Inaniwa T 2016 *Nuclear Instruments and Methods in Physics Research Section A* **831** 355–61.
50. Koide A et al. 2018 *Scientific Reports* **8** 8116.
51. Wrońska A, Hetzel R, Kasper J, Lalik R, Magiera A, Rusiecka K and Stahl A 2020 *Acta Physica Polonica B* **51** 17–25.
52. Niedźwiecki S et al. 2017 *Acta Physica Polonica B* **48** 1567–76.
53. Kasper J, Rusiecka K, Hetzel R, Kazemi Kozani M, Lalik R, Magiera A, Stahl A and Wrońska A 2020 *Physica Medica* **76** 317–25.
54. Pausch G, Golnik C, Schulz A and Enghardt W 2017 A novel scheme of Compton imaging for nuclear medicine. *2016 IEEE Nuclear Science Symposium, Medical Imaging Conference and Room-Temperature Semiconductor Detector Workshop, NSS/MIC/RTSD 2016* vol 2017-January (Institute of Electrical and Electronics Engineers Inc.) ISBN 9781509016426.
55. Livingstone J, Dauvergne D, Etxebeste A, Fontana M, Gallin-Martel M L, Huisman B, Létang J M, Marcatili S, Sarrut D, Testa É 2021 *Physics in Medicine and Biology* **66** 125012.
56. Muñoz E, Barrientos L, Bernabéu J, Borja-Lloret M, Llosá G, Ros A, Roser J and Oliver J F 2020 *Physics in Medicine and Biology* **65** 025011.
57. Fontana M, Testa E, Ley J L, Dauvergne D, Freud N, Krimmer J, Létang J M, Maxim V, Richard M H and Rinaldi I 2020 *IEEE Transactions on Radiation and Plasma Medical Sciences* **4** 218–32.
58. Ortega P G, Torres-Espallardo I, Böhlen T T, Cerutti F, Chin M P W, Ferrari A, Gillam J E, Lacasta C, Llosá G, Oliver J, Rafecas M, Sala P R and Solevi P 2013 Noise evaluation of prompt-gamma technique for proton-therapy range verification using a Compton Camera. *2013 IEEE Nuclear Science Symposium and Medical Imaging Conference (2013 NSS/MIC)* pp 1–7 ISSN: 1082-3654.
59. Rohling H, Priegnitz M, Schoene S, Schumann A, Enghardt W, Hueso-González F, Pausch G and Fiedler F 2017 *Physics in Medicine and Biology* **62** 2795.
60. Panaino C M V, Mackay R I, Kirkby K J and Taylor M J 2019 *Scientific Reports* **9** 18820
61. Golnik C et al. 2014 *Physics in Medicine and Biology* **59** 5399–422.

62. Testa M, Min C H, Verburg J M, Schümann J, Lu H M and Paganetti H 2014 *Physics in Medicine and Biology* **59** 4181–95.
63. Hueso-González F et al. 2015 *Physics in Medicine and Biology* **60** 6247–72.
64. Werner T et al. 2019 *Physics in Medicine and Biology* **64** 105023.
65. Marcatili S, Collot J, Curtoni S, Dauvergne D, Hostachy J Y, Koumeir C, Létang J M, Livingstone J, Metivier V, Gallin-Martel L, Gallin-Martel M L, Muraz J F, Servagent N, Testa E and Yamouni M 2020 *Physics in Medicine and Biology* **65** 245033.
66. Jacquet M, Marcatili S, Gallin-Martel M L, Bouly J L, Boursier Y, Dauvergne D, Dupont M, Gallin-Martel L, Hérault J, Létang J M, Manéval D, Morel C, Muraz J F, Testa É 2021 *Physics in Medicine and Biology* **66** 135003.
67. Verburg J M, Riley K, Bortfeld T and Seco J 2013 *Physics in Medicine and Biology* **58** L37–L49.
68. Verburg J M and Seco J 2014 *Physics in Medicine and Biology* **59** 7089–106.
69. Hueso-González F, Rabe M, Ruggieri T, Bortfeld T and Verburg J M 2018 *Physics in Medicine and Biology* **63** 185019.
70. Dal Bello R, Martins P M, Brons S, Hermann G, Kihm T, Seimetz M and Seco J 2020 *Physics in Medicine and Biology* **65** 95010.
71. Khamfongkhruea C, Berthold J, Janssens G, Petzoldt J, Smeets J, Pausch G and Richter C 2020 *Medical Physics* **47** 5102–5111.
72. Liu C C and Huang H M 2020 *Physica Medica* **69** 110–9.
73. Muñoz E. et al. 2021 *Nature Scientific Reports* **11** 9321–9339.
74. Basalyga J N, Barajas C A, Gobbert M K, Maggi P and Polf J 2021 *Proc. Appl. Math. Mech.* **20** e202000070.
75. Dauvergne D, Allegrini O, Caplan C, Chen X, Curtoni S, Etxebeste A, Gallin-Martel M L, Jacquet M, Létang J M, Livingstone J, Marcatili S, Morel C, Testa E and Zoccarato Y 2020 *Frontiers in Physics* **8** 434.

7

Compton Cameras and Their Applications

J. Roser, F. Hueso-González, A. Ros, and G. Llosá

CONTENTS

7.1 Introduction

Since the discovery of gamma-ray radiation by Paul Villard in 1900, instrumentation for its detection has experienced a permanent evolution, with new materials that allow for better resolutions and higher efficiencies. Usually, there is a need for retrieving spatial (sometimes only directional) and spectral information (i.e., the incident energy) from the incoming gamma-rays. On the one hand, such concern was soon solved for gamma-rays below hundreds of keV and above tens of MeV in a relatively easy way, by using passive collimators and photoelectric or pair production interactions undergone by the gamma-rays at these energies. On the other hand, gamma-rays in the few MeV range have posed a bigger problem. In this intermediate region, Compton interaction dominates over the two other processes, and physical collimation would only be feasible with large and dense passive systems.

It is in this context where Compton cameras were first proposed. The principles of the technique (which will be further explained in the Section 7.2.1) can be traced back to as early as 1950, when (Hofstadter and McIntyre 1950) proposed a setup for gamma-ray detection using two NaI(Tl) inorganic scintillation crystals[1]. The aim of the proposed setup was to retrieve spectral

DOI: 10.1201/9781003218364-7

information of the gamma-rays by requiring a Compton backscattering in the first scintillator followed by a further interaction in the second scintillator; the incoming direction of the gamma-ray was known beforehand. This *Compton spectrometer* concept was further developed in the literature; for example, (Weinzierl and Tisljar-Lentulis 1958; Kalish and Nardi 1964). In parallel, the idea of a *Compton telescope* as a device to retrieve spatial information of incoming gamma-rays was proposed in (Peterson and Howard 1961) and met a major breakthrough in the 1970s with the works at Max Planck *Institut für extraterrestrische Physik* (Schönfelder, Hirner, and Schneider 1973), the University of California, Riverside (Herzo et al. 1975) and the University of California, Berkeley (Dauber and Smith 1973). Schönfelder's proposal had been preceded by analogous experiments for non-relativistic solar and atmospheric neutrons with spark chambers (Pinkau 1966; Göllnitz et al. 1969; Heidbreder et al. 1971); similarly, at University of California, Riverside, (White 1968) had proposed an experiment to measure solar neutrons by taking advantage of the neutron-proton collisions on two plastic scintillator detectors, later developed and tested in (Grannan et al. 1972; Preszler, Simnett, and White 1972). Measurements with Compton telescopes in balloon flight experiments soon ensued (Lichti and Schönfelder 1974). Meanwhile, the application of this imaging system in other areas started to be recognized; for instance, in nuclear medicine imaging (Todd, Nightingale, and Everett 1974).

Compton cameras have received constant interest since then. A great variety of detector materials and geometries have been proposed, owing to the need of excellent energy resolution, accurate interaction position determination, improved efficiencies and fast readout systems. Initial scintillator-based proposals were soon replaced with semiconductor materials (e.g., Si, Ge, CZT, CdTe...), as a result of their unmatched energy and position resolutions. However, this trend has been contested with the recent discovery of new scintillator materials with improved capabilities (e.g., $LaBr_3$:Ce, GaGG:Ce...) and the commercial availability of silicon photomultipliers (SiPMs), allowing for the manufacturing of compact, easy operable Compton camera detectors (Llosá 2019). Nowadays, Compton cameras made with different types of detectors are commercially available for a variety of applications.

7.2 Theoretical Background

7.2.1 Working Principle

A Compton camera aims to determine the direction of incidence of a photon that undergoes incoherent scattering with an electron, see Figure 7.1 (left). The electron involved is considered to be at rest (free), as its kinetic and binding energy are usually much lower than the energy transferred by the

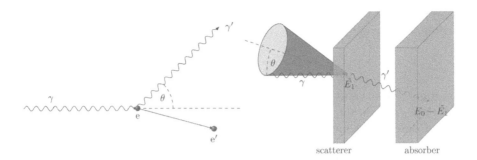

FIGURE 7.1
Sketch of a gamma-ray undergoing Compton scattering with a free electron (left) and of consecutive interactions of a gamma-ray in two parallel detection planes (scattering and absorption) of a Compton camera (right). The deflection angle θ of the gamma-ray (left) corresponds to the half-opening angle of the cone (right).

gamma-ray. Based on this approximation, the Compton equation (Compton 1923) relates the scattering angle θ with the initial (E_γ) and final ($E_{\gamma'}$) photon energies:

$$\frac{1-\cos\theta}{m_e c^2} = \frac{1}{E_{\gamma'}} - \frac{1}{E_\gamma}, \tag{7.1}$$

where m_e is the electron mass and c is the speed of light. For a gamma-ray to be detected as a signal event in a Compton camera, the gamma-ray must undergo a Compton interaction in one detector element, followed by a photonic interaction in another detector element. If the positions of interaction in both detector elements and the energy deposited in the first detector element are measured, the origin of the gamma-ray can be restricted to a conical surface, whose apex is given by the position of interaction in the first detector element; the axis is given by the line connecting both positions of interaction, see Figure 7.1 (right); and the aperture angle is given by Equation (7.1), which can be conveniently rewritten in the following way:

$$\cos\theta = 1 - \frac{m_e c^2 \tilde{E}_1}{E_0(E_0 - \tilde{E}_1)}, \tag{7.2}$$

where E_0 is equal to the incident gamma-ray energy E_γ and $\tilde{E}_1$ is the energy deposited in the first detector element; namely, $E_\gamma - E_{\gamma'}$.

It must be noted that knowledge over the incident gamma-ray energy is needed in order to determine correctly the aperture angle. In some applications, this knowledge cannot be obtained beforehand. Sometimes, the incident gamma-ray energy is estimated as the total energy deposited in the

Compton camera, which requires to measure the energy deposited in the second detector element, too:

$$E_0 \approx \tilde{E}_1 + \tilde{E}_2. \tag{7.3}$$

Equation (7.3) is valid only if the gamma-ray is completely absorbed in the Compton camera, for example, by undergoing a photoelectric effect in the second detector element. In this context, the first and second detector elements are usually referred to as *scatterer* and *absorber*, respectively. Moreover, it is common to combine this approximation with energetic cuts over the total deposited energy spectrum in order to select the photopeak, which in turn reduces the available statistics. It must be noted that the approach may be inappropriate for high energy gamma-rays and/or non-monochromatic sources, in particular if the absorber is not large enough.

A third approach is to estimate the incident energy together with the spatial distribution of incoming gamma-rays, which requires the use of a spectral image reconstruction algorithm (see Section 7.2.2).

Finally, the incident gamma-ray energy can be also estimated if the gamma-ray undergoes a further Compton interaction in the second detector element followed by a third photonic interaction in a third detector element (Dogan, Wehe, and Knoll 1990). We will refer to this sequence of interactions as a *three-interaction signal event*. If the positions of interaction and the energy deposited in the first and second detector elements are measured (e.g., using a multilayered Compton camera, see Figure 7.2), then the incident energy can be calculated by means of Compton kinematics (Kurfess et al. 2000):

$$E_0 = \tilde{E}_1 + \frac{1}{2}\left(\tilde{E}_2 + \sqrt{\tilde{E}_2^2 + \frac{4m_e c^2 \tilde{E}_2}{1 - \cos\theta_2}} \right), \tag{7.4}$$

where θ_2 is the angle between the line connecting the first and second positions of interaction and the second and third positions of interaction.

Some remarks can be added to the formalism described here:

- **Electron tracking**: Notably, the directions of the incoming gamma-ray, the scattered gamma-ray and the scattered electron must lie in the same plane. Thus, if the scattered electron direction is measured, the azimuthal redundancy in the determination of the incoming gamma-ray direction can be removed. This advantage was soon recognized (White, Zych, and Tumer 1990) and has motivated the development of electron tracking Compton cameras (ETCCs).

- **Polarization**: The Klein-Nishina differential cross-section for Compton scattering includes a term that depends on the angle between the photon polarization directions before and after the scattering (Klein and Nishina 1929). Consequently, an incoming

FIGURE 7.2
Image of the multilayer Compton camera MACACO. The three detectors (monolithic LaBr$_3$ scintillator crystals, coupled to SiPM and enclosed in plastic holders) are visible in the forefront. Such Compton camera geometry allows measuring three-interaction events.

unpolarized gamma-ray beam will become partially linearly polarized after Compton scattering in the first detector element; if the incoming gamma-ray beam is already polarized, an azimuthal asymmetry will be observed in the second detector element. Compton polarimetry can be useful in some Compton imaging applications (e.g., astrophysics). Moreover, three-interaction signal events may benefit from this property in order to reduce the incoming gamma-ray direction azimuthal redundancy (Dogan, Wehe, and Akcasu 1992; Boggs 2003).

- **Degradation effects**: The determination of the incoming gamma-ray direction is conditioned to the correct measurement of the interaction sites and deposited energies. For this reason, it is common to quantify the influence of degradation effects on Compton camera prototypes by means of the angular resolution measurement (ARM). The ARM is usually given as the full width at half maximum (FWHM) of the distribution of differences between the geometrical polar angle and the angle calculated by Equation (7.2). Other quantifiers are found in the literature; for example, the *point of closest approach* (Wilderman et al. 1997; Mackin et al. 2013). Both energy and spatial detector resolution are important degradation effects that can be related analytically to the angular resolution (Mihailescu et al. 2007) and that need to be minimized through careful detector design; as an example, increasing distances between detector elements improve the angular uncertainty but in turn reduce the overall efficiency of the imaging device. Numerous Compton camera optimization studies are found in the literature; for instance, (Chelikani, Gore, and Zubal 2004; Peterson, Robertson, and Polf 2010; Richard et al. 2011; Uche, Round, and Cree 2011; Muñoz et al. 2017). Finally, the derivation of Equation (7.1) ignores the fact that electrons are bound to atoms and have a non-zero moment distribution; the magnitude of this effect, known as Doppler broadening, depends on the detector material and the incoming gamma-ray energy (Ordonez, Bolozdynya, and Chang 1997).

7.2.2 Image Reconstruction

Compton imaging can be classified as an emission tomography reconstruction problem, where a photo-emitting object distribution $f(\vec{x})$ has to be estimated by using a discrete set of photo-counting projections $\vec{g}$ measured with the imaging system. A first approximation can be obtained with the simple backprojection in the image space of the cones given by the parameters of each measurement; this approach yields blurry, low-resolution images.

In the development of more sophisticated algorithms, a continuous-discrete linear model of the imaging process for signal data is generally assumed,

$$g_i = \int_V h_i(\vec{x})f(\vec{x})dV, \qquad (7.5)$$

where $h_i(\vec{x})$, often referred to as the point response function, represents the probability density of detecting an emission from $\vec{x}$ in terms of the measure i. Analytical methods aim to directly obtain the object distribution after reducing its relation with the photo-counting projections to a weighted integral over the surface of the cone. Several algorithms have been proposed (Cree and Bones 1994; Basko, Zeng, and Gullberg 1997, 1998; Parra 2000; Tomitani and Hirasawa 2002; Smith 2005; Maxim, Frandes, and Prost 2009; Maxim 2014; Jung and Moon 2015; Terzioglu 2015; Jung and Moon 2016; Kuchment and Terzioglu 2016; Maxim 2018; Kwon 2019; Moon and Kwon 2019; Terzioglu 2020; Zhang 2020), which differ in the geometries to which they are applicable, the weight assigned to the integral model, and the restrictions over the set of projections to be employed. A brief review is given in (Terzioglu, Kuchment, and Kunyansky 2018).

Currently, iterative algorithms are the most utilized in Compton imaging. They account for the randomness of the data (the photo-counting detections) as well as providing an easy framework to describe the imaging system; however, they are generally slower than analytical methods. Hereafter, we review the underlying theory of some iterative algorithms specifically employed in Compton imaging. The reader can find in (Qi and Leahy 2006) an extensive review on iterative reconstruction algorithms applied to emission tomography.

Photoemission and photodetection are stochastic processes that can be modeled with the binomial and Poisson-multinomial distribution, respectively (Sitek 2014). Under the proper conditions[2] these distributions can be approximated to a Poisson and a product of independent Poisson distributions, respectively; the latter is referred to as the likelihood probability distribution. The object distribution is often expressed in terms of a discrete basis (e.g., the voxel basis); as a consequence, Equation (7.5)[3] is rewritten as:

$$E[g_i] \approx \int_V h_i(\vec{x}) \left(\sum_v \lambda_v b_v(\vec{x}) \right) dV = \sum_v H_{iv} \lambda_v, \tag{7.6}$$

where λ_v is the number of emissions from voxel v, and the elements of the system matrix (SM) H_{iv} represent the probability of detecting an emission from voxel v in terms of the measurement element i. A precise estimation of the SM is crucial in the iterative reconstruction procedure and its calculation and/or storage is a major challenge. Given the large number of measurement elements[4] and the usually desired 3D (or even 4D, including the gamma-ray incident energy) imaging, Monte Carlo estimations can be very expensive in computation terms; similarly, direct estimation through measurements may not be practical. For this reason, several analytical models for the SM calculation have been proposed (Wilderman et al. 1998; Xu and He 2007; Wang et al. 2012; Maxim et al. 2015; Schoene et al. 2017; Muñoz et al. 2018, 2020; Roser et al. 2020), covering different Compton imaging scenarios.

In order to develop an iterative reconstruction algorithm, a criterion defining the best image estimate to which the algorithm should converge must be formulated first. From a Bayesian perspective, this best image is the point estimate obtained by maximizing the posterior expected loss, for which a cost function must be chosen. If the hit-or-miss cost function is chosen, then the best estimate for the reconstruction problem is the mode of the posterior distribution or maximum a posteriori (MAP) criterion, which reduces to the maximum likelihood (ML) criterion when the uniform prior is assumed.

The maximum-likelihood expectation-maximization algorithm (ML-EM) (Dempster, Laird, and Rubin 1977) was demonstrated for emission tomography in (Shepp and Vardi 1982; Lange and Carson 1984). In its List Mode version (Wilderman et al. 1998), it is currently one of the most employed algorithms in Compton imaging. The recursion formula

$$\lambda_v^{(n+1)} = \frac{\lambda_v^{(n)}}{s_v} \sum_i \frac{H_{iv}}{\sum_k H_{ik} \lambda_k^{(n)}} \tag{7.7}$$

converges to the image that maximizes the Poisson likelihood. It has the desirable properties of non-negativity and monotonicity, aside from the asymptotic unbiasedness and efficiency properties of the ML estimate. However, given a finite set of realistic projection data, the image to which ML-EM converges is highly noisy. For this reason, the iteration process is usually stopped before convergence and/or post-filtering is applied. With the proper prior selection, algorithms driven by the MAP criterion can converge to smoother estimates. Among these, we can cite the median root prior (Alenius and Ruotsalainen 1997) and the total variation (Rudin, Osher, and Fatemi 1992), which were applied in the Compton imaging context with promising results in (Sakai et al. 2019, 2020) and (Feng et al. 2018) respectively. One issue concerning MAP algorithms is the need of choosing the correct value for the (usually object-dependent) hyperparameters, which regulate the influence of the prior on the final image estimate.

If the quadratic cost function is chosen, then the best estimate is given by the mean of the posterior distribution. This is the criterion followed by the origin ensembles (OE) algorithm. OE was proposed for emission tomography in (Sitek 2008, 2011) and specifically formulated for Compton imaging in (Andreyev, Sitek, and Celler 2011). The algorithm relies on a Markov chain formalism to stochastically obtain samples of the posterior distribution[5] via the Metropolis-Hastings algorithm; some acceptance ratios are discussed in (Sitek 2012, 2014). The point estimate is calculated by averaging the obtained samples once the Markov Chain has reached equilibrium; the number of iterations should be large enough for the Markov Chain to explore the entire posterior distribution. OE-based algorithms have been proposed and applied in Compton imaging both with simulations (Mackin et al. 2012; Calderón et al. 2014; Andreyev et al. 2016; Yao, Xiao, and Chen 2017; Yao et al. 2019; Zheng, Yao, and Xiao 2020) and experimental data (Polf et al. 2015; Chen et al. 2018; Draeger et al. 2018; Gutierrez et al. 2018).

The OE algorithm is not affected by the noisy convergence problem of the ML-EM and provides an easy framework to include resolution recovery; however, it features other issues such as the unknown extent of the burn-up region (the number of iterations needed to reach the equilibrium) or the highly blurred images obtained in low statistics scenarios. An insightful comparison with focus in the Compton camera treatment monitoring application was recently driven in (Kohlhase et al. 2020). So far, no consensus has been reached as for an optimal image reconstruction algorithm for Compton imaging, and research literature on this subject is still abundant.

7.3 Applications

7.3.1 Astrophysics

Compton cameras (hereafter in this section Compton telescopes) were first developed in the astronomy field as a response to the lack of imaging detectors for gamma-ray sources in the few MeV region. Examples of such sources are gamma-ray bursts, the positron annihilation and neutron capture lines, unstable isotopes like ^{26}Al, ^{60}Fe and ^{44}Ti and the nuclear de-excitation emission (which is also relevant in the context of medical applications). Their detection is crucial to improve the understanding of a variety of topics like supernovae formation, neutron star features, the existence of dark matter, solar flares, etc[6].

Following the first steps in the Compton telescope development, which crystallized in first balloon flight measurements during the 1970s, the Imaging Gamma-Ray Telescope COMPTEL aboard the Compton Gamma-Ray Observatory (Schönfelder et al. 1993) became the first Compton telescope flown in orbit. COMPTEL featured a scatterer consisting of seven modules of liquid scintillator NE 213A and an absorber containing fourteen modules of NaI(Tl); all coupled to photomultiplier tubes (PMTs) and surrounded by veto-domes of 1.5 cm thick plastic scintillator to reject background; in addition, time of flight (ToF) background rejection was applied. The detectors featured an energy resolution of 8.8% FWHM and an ARM of 4.74° FWHM for 1.27 MeV gamma-rays (Schönfelder 1991). Some of its numerous achievements were the first all-sky image of 1.8 MeV ^{26}Al emission from the galactic plane (Oberlack et al. 1996), the observation of ^{44}Ti gamma-ray line emission from Cas A (Iyudin et al. 1994), the ^{56}Co gamma-ray line from SN 1991T (Morris et al. 1995), the measurement of ^{12}C and ^{16}O nuclear de-excitation lines in the Orion region (Bloemen et al. 1994), the observation of several pulsars and X-ray binaries (Dijk 1996; Kuiper et al. 2001), and the first source catalogue in the energy range 0.75–30 MeV (Schönfelder et al. 2000). Many of these results are reviewed in (Schönfelder et al. 1996).

Despite its unprecedented success, COMPTEL had limited sensitivity and was subject to important instrumental background (Schönfelder 2004). This sparked a variety of new proposed Compton telescope designs that take advantage of the recent improvements in gamma-ray detection materials and techniques, in what has been called the *second generation* of Compton telescopes. Many of them are based on semiconductor detectors, owing to their excellent energy and position resolution. The NCT balloon-borne Compton telescope (Boggs et al. 2001), later upgraded and renamed as COSI, comprises high-purity cross-strip germanium detectors in a multilayered geometry. It has detected and imaged several compact gamma-ray sources, a long duration gamma-ray burst and the positron annihilation emission (Bandstra et al. 2011; Kierans et al. 2017, 2020); currently, the prototype is under Phase A study after being selected in the 2019 NASA small explorer (SMEX) call. Another semiconductor-based example was the Si/CdTe Compton telescope proposed in (Tanaka et al. 2004) and implemented in ASTRO-H (Watanabe et al. 2016; Hitomi Collaboration et al. 2018; Tajima et al. 2018) for sub-MeV, narrow field of view (FOV) imaging. It featured 32 layers of Si pad detectors and 8 layers of CdTe pad detectors, mounted inside the bottom of a well-type active shield made of BGO scintillators and cooled to –25°C.

Scintillator-based Compton telescopes are being proposed again as a consequence of the improved performance of recently discovered scintillation materials and compact light readout devices like SiPMs. Balloon-born examples are the FACTEL, SolCompT and ASCOT prototypes (Julien et al. 2012; Bloser et al. 2016, 2018). The latter features $CeBr_3$ and p-terphenyl scintillators coupled to SiPMs and uses ToF background rejection; recent balloon measurements of the Crab nebula have been reported (Sharma et al. 2020). Furthermore, the combination of $CeBr_3$ scintillators with artificial single-crystal diamond detectors is being investigated (Poulson et al. 2020). A ground-based, scintillation-based Compton telescope example is the i-TED prototype (Domingo-Pardo 2016; Babiano-Suárez et al. 2020), based on $LaCl_3$ crystals coupled to SiPMs and aimed at the measurement of neutron capture cross sections for stellar nucleosynthesis studies (Figure 7.3).

A different approach is represented by LXeGRIT, a Compton telescope based on a fully active liquid xenon ionization chamber, operated as a time projection chamber (TPC) (Aprile, Mukherjee, and Suzuki 1989), and tested in balloon flight experiments, where it measured the internal and atmospheric gamma-ray background spectrum (Aprile et al. 2003). An ARM of about 4° at 1.8 MeV was reported (Aprile et al. 2008). A similar imaging system based on liquid argon with plastic scintillators read by SiPMs has been recently proposed (Aramaki et al. 2020).

FIGURE 7.3
Image of i-TED and associated electronics. (Courtesy of C. Domingo-Pardo.)

Some prototypes use the electron tracking technique to improve the Compton imaging and background suppression capabilities, as mentioned earlier. Examples using a stack of silicon strip detectors surrounded by scintillation absorbers are TIGRE (Tumer et al. 1995; Zych et al. 2006), MEGA (Kanbach et al. 2003; Andritschke et al. 2005), the GRIPS Gamma-Ray Monitor (Zoglauer et al. 2008), and more recently the proposed concept of a nano-satellite Compton telescope (Lucchetta et al. 2017; Rando et al. 2019). A different technology is employed in the SMILE/ETCC prototype (Orito et al. 2003), which features a gaseous argon filled TPC scatterer based on a micro-pixel gas chamber with electron tracking capabilities, surrounded with GSO scintillator absorbers read by PMTs. An efficient background suppression was achieved with the first prototype (SMILE-I) observing diffuse cosmic and atmospheric gamma-ray spectra (Takada et al. 2011), and further improvement has been reported in the preliminary results obtained with a second prototype (SMILE-2+) with an ARM of 5.3° at 662 keV (Tanimori et al. 2017; Tanimori 2020). A third improved prototype (SMILE-3) is under development (Takada et al. 2020).

The astrophysical application of Compton imaging is a field that remains active and with a strong scientific case. Future steps will likely involve the development of a space-based mission with a second generation Compton telescope, following the COMPTEL successful example. Recently, the e-ASTROGAM (De Angelis et al. 2017) and the AMEGO (Rando 2017) missions proposed the launching of Compton telescopes with similar features, including trackers with 50–60 double-sided Si strip detectors (DSSDs) followed by absorbers (an array of CsI(Tl) and segmented CdZnTe (CZT), respectively) and anticoincidence systems. Next generation space-based missions present elevated costs and payloads; for illustration, the e-ASTROGRAM scatterers have a total area of around 1 m^2 each plane, and the dimensions of each absorber bar are $5 \times 5 \times 80$ mm^3. In return, unprecedented performance and measurement capabilities with enhanced sensitivity can be expected.

7.3.2 Environmental Radiation Monitoring

One of the main uses of Compton cameras is in the fields of safety and security, for the detection of hotspots of contaminated areas due to human-driven (Chernobyl) or nature-driven (Fukushima) disasters. The correct localization of such hotspots is necessary in order to decommission and decontaminate the affected areas. Compton cameras are also used in nuclear warhead detection, in the characterization of shielding around special nuclear materials (SNMs) for homeland security and potential nuclear disarmament. Compton cameras have also been widely used for homeland security in airports and borders.

Regarding hotspot detection, in recent years, since the Fukushima Daiichi Nuclear Power Station (FDNPS) disaster in Japan on 11 March 2011, the development of new approaches for the use of Compton cameras in potentially hazardous areas has seen an exponential increase.

Portable Compton cameras are often used to obtain a dose map of contaminated areas and to identify hotspots. This kind of detector is commercially available for industrial use: Fulcrum and GeGI[7] (Figure 7.4), Polaris-H[8] (Wahl et al. 2015), ASTROCAM 7000HS[9] (Takeda et al. 2015) or NuVISION[10] (Montémont et al. 2017). Different materials are employed in these types of devices in which size, weight and energy resolution are the main constraints. NuVISION, Polaris-H and ASTROCAM 7000HS use semiconductor detectors for the Compton camera, CZT for the first two and Si/CdTe for the latter. The use of semiconductor detectors allows its size to be minimized while providing very good energy resolution. Fulcrum and GeGI use HPGe for the detector because of its excellent energy resolution, although HPGe detectors need to be cooled, which can take 1.5–4 hours, in order to be able to measure. Although currently available commercial portable cameras have excellent performance, research in this field is still ongoing. Inorganic scintillation-based Compton cameras are also being proposed.

A portable Compton camera with CeBr$_3$ scintillation crystals and digital photon counter detectors with ToF capability (Hmissi et al. 2018) has been tested in the laboratory giving good results for energy and angular resolutions. Another example is SCoTSS, which has been designed for use in radiological and nuclear incident investigations, nuclear non-proliferation

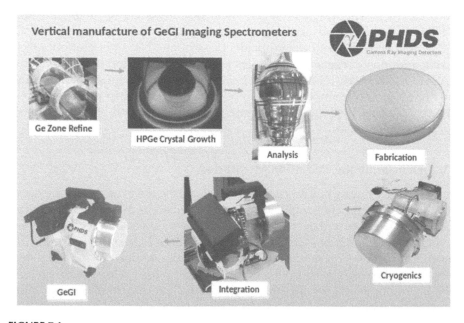

FIGURE 7.4
Image of the manufacturing process of a GeGI portable Compton camera. (Courtesy of PHDS Co.)

applications, and to support safety and security activities in Canada (Sinclair et al. 2020). This prototype uses CsI(Tl) and SiPMs in a modular way, which allows the number of modules per layer and the distance between them to be changed. This means that both images with a high resolution in a narrow field and images with low resolution in a wide-angle field can be reconstructed. It also performs isotope identification and can act as a directional spectrometer. Depending on the required capabilities, the number of modules can vary so that SCoTSS can be used handheld or in a backpack, or for the bigger configurations that could be deployed in an unmanned ground vehicle (UGV). In order to improve the results obtained with Compton cameras, hybrid systems are also under study. A hybrid system based on GAMPIX (Carrel et al. 2011) with a MURA coded-aperture imaging spectrometer has been developed for nuclear industry applications (Amoyal et al. 2021). The Compton camera uses a TimePix3 readout and a 1-mm thick CdTe detector. A characterization of the prototype with different radioactive sources in the laboratory showed an improvement in the results when measuring with the hybrid system compared to the Compton camera.

Scene-data fusion (SDF) is a technique that combines gamma-ray cameras and sensors such as visual cameras, structured light or LiDAR (light detection and ranging). It was applied in (Vetter et al. 2018) with a system comprising the high-efficiency multimode imager (HEMI) and a Microsoft Kinect sensor (Zhang 2012). HEMI is a hand-portable 4.5 kg gamma-ray imager with CZT detectors that provides coded aperture and Compton imaging capabilities. It has been used successfully to detect single radioactive sources by walking around the area of interest with the detector system. Efforts have been made to minimize the exposure of human workers by sending UGVs into contaminated areas, although these vehicles are limited to certain types of terrain. For this reason, the use of unmanned aerial vehicles (UAVs) has been widely explored in recent years. The feasibility of mounting HEMI in a UAV in order to detect and localize hotspots at ground level in a contaminated area in Fukushima Prefecture has been evaluated. The results obtained in Fukushima were promising although it would be desirable to develop lighter cameras. For this kind of device, the challenge is to develop a Compton camera light enough to be mounted on an UAV. It should be light, not only to allow it to be mounted on a smaller aerial vehicle and thus permit access to hidden areas, but also to allow longer flight times and thus a larger measured area.

SDF is extensively used in UAVs where optical cameras are employed in addition to the Compton camera. A Compton camera designed to reduce noise on gamma-ray images by adding a Pb layer to the rear-panel has been designed and tested in Fukushima (Nishiyama et al. 2014). The camera consists of Ce:GAGG scintillator arrays and SiPMs and weighs 1.9 kg. Studies have been carried out with the same camera without the rear-panel, mounted on a drone (Spreading Wings S1000+) with a diagonal wheelbase of approximately 1 m (Mochizuki et al. 2017). To test the device, an air dose map at ground level was measured in an area of 80×80 m². In-flight measurements have also been performed, in which studies at different heights showed that measuring at an elevation of 20 m from the ground improved the measuring time by ten times with respect to the same measurement at a 10-m elevation.

A similar approach has been published by (Sato et al. 2020), in which a Compton camera (1.5 kg) was mounted on a drone system with a diagonal base of approximately 2 m. The prototype was tested in an area of Fukushima of about 7,000 m² comprising three differentiated dose areas. The flight altitude was approximately 11 m from the ground. The intensity map for the three areas was reconstructed, showing the same dose distribution as that measured with a survey meter. This prototype lacks the capability to determine the intensity of the contamination level for each hotspot and only reflects the general behavior. In relation to both Mochizuki's and Sato's studies, it will be necessary to improve in future work the precision of the determination of the altitude of the drone, in order to estimate the dose correctly.

Small and light Compton cameras are desirable for use in UAVs. A first prototype of a Compton camera in millimetric scale based on a single CZT detector with 4×4 pixels has been presented (Liu et al. 2018) and it was able to localize and identify radioactive sources. One of the smallest Compton cameras assembled in a UAV to date (Baca et al. 2020) weighs 40 g and has dimensions of $80 \times 20 \times 15$ mm³. This means that it could be mounted in a micro aerial vehicle (MAV), defined as a sub-250 g UAV. Nevertheless, as the Compton camera was not the only sensor mounted in the aerial vehicle, the authors of this work used a UAV with a diagonal dimension of 650 mm. The Compton camera employs a single sensor with a 2-mm thick CdTe detector and TimePix3 for the readout (Poikela et al. 2014). The thickness, material and bias voltage employed in the detector can be changed depending on the application. This prototype was able to localize and track a ^{137}Cs source autonomously in an area of 40×20 m². Figure 7.5 shows a graphic of the weight versus date for selected portable Compton cameras mentioned in the text.

Using UAVs for hotspot positioning allows large areas to be measured, while reducing the exposure to human workers. However, the Compton cameras used for this purpose do not provide a measure of the dose at the ground level. An ETCC, based on the aforementioned SMILE-I telescope (the only difference being that the absorber plane is formed by a GSO pixelated crystal coupled to PMTs), has been developed in order to obtain a dose map at ground level (Tomono et al. 2017). It was also utilized to successfully identify a micro-hotspot in a previously decontaminated area. The main drawbacks of this system are the weight (40–50 kg) and the fact that the gas chamber must be refilled every 3 weeks.

One of the main drawbacks of Compton cameras is the directional variance of its detection efficiency. In the case of Compton cameras mounted on UAVs, this problem can be partially compensated for by the movement of the camera, allowing measurements to be taken from different points of

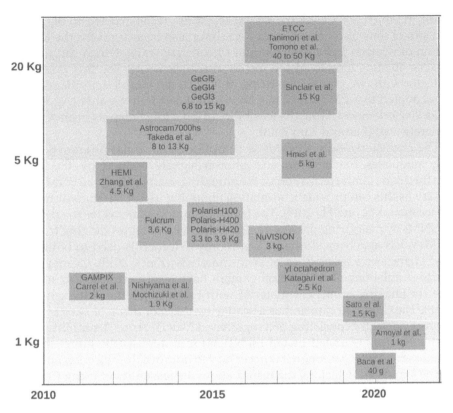

FIGURE 7.5
Weight versus date for selected portable Compton cameras. The blue boxes correspond to commercial cameras and the green boxes to cameras used for research purposes.

view with respect to the area of interest. Even so, a Compton camera with homogeneous efficiency would be desirable. An all-sky Compton camera (γI octahedron) with almost isotropic acceptance, the directional variance of its acceptance being about 23% (full width) of the average value, has been developed by (Katagiri et al. 2018). The detector (2.5 kg) is formed of six NaI(Tl) scintillator cylinders placed at the vertices of an octahedron. This geometry is suitable for detectors in which uniform acceptance and angular resolution in all directions are necessary. The detector was tested in the Fukushima Daiichi nuclear power plant, demonstrating its capability to visualize gamma-rays in environments with multiple and/or extended sources.

At the other end of the spectrum, there are situations in which the area under study is well known, delimited and easy to access, such as the decommissioning of large facilities like cyclotrons or nuclear power plants. In these cases, a large-area Compton camera (LACC) would not only increase the sensitivity, but also allow reconstruction in 3D. A prototype LACC system with two planes with NaI(Tl) crystals and PMTs has been assembled (Kim et al.

2018), in which a ^{137}Cs source was successfully reconstructed in 3D within a depth of tens of centimeters. LACCs have also been used for the identification of neutron induced activation in concrete walls, which can often be misclassified as long-lived low or intermediate waste. For this purpose, a detector with two planes consisting of NaI(Tl) monolithic scintillation crystals coupled to a PMT array has been simulated with Geant4 (Agostinelli et al. 2003), demonstrating the feasibility of detecting hotspots in an activated concrete wall (Lee and Kim 2018).

The characterization of SNMs is complicated since the unknown shielding modulates the neutron and photon signals emitted by the source. Furthermore, characterization of the shielding surrounding the SNM is necessary as this can provide information about the use of the nuclear material (Goodman, Xia, and He 2019). The use of Compton cameras for the detection of SNMs by measuring passive gamma-rays, with energies ranging from keV to MeV, has also been studied. Gamma-ray detection is used to both detect and characterize assemblies of such materials (Ziock 2018). Geant4 Monte Carlo simulations of a Compton camera have been performed to study its use for imaging of nuclear material with heavy shielding (Yingzeng et al. 2018). The Compton camera has a scattering detector made of DSSDs and an absorber detector consisting of a segmented NaI(Tl) array. A simplified Steve Fetter Nuclear Warhead model (SFNWM) was used as the target material (Fetter et al. 1990). Although it is necessary to improve the detection efficiency of the simulated detectors, the results show the feasibility of using Compton cameras for detecting nuclear warheads. The passive characterization of nuclear materials can be performed by detecting neutron-induced gamma-rays with CZT detectors (Goodman, Xia, and He 2019). Measurements with a ^{238}PuBe source, two targets (PVC and polyethylene) and CZT detectors have shown that it is possible to detect and spatially localize the presence of certain shielding isotopes in neutron fields.

7.3.3 Medical Applications

While not employed yet clinically in medical imaging, Compton cameras could potentially offer several advantages as compared to conventional gamma cameras. Avoiding the use of mechanical collimators overcomes the resolution-efficiency tradeoff associated with the latter, improving both parameters. An enhanced efficiency at comparable resolution would enable dose or imaging time reduction (Fontana et al. 2017).

Compton cameras can image a wide range of radioisotope energies (typically from a few hundred keV to above 1 MeV), and their performance improves with energy, showing significant advantage over collimated cameras at high energies. This also opens the possibility to the use of higher energy radiotracers with reduced scattering and lower effective dose to the tissues. The simultaneous visualization of different radiotracers could lead to more advanced and accurate diagnosis. In addition, the FOV covers an area that is bigger than the first detector, allowing to image large FOVs with compact devices.

Their main limitations are their higher technological complexity than gamma cameras and the fact that at the relatively low radiotracer energies suited for gamma cameras, Compton cameras need to have excellent energy resolution in order to achieve similar spatial resolution. In addition, image reconstruction is more complex and the gain in efficiency does not directly translate into the reconstructed images. Although these facts still hinder their clinical use, steady progress has been made and Compton cameras made of different materials and following diverse design approaches, some of them developed initially for other applications, are being evaluated for medical imaging.

After their proposal by (Todd, Nightingale, and Everett 1974), the application of Compton cameras to medical imaging was pioneered by Singh et al., between 1977 and 1990. The first experimental prototype (Singh and Doria 1983), constructed with an HPGe detector as scatterer and a NaI absorber behind it, demonstrated the possibility of imaging ^{99m}Tc and ^{137}Cs sources. Later experiments (Singh and Brechner 1990) allowed imaging 3D cylindrical test phantoms. Extrapolating to a larger system, the results showed a sensitivity gain of an order of magnitude over a collimated camera at the same spatial resolution. The same scatter detector was employed by (Martin et al. 1993) with a ring-shaped NaI second detector, which avoided direct illumination by the source but reduced the solid angle and thus the efficiency. This system was not intended for medical imaging, but for industrial use.

The advances in the development of silicon detectors made them the detector of choice in subsequent efforts, given their simpler operation, low Doppler broadening, excellent energy and spatial resolution, and capability of operating at room temperature with no external cooling requirements. The first experiments were carried out by (LeBlanc et al. 1998) with C-SPRINT, employing a silicon pad detector as scatterer and an existing SPECT NaI ring with its lead collimators removed as absorber. Further work with silicon detectors included the use of a stack of silicon pad detectors to increase the efficiency (Studen et al. 2004; Llosá et al. 2008), or the use of silicon drift detectors (Conka-Nurdan et al. 2005).

Some of the developed systems reached the necessary performance for in vivo imaging. The development of a Compton camera composed of two double-sided strip Ge detectors in (Motomura et al. 2008) enabled imaging of three radiotracers (^{65}ZnCl$_2$, ^{85}SrCl$_2$ and iodinated (^{131}I) methyl-norcholestenol) injected to a living mouse and measured simultaneously. Data taking for 12 hours with the mouse under anesthesia allowed for the acquisition of 2D and 3D images of the tracers simultaneously and to study their in vivo behavior. Further work with the Gamma-ray emission imaging (GREI) system enabled the visualization of different distributions of several ^{65}Zn-labeled compounds after their intravenous injection in live mice (Munekane et al. 2016).

The Si/CdTe Compton camera developed originally for the Japanese X-ray astronomy mission ASTRO-H mentioned earlier was also employed for obtaining 2D (Takeda et al. 2012) and 3D (Suzuki et al. 2013) rat images. The camera was composed of a single 2.56×2.56 cm^2 DSSD and four layers of pixelated CdTe detectors, installed in a chamber and cooled to $-20°C$. A spatial resolution similar to SPECT (~10 mm) was achieved, but efficiency was substantially lower. Although the system was able to distinguish different isotopes simultaneously, further improvement was required for imaging applications. Recent work of the group (Sakai et al. 2018) has allowed for in vivo imaging with ^{99m}Tc-DMSA and ^{18}F-FDG in rats, both separated and simultaneously, with an improved system achieving 4.9° ARM at 511 keV. Accumulations of ^{99m}Tc in the kidneys and ^{18}F in the bladder were observed in the reconstructed images and the results were consistent with PET images and well counter measurements.

The study was also extended to a human torso phantom including several organs filling the kidneys with a ^{99m}Tc solution (Sakai et al. 2019), and to a human volunteer who was injected intravenously 30 MBq ^{99m}Tc-DMSA and 150 MBq ^{18}F-FDG simultaneously (Nakano et al. 2020). In the latter work, the Si/CdTe Compton camera achieved an ARM of 11.4° FWHM at 122 keV, and 4.2° at 511 keV. The efficiency was 3.4×10^{-6} and 1.3×10^{-6} respectively, both at 10 cm. Images were taken at an estimated distance of 30 cm and could simultaneously image the distributions of the two radiopharmaceuticals accumulated in different organs as expected from SPECT and PET studies. The imaging resolution of the system is not yet sufficient for clinical use and does not surpass that of SPECT and PET. However, the results indicate high potential for the application of Compton cameras in the field of multi-energy radioisotope tomography.

The technical complexity of operating solid state detectors in clinical environments, the need to refrigerate them and their high cost can be overcome by the use of scintillator detectors also as scatterers, at the expense of a reduced energy resolution and thus worse imaging performance. This has been possible only recently through the introduction of SiPMs as photodetectors. With this approach, different Compton camera designs are under study.

A laparoscopic Compton camera of 11.8 mm diameter for radio-guided surgery (RGS) was developed in (Nakamura et al. 2016) with four layers of gadolinium fine aluminum gallate (GFAG) cross-shaped pixel arrays of $2 \times 2 \times 3$ mm^3 crystals coupled to SiPMs. The system had an ARM of 17° FWHM at 511 keV, 4 mm resolution FWHM at 10 mm and an efficiency of 0.11 cps kBq^{-1}, comparable to the performance of commercial gamma cameras used for RGS with a low-energy gamma-ray source such as ^{99m}Tc. The system allowed obtaining images of a resected lymph node and a part of primary tumor ex vivo after FDG administration to the patient. Another version of the system (Koyama et al. 2017) with two detectors made of GAGG crystals includes a position tracking system for performing real-time radiation-guided surgery.

In vivo 3D imaging of a mouse with three different radioisotopes (^{131}I, ^{85}Sr and ^{65}Zn) was also possible employing a compact and light Compton camera made of two layers of GAGG scintillator crystal arrays coupled to SiPM arrays. The energy resolution was 7.4% (FWHM) and the angular resolution was 4.5° at 662 keV. The system was placed at ten different positions around the mouse for tomographic acquisition, and data were taken for 10 minutes at each angle. The reconstructed images showed the radiotracers concentrated in their specific target organs.

Alternative detector types are also employed. The ETCC system mentioned earlier was employed to obtain images with 204 keV, 582 keV and 835 keV photons from ^{95m}Tc (Hatsukawa et al. 2018). This isotope could be employed to replace ^{99m}Tc, the most widely adopted radioisotope for medical diagnostic scans, reducing scattering in the patient.

Compton cameras have also been proposed for monitoring drug delivery and radionuclide therapy with alpha particles, for which current imaging methods are suboptimal since high energies are involved (Nurdan et al. 2015; Seo 2019). With this aim, a system has been developed and tested for the detection of ^{211}At emission photons of 570, 687 and 898 keV (Nagao et al. 2018). The system employed consists of a scatterer made of a $20.8 \times 20.8 \times 5$ mm^3 GAGG scintillator block coupled to a SiPM array, and an absorber as a $41.7 \times 41.7 \times 10$ mm^3 GAGG array block coupled to a multianode PMT.

In the same application area, a system combining a pinhole and a Compton camera was developed to image both X-rays and gamma-rays (Omata et al. 2020). The system, composed of two-layers of GAGG crystals coupled to SiPM arrays, has a 5 mm $\times$ 5 mm hole in the first detector, acting as a pinhole for low energy events interacting in the second detector. Images of a ^{211}At source in a bottle were reconstructed using 79 keV X-rays in the pinhole mode and 570 keV gamma-rays in the Compton mode. In addition, a mouse was injected with ^{211}At (971 kBq) and imaged with both modalities taking data for 11 hours after being euthanized. While the pinhole images showed accumulation in the thyroid, stomach and bladder, Compton events were too scarce for a precise image. Imaging of ^{223}Ra, widely used for treating bone metastasis of prostate cancer with radionuclide therapy, has also been tested with an unoptimized Compton camera developed for environmental survey in Fukushima (Fujieda et al. 2020). 3D imaging of a complex phantom filled with a ^{223}Ra solution was performed successfully, and imaging of the tracer in the body of a patient was also possible in a shorter time than that using a SPECT system, but with worse spatial resolution.

The combination of Compton and PET imaging modalities in a single device has also been proposed for simultaneous imaging of both modalities or for triple gamma imaging of non-pure positron emitters (e.g., ^{44}Sc), which emit an additional gamma-ray almost at the same time. An additional detector ring, which is used as the scatterer, is inserted in the FOV of a PET ring, and the performance improvement is being investigated (Yoshida et al. 2020).

While performance and cost issues still need to be further addressed, studies (Zhang, Rogers, and Clinthorne 2004; Han et al. 2008; Nurdan et al. 2015; Fontana et al. 2017) keep showing the potential advantage of employing Compton cameras, and technological advances get them closer to their medical application.

A further field for medical application of Compton cameras has emerged recently in the context of radiation oncology with accelerated proton or ion beams. These beams have a characteristic dose deposition profile: the Bragg curve (Wilson 1946). Compared to conventional radiotherapy with electron or photon beams, proton or heavier ions have a finite range in the body, and exhibit an increased ionization density close to that stopping point, with no dose deposition beyond.

Nevertheless, due to uncertainties in patient imaging and tissue composition, the particle range cannot be predicted with high accuracy (Andreo 2009). To tackle this shortcoming, research groups around the world are trying for several decades to develop methods to measure the particle range in parallel to the treatment (Knopf and Lomax 2013). One strategy is to measure prompt gamma-rays emitted in nuclear interactions between the accelerated charged particles and nuclei of the tissue (Krimmer et al. 2018). This secondary radiation has energies between 1 MeV and 6 MeV. As the emission distribution of these gamma-rays is correlated to the range of the ion beam, the detection system should reconstruct their origin with as high certainty as possible.

A natural candidate for this task is the Compton camera. In theory, this technology is more compact than passively collimated cameras, which require very dense and thick collimators to block gamma-rays of up to 6 MeV. Figure 7.6 shows an experimental measurement of a BGO-based Compton camera and the resulting 2D backprojection image superimposed with the expected proton range in a homogeneous target.

The use of two-stage and multi-stage Compton cameras for application in ion beam therapy has been pursued by numerous research groups based on scintillation as well as semiconductor detectors: a gaseous micro-pixel chamber (Kabuki et al. 2009; Kang and Kim 2009), CZT cross strip detectors (Golnik et al. 2016), the CZT-based Polaris-J camera (Polf et al. 2015), DSSDs for electron tracking (Frandes et al. 2010; Seo et al. 2011), LYSO (Richard et al. 2009), BGO and LSO block crystals coupled to PMT tubes (Golnik et al. 2016; Hueso-Gonzalez et al. 2017), LaBr$_3$ pixelated (Aldawood et al. 2017) or monolithic (Frandes et al. 2010; Llosá et al. 2013) scintillation crystals coupled to multi-anode PMTs or SiPMs, silicon drift detectors coupled to LYSO, CsI (Peloso et al. 2011) or GSO pixels (Kabuki et al. 2009) on a multi-anode PMT, GAGG pixels coupled to MPPCs (Taya et al. 2016) or multi-anode PMTs (Koide et al. 2018) and a NaI cylindrical crystal coupled to a PMT (Seo et al. 2011).

The harsh radiation conditions in a clinical treatment room, especially high-count rate and neutron background, impose strong prerequisites for the detectors and electronics design, which can significantly increase the cost of the system and cast doubts on its applicability (Rohling et al. 2017;

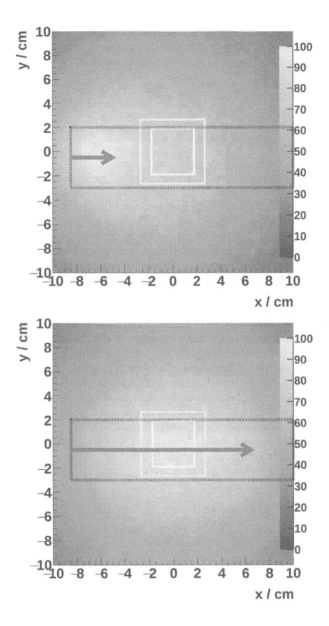

FIGURE 7.6

Experimental images obtained in a clinical proton therapy center with a two-layer Compton camera using simple backprojection, adapted from (Hueso-González et al. 2017). The scale of the color palette is arbitrary. The gray rectangles correspond to the projection on the imaging plane of the two BGO block detectors (refurbished from PET scanners) deployed in this Compton camera. The black dotted rectangle indicates the acrylic glass target. The proton beam incidence is horizontal from the left, and the theoretical proton range is marked by the tip of the red arrow (3.5 cm for 70 MeV protons on the top, 15.1 cm for 160 MeV protons on the bottom). The top and bottom images contain 40,000 and 85,000 coincidences, approximately, measured with a low-proton beam current of 40 pA incident on the target.

Pausch et al. 2020). Consequently, new technologies are being developed to reduce costs, for example, using scintillation fibers (Kasper et al. 2020), and to facilitate a future clinical translation of Compton cameras for treatment monitoring.

Notes

1. R. Hofstadter had discovered the scintillation properties of NaI(Tl) just two years before (Hofstadter 1948). A review is provided by the same author in (Hofstadter and Stein 1975).
2. Namely, the average number of emissions being small compared to the number of possible emitters, and the probability of detection being low. These conditions are usually met in emission tomography imaging. On the limitations of the generally assumed Poisson model we can also cite (Sitek and Celler 2015).
3. Note that Equation (7.5) is also reinterpreted as describing the expected value of the photo-counting detections.
4. As an example, (Wilderman et al. 1998) estimates around 2.3×10^{10} measurement elements for the described prototype. More complex (e.g., multilayered) Compton camera geometries could yield higher numbers.
5. The Poisson likelihood approximation is not applied; instead, the formalism relies on the more general Poisson-multinomial photo-detection model. The reader can find a comprehensive derivation of the algorithm in (Sitek, 2014).
6. Comprehensive descriptions of astronomical gamma-ray sources can be found in dedicated publications, such as (Takada et al. 2011) or (De Angelis et al. 2017).
7. PHDS Co. 3011 Amherst Road, Knoxville, TN 37921, USA.
8. H3D Inc. 812 Avis Dr. Ann Arbor, MI 48108, USA.
9. Mitsubishi Heavy Industries, Ltd. (MHI).
10. NUVIATech Instruments.

References

Agostinelli, S., J. Allison, K. Amako, J. Apostolakis, H. Araujo, P. Arce, M. Asai, et al. 2003. "Geant4—a Simulation Toolkit." Nuclear Instruments and Methods in Physics Research Section A 506 (3): 250–303. doi:10.1016/S0168-9002(03)01368-8.

Aldawood, S., P. G. Thirolf, A. Miani, M. Böhmer, G. Dedes, R. Gernhäuser, C. Lang, et al. 2017. "Development of a Compton Camera for Prompt-Gamma Medical Imaging." Radiation Physics and Chemistry 140 (November): 190–7. doi:10.1016/j.radphyschem.2017.01.024.

Alenius, S., and U. Ruotsalainen. 1997. "Bayesian Image Reconstruction for Emission Tomography Based on Median Root Prior." European Journal of Nuclear Medicine 24 (3): 258–65. doi:10.1007/BF01728761.

Amoyal, G., V. Schoepff, F. Carrel, M. Michel, N. Blanc de Lanaute, and J. C. Angélique. 2021. "Development of a Hybrid Gamma Camera Based on Timepix3 for Nuclear Industry Applications." Nuclear Instruments and Methods in Physics Research Section A 987 (January): 164838. doi:10.1016/j.nima.2020.164838.

Andreo, P. 2009. "On the Clinical Spatial Resolution Achievable with Protons and Heavier Charged Particle Radiotherapy Beams." Physics in Medicine and Biology 54 (11): N205–15. IOP Publishing. doi:10.1088/0031-9155/54/11/N01.

Andreyev, A., A. Celler, I. Ozsahin, and A. Sitek. 2016. "Resolution Recovery for Compton Camera Using Origin Ensemble Algorithm." Medical Physics 43 (8 Part 1): 4866–76. doi:10.1118/1.4959551.

Andreyev, A., A. Sitek, and A. Celler. 2011. "Fast Image Reconstruction for Compton Camera Using Stochastic Origin Ensemble Approach." Medical Physics 38 (1): 429–38. doi:10.1118/1.3528170.

Andritschke, R., A. Zoglauer, G. Kanbach, P. F. Bloser, and F. Schopper. 2005. "The Compton and Pair Creation Telescope MEGA." Experimental Astronomy 20 (1): 395–403. doi:10.1007/s10686-006-9040-7.

Aprile, E., A. Curioni, K. L. Giboni, M. Kobayashi, U. G. Oberlack, and S. Zhang. 2008. "Compton Imaging of MeV Gamma-Rays with the Liquid Xenon Gamma-Ray Imaging Telescope (LXeGRIT)." Nuclear Instruments and Methods in Physics Research Section A 593 (3): 414–25. doi:10.1016/j.nima.2008.05.039.

Aprile, E., A. Curioni, K. L. Giboni, M. Kobayashi, U. G. Oberlack, E. I.. Chupp, P. P. Dunphy, T. Doke, J. Kikuchi, and S. Ventura. 2003. "LXeGRIT Compton Telescope Prototype: Current Status and Future Prospects." X-Ray and Gamma-Ray Telescopes and Instruments for Astronomy, 4851: 1196–208. International Society for Optics and Photonics. doi:10.1117/12.461567.

Aprile, E., R. Mukherjee, and M. Suzuki. 1989. "A Liquid Xenon Imaging Telescope for 1-30 MeV Gamma-Ray Astrophysics." EUV, X-Ray, and Gamma-Ray Instrumentation for Astronomy and Atomic Physics, 1159: 295–305. International Society for Optics and Photonics. doi:10.1117/12.962588.

Aramaki, T., P. O. H. Adrian, G. Karagiorgi, and H. Odaka. 2020. "Dual MeV Gamma-Ray and Dark Matter Observatory—GRAMS Project." Astroparticle Physics 114 (January): 107–14. doi:10.1016/j.astropartphys.2019.07.002.

Babiano-Suárez, V., J. Lerendegui-Marco, J. Balibrea-Correa, L. Caballero, D. Calvo, I. Ladarescu, C. Domingo-Pardo, et al. 2020. "Imaging Neutron Capture Cross Sections: I-TED Proof-of-Concept and Future Prospects Based on Machine-Learning Techniques." ArXiv:2012.10374 [Astro-Ph, Physics:Nucl-Ex, Physics:Physics], December. http://arxiv.org/abs/2012.10374.

Baca, T., P. Stibinger, D. Doubravova, D. Turecek, J. Solc, J. Rusnak, M. Saska, and J. Jakubek. 2020. "Gamma Radiation Source Localization for Micro Aerial Vehicles with a Miniature Single-Detector Compton Event Camera." ArXiv:2011.03356 [Physics], November. http://arxiv.org/abs/2011.03356.

Bandstra, M. S., E. C. Bellm, S. E. Boggs, D. Perez-Becker, A. Zoglauer, H.-K. Chang, J.-L. Chiu, et al. 2011. "Detection and Imaging of the Crab Nebula with the Nuclear Compton Telescope." The Astrophysical Journal 738 (1): 8. American Astronomical Society. doi:10.1088/0004-637X/738/1/8.

Basko, R., G. L. Zeng, and G. T. Gullberg. 1997. "Analytical Reconstruction Formula for One-Dimensional Compton Camera." IEEE Transactions on Nuclear Science 44 (3): 1342–6. doi:10.1109/23.597011.

————. 1998. "Application of Spherical Harmonics to Image Reconstruction for the Compton Camera." Physics in Medicine and Biology 43 (4): 887–94. IOP Publishing. doi:10.1088/0031-9155/43/4/016.

Bloemen, H., R. Wijnands, K. Bennett, R. Diehl, W. Hermsen, G. Lichti, D. Morris, et al. 1994. "COMPTEL Observations of the Orion Complex: Evidence for Cosmic-Ray Induced Gamma-Ray Lines." Astronomy and Astrophysics 281 (January): L5–8.

Bloser, P. F., J. S. Legere, C. M. Bancroft, J. M. Ryan, and M. L. McConnell. 2016. "Balloon Flight Test of a Compton Telescope Based on Scintillators with Silicon Photomultiplier Readouts." Nuclear Instruments and Methods in Physics Research Section A 812 (March): 92–103. doi:10.1016/j.nima.2015.12.047.

Bloser, P. F., T. Sharma, J. S. Legere, C. M. Bancroft, M. L. McConnell, J. M. Ryan, and A. M. Wright. 2018. "The Advanced Scintillator Compton Telescope (ASCOT) Balloon Payload." Space Telescopes and Instrumentation 2018: Ultraviolet to Gamma Ray, 10699: 106995X. International Society for Optics and Photonics. doi:10.1117/12.2312150.

Boggs, S. E. 2003. "Polarization Constraints on Gamma-Ray Event Circles in Compton Scatter Instruments." Nuclear Instruments and Methods in Physics Research Section A 503 (3): 562–66. doi:10.1016/S0168-9002(03)00989-6.

Boggs, S. E., P. Jean, R. P. Lin, D. M. Smith, P. vonBallmoos, N. W. Madden, P. N. Luke, et al. 2001. "The Nuclear Compton Telescope: A Balloon-Borne Soft γ-Ray Spectrometer, Polarimeter, and Imager." AIP Conference Proceedings, 587 (1): 877–81. American Institute of Physics. doi:10.1063/1.1419514.

Calderón, Y., M. Chmeissani, M. Kolstein, and G. De Lorenzo. 2014. "Evaluation of Compton Gamma Camera Prototype Based on Pixelated CdTe Detectors." Journal of Instrumentation 9 (06): C06003–C06003. IOP Publishing. doi:10.1088/1748-0221/9/06/C06003.

Carrel, F., R. A. Khalil, S. Colas, D. de Toro, G. Ferrand, E. Gaillard-Lecanu, M. Gmar, et al. 2011. "GAMPIX: A New Gamma Imaging System for Radiological Safety and Homeland Security Purposes." In 2011 IEEE Nuclear Science Symposium Conference Record, 4739–44. doi:10.1109/NSSMIC.2011.6154706.

Chelikani, S., J. Gore, and G. Zubal. 2004. "Optimizing Compton Camera Geometries." Physics in Medicine and Biology 49 (8): 1387–408. IOP Publishing. doi:10.1088/0031-9155/49/8/002.

Chen, H., H. H. Chen-Mayer, D. J. Turkoglu, B. K. Riley, E. Draeger, and J. P. Polf. 2018. "Spectroscopic Compton Imaging of Prompt Gamma Emission at the MeV Energy Range." Journal of Radioanalytical and Nuclear Chemistry 318 (1): 241–6. doi:10.1007/s10967-018-6070-3.

Compton, A. H. 1923. "A Quantum Theory of the Scattering of X-Rays by Light Elements." Physical Review 21 (5): 483–502. doi:10.1103/PhysRev.21.483.

Conka-Nurdan, T., K. Nurdan, A. H. Walenta, I. Chiosa, B. Freisleben, N. A. Pavel, and L. Struder. 2005. "First Results on Compton Camera Coincidences with the Silicon Drift Detector." IEEE Transactions on Nuclear Science 52 (5): 1381–5. doi:10.1109/TNS.2005.858211.

Cree, M. J., and P. J. Bones. 1994. "Towards Direct Reconstruction from a Gamma Camera Based on Compton Scattering." IEEE Transactions on Medical Imaging 13 (2): 398–407. doi:10.1109/42.293932.

Dauber, P. M., and L. H. Smith. 1973. "The Liquid-Xenon Compton Telescope: A New Technique for Gamma-Ray Astronomy." Proceedings of the 13th International Conference on Cosmic Rays, 4: 2716–21.

De Angelis, A., V. Tatischeff, M. Tavani, U. Oberlack, I. Grenier, L. Hanlon, R. Walter, et al. 2017. "The E-ASTROGAM Mission." Experimental Astronomy 44 (1): 25–82. doi:10.1007/s10686-017-9533-6.

Dempster, A. P., N. M. Laird, and D. B. Rubin. 1977. "Maximum Likelihood from Incomplete Data via the EM Algorithm." Journal of the Royal Statistical Society 39 (1): 1–22. doi:10.1111/j.2517-6161.1977.tb01600.x.

Dijk, R. van. 1996. "Gamma-Ray Observations of X-Ray Binaries with COMPTEL." Ph. D. dissertation, University of Amsterdam.

Dogan, N., D. K. Wehe, and A. Z. Akcasu. 1992. "A Source Reconstruction Method for Multiple Scatter Compton Cameras." IEEE Transactions on Nuclear Science 39 (5): 1427–30. doi:10.1109/23.173219.

Dogan, N., D. K. Wehe, and G. F. Knoll. 1990. "Multiple Compton Scattering Gamma Ray Imaging Camera." Nuclear Instruments and Methods in Physics Research Section A 299 (1): 501–6. doi:10.1016/0168-9002(90)90832-Q.

Domingo-Pardo, C. 2016. "I-TED: A Novel Concept for High-Sensitivity (n,γ) Cross-Section Measurements." Nuclear Instruments and Methods in Physics Research Section A 825 (July): 78–86. doi:10.1016/j.nima.2016.04.002.

Draeger, E., D. Mackin, S. Peterson, H. Chen, S. Avery, S. Beddar, and J. C. Polf. 2018. "3D Prompt Gamma Imaging for Proton Beam Range Verification." Physics in Medicine and Biology 63 (3): 035019. doi:10.1088/1361-6560/aaa203.

Feng, Y., A. Etxebeste, J. M. Létang, D. Sarrut, and V. Maxim. 2018. "Total Variation and Point Spread Function Priors for MLEM Reconstruction in Compton Camera Imaging." In 2018 IEEE Nuclear Science Symposium and Medical Imaging Conference Proceedings (NSS/MIC), 1–3. doi:10.1109/NSSMIC.2018.8824767.

Fetter, S., V. A. Frolov, M. Miller, R. Mozley, O. F. Prilutsky, S. N. Rodionov, and R. Z. Sagdeev. 1990. "Detecting Nuclear Warheads." Science & Global Security 1 (3–4): 225–53. Routledge. doi:10.1080/08929889008426333.

Fontana, M., D. Dauvergne, J. M. Létang, J.-L. Ley, and É. Testa. 2017. "Compton Camera Study for High Efficiency SPECT and Benchmark with Anger System." Physics in Medicine and Biology 62 (23): 8794–812. IOP Publishing. doi:10.1088/1361-6560/aa926a.

Frandes, M., A. Zoglauer, V. Maxim, and R. Prost. 2010. "A Tracking Compton-Scattering Imaging System for Hadron Therapy Monitoring." IEEE Transactions on Nuclear Science 57 (1): 144–50. doi:10.1109/TNS.2009.2031679.

Fujieda, K., J. Kataoka, S. Mochizuki, L. Tagawa, S. Sato, R. Tanaka, K. Matsunaga, et al. 2020. "First Demonstration of Portable Compton Camera to Visualize 223-Ra Concentration for Radionuclide Therapy." Nuclear Instruments and Methods in Physics Research Section A, Proceedings of the Vienna Conference on Instrumentation 2019, 958 (April): 162802. doi:10.1016/j.nima.2019.162802.

Göllnitz, H., E. Heidbreder, K. Pinkau, C. Reppin, V. Schönfelder, and R. Gorenflo. 1969. "Design of a Neutron Scattering Chamber Using Monte Carlo Calculations." Nuclear Instruments and Methods 74 (1): 109–22. doi:10.1016/0029-554X(69)90498-4.

Golnik, C., D. Bemmerer, W. Enghardt, F. Fiedler, F. Hueso-González, G. Pausch, K. Römer, H. Rohling, S. Schöne, and L. Wagner. 2016. "Tests of a Compton Imaging Prototype in a Monoenergetic 4.44 MeV Photon Field—A Benchmark Setup for Prompt Gamma-Ray Imaging Devices." Journal of Instrumentation 11 (06): P06009. IOP Publishing.

Goodman, D., J. Xia, and Z. He. 2019. "Qualitative Measurement of Spatial Shielding Isotopics via Compton Imaging Neutron-Induced Gamma Rays Using 3-D CdZnTe Detectors." Nuclear Instruments and Methods in Physics Research Section A 935 (August): 214–21. doi:10.1016/j.nima.2019.04.026.

Grannan, R. T., R. Koga, W. A. Millard, A. M. Preszler, G. M. Simnett, and R. S. White. 1972. "A Large Area Detector for Neutrons between 2 and 100 MeV." Nuclear Instruments and Methods 103 (1): 99–108. doi:10.1016/0029-554X(72)90465-X.

Gutierrez, A., C. Baker, H. Boston, S. Chung, D. S. Judson, A. Kacperek, B. Le Crom, et al. 2018. "Progress towards a Semiconductor Compton Camera for Prompt Gamma Imaging during Proton Beam Therapy for Range and Dose Verification." Journal of Instrumentation 13 (01): C01036–C01036. IOP Publishing. doi:10.1088/1748-0221/13/01/C01036.

Han, L., W. L. Rogers, S. S. Huh, and N. Clinthorne. 2008. "Statistical Performance Evaluation and Comparison of a Compton Medical Imaging System and a Collimated Anger Camera for Higher Energy Photon Imaging." Physics in Medicine and Biology 53 (24): 7029–45. IOP Publishing. doi:10.1088/0031-9155/53/24/002.

Hatsukawa, Y., T. Hayakawa, K. Tsukada, K. Hashimoto, T. Sato, et al. 2018. "Electron-Tracking Compton Camera Imaging of Technetium-95m". PLOS ONE 13 (12): e0208909. Public Library of Science. doi:10.1371/journal.pone.0208909.

Heidbreder, E., K. Pinkau, C. Reppin, and V. Schönfelder. 1971. "Measurements of the Distribution in Energy and Angle of High-Energy Neutrons in the Lower Atmosphere." Journal of Geophysical Research (1896-1977) 76 (13): 2905–16. doi:10.1029/JA076i013p02905.

Herzo, D., R. Koga, W. A. Millard, S. Moon, J. Ryan, R. Wilson, A. D. Zych, and R. S. White. 1975. "A Large Double Scatter Telescope for Gamma Rays and Neutrons." Nuclear Instruments and Methods 123 (3): 583–97. doi:10.1016/0029-554X(75)90215-3.

Hitomi Collaboration, F. Aharonian, H. Akamatsu, F. Akimoto, S. W. Allen, L. Angelini, M. Audard, et al. 2018. "Detection of Polarized Gamma-Ray Emission from the Crab Nebula with the Hitomi Soft Gamma-Ray Detector." Publications of the Astronomical Society of Japan 70 (113). doi:10.1093/pasj/psy118.

Hmissi, M. Z., A. Iltis, C. Tata, G. Zeufack, L. Rodrigues, B. Mehadji, C. Morel, and H. Snoussi. 2018. "First Images from a CeBr3/LYSO:Ce Temporal Imaging Portable Compton Camera at 1.3 MeV." In 2018 IEEE Nuclear Science Symposium and Medical Imaging Conference Proceedings (NSS/MIC), 1–3. doi:10.1109/NSSMIC.2018.8824429.

Hofstadter, R., and J. A. McIntyre. 1950. "Measurement of Gamma-Ray Energies with Two Crystals in Coincidence." Physical Review 78 (5): 619–20. American Physical Society. doi:10.1103/PhysRev.78.619.

Hueso-González, F., G. Pausch, J. Petzoldt, K. E. Römer, and W. Enghardt. 2017. "Prompt Gamma Rays Detected with a BGO Block Compton Camera Reveal Range Deviations of Therapeutic Proton Beams." IEEE Transactions on Radiation and Plasma Medical Sciences 1 (1): 76–86. doi:10.1109/TNS.2016.2622162.

Iyudin, A. F., R. Diehl, H. Bloemen, W. Hermsen, G. G. Lichti, D. Morris, J. Ryan, et al. 1994. "COMPTEL Observations of Ti-44 Gamma-Ray Line Emission from CAS A." Astronomy and Astrophysics 284 (April): L1–4.

Julien, M., J. M. Ryan, P. F. Bloser, J. S. Legere, C. M. Bancroft, M. L. McConnell, R. M. Kippen, and S. Tornga. 2012. "Balloon Flight Results of a FAst Compton Telescope (FACTEL)." In 2012 IEEE Nuclear Science Symposium and Medical Imaging Conference Record (NSS/MIC), 1893–1900. doi:10.1109/NSSMIC.2012.6551439.

Jung, C. Y., and S. Moon. 2015. "Inversion Formulas for Cone Transforms Arising in Application of Compton Cameras." Inverse Problems 31 (1): 015006. IOP Publishing. doi:10.1088/0266-5611/31/1/015006.

————. 2016. "Exact Inversion of the Cone Transform Arising in an Application of a Compton Camera Consisting of Line Detectors." SIAM Journal on Imaging Sciences 9 (2): 520–36. Society for Industrial and Applied Mathematics. doi:10.1137/15M1033617.

Kabuki, S., K. Ueno, S. Kurosawa, S. Iwaki, H. Kubo, K. Miuchi, Y. Fujii, et al. 2009. "Study on the Use of Electron-Tracking Compton Gamma-Ray Camera to Monitor the Therapeutic Proton Dose Distribution in Real Time." In 2009 IEEE Nuclear Science Symposium Conference Record (NSS/MIC), 2437–40. doi:10.1109/NSSMIC.2009.5402130.

Kalish, Y., and E. Nardi. 1964. "A Three Crystal NaI(Tl) Gamma-Ray Scintillation Spectrometer." Nuclear Instruments and Methods 26 (February): 329–32. doi:10.1016/0029-554X(64)90099-0.

Kanbach, G., R. Andritschke, P. F. Bloser, F. Schopper, V. Schönfelder, and A. Zoglauer. 2003. "Concept Study for the Next Generation Medium-Energy Gamma-Ray Astronomy Mission: MEGA." X-Ray and Gamma-Ray Telescopes and Instruments for Astronomy, 4851: 1209–20. International Society for Optics and Photonics. doi:10.1117/12.461348.

Kang, B., and J. Kim. 2009. "Monte Carlo Design Study of a Gamma Detector System to Locate Distal Dose Falloff in Proton Therapy." IEEE Transactions on Nuclear Science 56 (1): 46–50. doi:10.1109/TNS.2008.2005189.

Kasper, J., K. Rusiecka, R. Hetzel, M. K. Kozani, R. Lalik, A. Magiera, A. Stahl, and A. Wrońska. 2020. "The SiFi-CC Project—Feasibility Study of a Scintillation-Fiber-Based Compton Camera for Proton Therapy Monitoring." Physica Medica 76 (August): 317–25. doi:10.1016/j.ejmp.2020.07.013.

Katagiri, H., W. Satoh, R. Enomoto, R. Wakamatsu, T. Watanabe, H. Muraishi, M. Kagaya, et al. 2018. "Development of an All-Sky Gamma-Ray Compton Camera Based on Scintillators for High-Dose Environments." Journal of Nuclear Science and Technology 55 (10): 1172–9. Taylor & Francis. doi:10.1080/00223131.2018.1485598.

Kierans, C. A., S. E. Boggs, A. Zoglauer, A. W. Lowell, C. Sleator, J. Beechert, T. J. Brandt, et al. 2020. "Detection of the 511 KeV Galactic Positron Annihilation Line with COSI." The Astrophysical Journal 895 (1): 44. American Astronomical Society. doi:10.3847/1538-4357/ab89a9.

Kierans, C. A., S. E. Boggs, J. L. Chiu, A. Lowell, C. Sleator, J. A. Tomsick, A. Zoglauer, et al. 2017. "The 2016 Super Pressure Balloon Flight of the Compton Spectrometer and Imager." ArXiv, 1701: 05558 [Astro-Ph], January. http://arxiv.org/abs/1701.05558.

Kim, Y., J. H. Kim, J. Lee, and C. H. Kim. 2018. "Large-Area Compton Camera for High-Speed and 3-D Imaging." IEEE Transactions on Nuclear Science 65 (11): 2817–22. doi:10.1109/TNS.2018.2874890.

Klein, O., and Y. Nishina. 1929. "Über die Streuung von Strahlung durch freie Elektronen nach der neuen relativistischen Quantendynamik von Dirac." Zeitschrift für Physik 52 (11): 853–68. doi:10.1007/BF01366453.

Knopf, A. C., and A. Lomax. 2013. "In Vivo proton Range Verification: A Review." Physics in Medicine and Biology 58 (15): R131–60. IOP Publishing. doi:10.1088/0031-9155/58/15/R131.

Kohlhase, N., T. Wegener, M. Schaar, A. Bolke, A. Etxebeste, D. Sarrut, and M. Rafecas. 2020. "Capability of MLEM and OE to Detect Range Shifts with a Compton Camera in Particle Therapy." IEEE Transactions on Radiation and Plasma Medical Sciences 4 (2): 233–42. doi:10.1109/TRPMS.2019.2937675.

Koide, A., J. Kataoka, T. Masuda, S. Mochizuki, T. Taya, K. Sueoka, L. Tagawa, et al. 2018. "Precision Imaging of 4.4 MeV Gamma Rays Using a 3-D Position Sensitive Compton Camera." Scientific Reports 8 (1): 8116. Nature Publishing Group. doi:10.1038/s41598-018-26591-2.

Koyama, A., Y. Nakamura, K. Shimazoe, H. Takahashi, and I. Sakuma. 2017. "Prototype of a Single Probe Compton Camera for Laparoscopic Surgery." Nuclear Instruments and Methods in Physics Research Section A, Proceedings. of the Vienna Conference on Instrumentation 2016, 845 (February): 660–3. doi:10.1016/j.nima.2016.06.071.

Krimmer, J., D. Dauvergne, J. M. Létang, and É. Testa. 2018. "Prompt-Gamma Monitoring in Hadrontherapy: A Review." Nuclear Instruments and Methods in Physics Research Section A 878 (January): 58–73. doi:10.1016/j.nima.2017.07.063.

Kuchment, P., and F. Terzioglu. 2016. "Three-Dimensional Image Reconstruction from Compton Camera Data." SIAM Journal on Imaging Sciences 9 (4): 1708–25. Society for Industrial and Applied Mathematics. doi:10.1137/16M107476X.

Kuiper, L., W. Hermsen, G. Cusumano, R. Diehl, V. Schönfelder, A. Strong, K. Bennett, and M. L. McConnell. 2001. "The Crab Pulsar in the 0.75-30 MeV Range as Seen by CGRO COMPTEL—A Coherent High-Energy Picture from Soft X-Rays up to High-Energy γ-Rays." Astronomy & Astrophysics 378 (3): 918–35. EDP Sciences. doi:10.1051/0004-6361:20011256.

Kurfess, J. D., W. N. Johnson, R. A. Kroeger, and B. F. Phlips. 2000. "Considerations for the next Compton Telescope Mission." AIP Conference Proceedings, 510 (1): 789–93. American Institute of Physics. doi:10.1063/1.1303306.

Kwon, K. 2019. "An Inversion of the Conical Radon Transform Arising in the Compton Camera with Helical Movement." Biomedical Engineering Letters 9 (2): 233–43. doi:10.1007/s13534-019-00106-y.

Lange, K., and R. Carson. 1984. "EM Reconstruction Algorithms for Emission and Transmission Tomography." Journal of Computer Assisted Tomography 8 (2): 306–16.

LeBlanc, J. W., N. H. Clinthorne, C. Hua, E. Nygard, W. L. Rogers, D. K. Wehe, P. Weilhammer, and S. J. Wilderman. 1998. "C-SPRINT: A Prototype Compton Camera System for Low Energy Gamma Ray Imaging." IEEE Transactions on Nuclear Science 45 (3): 943–9. doi:10.1109/23.682679.

Lee, J., and C. H. Kim. 2018. "Feasibility Study for 3D Imaging of Hot Spot in Activated Concrete Wall Using Large-Area Compton Camera (LACC)."

Lichti, G., and V. Schönfelder. 1974. "Measurement of the Spectrum of Diffuse Gamma Quanta in the Energy Range from 1 to 10 MeV with the Aid of a Double-Compton Telescope." Mitteilungen Der Astronomischen Gesellschaft Hamburg 35: 228–33.

Liu, Y. L., J. Q. Fu, Y. L. Li, Y. J. Li, X. M. Ma, and L. Zhang. 2018. "Preliminary Results of a Compton Camera Based on a Single 3D Position-Sensitive CZT Detector." Nuclear Science and Techniques 29 (10): 145. doi:10.1007/s41365-018-0483-0.

Llosá, G. 2019. "SiPM-Based Compton Cameras." Nuclear Instruments and Methods in Physics Research Section A 926 (May): 148–52. doi:10.1016/j.nima.2018.09.053.

Llosá, G., J. Bernabeu, D. Burdette, E. Chesi, N. H. Clinthorne, K. Honscheid, H. Kagan, et al. 2008. "Last Results of a First Compton Probe Demonstrator." IEEE Transactions on Nuclear Science 55 (3): 936–41. doi:10.1109/TNS.2008.922817.

Llosá, G., J. Cabello, S. Callier, J. E. Gillam, C. Lacasta, M. Rafecas, L. Raux, et al. 2013. "First Compton Telescope Prototype Based on Continuous LaBr3-SiPM Detectors." Nuclear Instruments and Methods in Physics Research Section A, Proceedings of the 12th Pisa Meeting on Advanced Detectors, 718 (August): 130–3. doi:10.1016/j.nima.2012.08.074.

Lucchetta, G., F. Berlato, R. Rando, D. Bastieri, and G. Urso. 2017. "Scientific Performance of a Nano-Satellite MeV Telescope." The Astronomical Journal 153 (5): 237. American Astronomical Society. doi:10.3847/1538-3881/aa6a1b.

Mackin, D., S. Peterson, S. Beddar, and J. Polf. 2012. "Evaluation of a Stochastic Reconstruction Algorithm for Use in Compton Camera Imaging and Beam Range Verification from Secondary Gamma Emission during Proton Therapy." Physics in Medicine and Biology 57 (11): 3537–53. IOP Publishing. doi:10.1088/0031-9155/57/11/3537.

Martin, J. B., G. F. Knoll, D. K. Wehe, N. Dogan, V. Jordanov, N. Petrick, and M. Singh. 1993. "A Ring Compton Scatter Camera for Imaging Medium Energy Gamma Rays." IEEE Transactions on Nuclear Science 40 (4): 972–8. doi:10.1109/23.256695.

Maxim, V. 2014. "Filtered Backprojection Reconstruction and Redundancy in Compton Camera Imaging." IEEE Transactions on Image Processing 23 (1): 332–41. doi:10.1109/TIP.2013.2288143.

———. 2018. "Enhancement of Compton Camera Images Reconstructed by Inversion of a Conical Radon Transform." Inverse Problems 35 (1): 014001. IOP Publishing. doi:10.1088/1361-6420/aaecdb.

Maxim, V., M. Frandes, and R. Prost. 2009. "Analytical Inversion of the Compton Transform Using the Full Set of Available Projections." Inverse Problems 25 (9): 095001. IOP Publishing. doi:10.1088/0266-5611/25/9/095001.

Maxim, V., X. Lojacono, E. Hilaire, J. Krimmer, E. Testa, D. Dauvergne, I. Magnin, and R. Prost. 2015. "Probabilistic Models and Numerical Calculation of System Matrix and Sensitivity in List-Mode MLEM 3D Reconstruction of Compton Camera Images." Physics in Medicine and Biology 61 (1): 243–64. IOP Publishing. doi:10.1088/0031-9155/61/1/243.

Mihailescu, L., K. M. Vetter, M. T. Burks, E. L. Hull, and W. W. Craig. 2007. "SPEIR: A Ge Compton Camera." Nuclear Instruments and Methods in Physics Research Section A 570 (1): 89–100. doi:10.1016/j.nima.2006.09.111.

Mochizuki, S., J. Kataoka, L. Tagawa, Y. Iwamoto, H. Okochi, N. Katsumi, S. Kinno, et al. 2017. "First Demonstration of Aerial Gamma-Ray Imaging Using Drone for Prompt Radiation Survey in Fukushima." Journal of Instrumentation 12 (11): P11014. IOP Publishing. doi:10.1088/1748-0221/12/11/P11014.

Montémont, G., P. Bohuslav, J. Dubosq, B. Feret, O. Monnet, O. Oehling, L. Skala, S. Stanchina, L. Verger, and G. Werthmann. 2017. "NuVISION: A Portable Multimode Gamma Camera Based on HiSPECT Imaging Module." In 2017 IEEE Nuclear Science Symposium and Medical Imaging Conference (NSS/MIC), 1–3. doi:10.1109/NSSMIC.2017.8532713.

Moon, S., and K. Kwon. 2019. "Inversion Formula for the Conical Radon Transform Arising in a Single First Semicircle Second Compton Camera with Rotation." Japan Journal of Industrial and Applied Mathematics 36 (3): 989–1004. doi:10.1007/s13160-019-00379-x.

Morris, D. J., K. Bennett, H. Bloemen, W. Hermsen, G. G. Lichti, M. L. McConnell, J. M. Ryan, and V. Schönfelder. 1995. "Evidence for 56Co Line Emission from the Type Ia Supernova 1991T Using COMPTEL[a]." Annals of the New York Academy of Sciences 759 (1): 397–400. doi:10.1111/j.1749-6632.1995.tb17571.x.

Motomura, S., Y. Kanayama, H. Haba, Y. Watanabe, and S. Enomoto. 2008. "Multiple Molecular Simultaneous Imaging in a Live Mouse Using Semiconductor Compton Camera." Journal of Analytical Atomic Spectrometry 23 (8): 1089–92. Royal Society of Chemistry. doi:10.1039/B802964D.

Munekane, M., S. Motomura, S. Kamino, M. Ueda, H. Haba, Y. Yoshikawa, H. Yasui, M. Hiromura, and S. Enomoto. 2016. "Visualization of Biodistribution of Zn Complex with Antidiabetic Activity Using Semiconductor Compton Camera GREI." Biochemistry and Biophysics Reports 5 (March): 211–5. doi:10.1016/j.bbrep.2015.12.004.

Muñoz, E., J. Barrio, A. Etxebeste, P. G. Ortega, C. Lacasta, J. F. Oliver, C. Solaz, and G. Llosá. 2017. "Performance Evaluation of MACACO: A Multilayer Compton Camera." Physics in Medicine and Biology 62 (18): 7321. IOP Publishing. doi:10.1088/1361-6560/aa8070.

Muñoz, E., J. Barrio, J. Bernabéu, A. Etxebeste, C. Lacasta, G. Llosá, A. Ros, J. Roser, and J. F. Oliver. 2018. "Study and Comparison of Different Sensitivity Models for a Two-Plane Compton Camera." Physics in Medicine and Biology 63 (13): 135004. IOP Publishing. doi:10.1088/1361-6560/aac8cd.

Muñoz, E., L. Barrientos, J. Bernabéu, M. Borja-Lloret, G. Llosá, A. Ros, J. Roser, and J. F. Oliver. 2020. "A Spectral Reconstruction Algorithm for Two-Plane Compton Cameras." Physics in Medicine and Biology 65 (2): 025011. IOP Publishing. doi:10.1088/1361-6560/ab58ad.

Nagao, Y., M. Yamaguchi, S. Watanabe, N. S. Ishioka, N. Kawachi, and H. Watabe. 2018. "Astatine-211 Imaging by a Compton Camera for Targeted Radiotherapy." Applied Radiation and Isotopes 139 (September): 238–43. doi:10.1016/j.apradiso.2018.05.022.

Nakamura, Y., K. Shimazoe, H. Takahashi, S. Yoshimura, Y. Seto, S. Kato, M. Takahashi, and T. Momose. 2016. "Development of a Novel Handheld Intra-Operative Laparoscopic Compton Camera For 18F-Fluoro-2-Deoxy-2-D-Glucose-Guided Surgery." Physics in Medicine and Biology 61 (15): 5837–50. IOP Publishing. doi:10.1088/0031-9155/61/15/5837.

Nakano, T., M. Sakai, K. Torikai, Y. Suzuki, S. Takeda, S. Noda, M. Yamaguchi, et al. 2020. "Imaging of 99mTc-DMSA and 18F-FDG in Humans Using a Si/CdTe Compton Camera." Physics in Medicine and Biology 65 (5): 05LT01. IOP Publishing. doi:10.1088/1361-6560/ab33d8.

Nishiyama, T., J. Kataoka, A. Kishimoto, T. Fujita, Y. Iwamoto, T. Taya, S. Ohsuka, et al. 2014. "A Novel Compton Camera Design Featuring a Rear-Panel Shield for Substantial Noise Reduction in Gamma-Ray Images." Journal of Instrumentation 9 (12): C12031. IOP Publishing. doi:10.1088/1748-0221/9/12/C12031.

Nurdan, T. C., K. Nurdan, A. B. Brill, and A. H. Walenta. 2015. "Design Criteria for a High Energy Compton Camera and Possible Application to Targeted Cancer Therapy." Journal of Instrumentation 10 (07): C07018–C07018. IOP Publishing. doi:10.1088/1748-0221/10/07/C07018.

Oberlack, U., K. Bennett, H. Bloemen, R. Diehl, C. Dupraz, W. Hermsen, J. Knoedlseder, et al. 1996. "The COMPTEL 1.809MeV All-Sky Image." Astronomy and Astrophysics Supplement Series 120 (December): 311–4.

Omata, A., J. Kataoka, K. Fujieda, S. Sato, E. Kuriyama, H. Kato, A. Toyoshima, et al. 2020. "Performance Demonstration of a Hybrid Compton Camera with an Active Pinhole for Wide-Band X-Ray and Gamma-Ray Imaging." Scientific Reports 10 (1): 14064. Nature Publishing Group. doi:10.1038/s41598-020-71019-5.

Ordonez, C. E., A. Bolozdynya, and W. Chang. 1997. "Doppler Broadening of Energy Spectra in Compton Cameras." In 1997 IEEE Nuclear Science Symposium Conference Record, 2: 1361–5. doi:10.1109/NSSMIC.1997.670574.

Orito, R., H. Kubo, K. Miuchi, T. Nagayoshi, A. Takada, T. Tanimori, and M. Ueno. 2003. "A Novel Design of the MeV Gamma-Ray Imaging Detector with Micro-TPC." Nuclear Instruments and Methods in Physics Research Section A, Proceedings of the 6th International Conference on Position-Sensitive Detectors, 513 (1): 408–12. doi:10.1016/j.nima.2003.08.071.

Parra, L. C. 2000. "Reconstruction of Cone-Beam Projections from Compton Scattered Data." IEEE Transactions on Nuclear Science 47 (4): 1543–50. doi:10.1109/23.873014.

Pausch, G., J. Berthold, W. Enghardt, K. Römer, A. Straessner, A. Wagner, T. Werner, and T. Kögler. 2020. "Detection Systems for Range Monitoring in Proton Therapy: Needs and Challenges." Nuclear Instruments and Methods in Physics Research Section A, Symposium on Radiation Measurements and Applications XVII, 954(February): 161227. doi:10.1016/j.nima.2018.09.062.

Peloso, R., P. Busca, C. Fiorini, M. Basilavecchia, T. Frizzi, J. Smeets, F. Roellinghoff, D. Prieels, F. Stichelbaut, and A. Benilov. 2011. "Application of the HICAM Camera for Imaging of Prompt Gamma Rays in Measurements of Proton Beam Range." In 2011 IEEE Nuclear Science Symposium Conference Record, 2285–89. doi:10.1109/NSSMIC.2011.6153863.

Peterson, L. E., and R. L. Howard. 1961. "Gamma-Ray Astronomy in Space in the 50-KEV to 3-MEV Region." IRE Transactions on Nuclear Science 8 (4): 21–9. doi:10.1109/TNS2.1961.4315853.

Peterson, S. W., D. Robertson, and J. Polf. 2010. "Optimizing a Three-Stage Compton Camera for Measuring Prompt Gamma Rays Emitted during Proton Radiotherapy." Physics in Medicine and Biology 55 (22): 6841–56. IOP Publishing. doi:10.1088/0031-9155/55/22/015.

Pinkau, K. 1966. "Notizen: Die Messung Solarer Und Atmosphärischer Neutronen." Zeitschrift Für Naturforschung A 21 (12): 2100–1. De Gruyter. doi:10.1515/zna-1966-1216.

Poikela, T., J. Plosila, T. Westerlund, M. Campbell, M. D. Gaspari, X. Llopart, V. Gromov, et al. 2014. "Timepix3: A 65K Channel Hybrid Pixel Readout Chip with Simultaneous ToA/ToT and Sparse Readout." Journal of Instrumentation 9 (05): C05013–C05013. IOP Publishing. doi:10.1088/1748-0221/9/05/C05013.

Polf, J. C., S. Avery, D. S. Mackin, and S. Beddar. 2015. "Imaging of Prompt Gamma Rays Emitted during Delivery of Clinical Proton Beams with a Compton Camera: Feasibility Studies for Range Verification." Physics in Medicine and Biology 60 (18): 7085–99. IOP Publishing. doi:10.1088/0031-9155/60/18/7085.

Poulson, D., P. F. Bloser, K. Ogasawara, J. A. Trevino, J. S. Legere, J. M. Ryan, and M. L. McConnell. 2020. "Development of a Compton Telescope Based on Single-Crystal Diamond Detectors and Fast Scintillators." In Space Telescopes and Instrumentation 2020: Ultraviolet to Gamma Ray, 11444: 114446D. International Society for Optics and Photonics. doi:10.1117/12.2576091.

Preszler, A. M., G. M. Simnett, and R. S. White. 1972. "Earth Albedo Neutrons from 10 to 100 MeV." Physical Review Letters 28 (15): 982–5. American Physical Society. doi:10.1103/PhysRevLett.28.982.

Qi, J., and R. M. Leahy. 2006. "Iterative Reconstruction Techniques in Emission Computed Tomography." Physics in Medicine and Biology 51 (15): R541–78. IOP Publishing. doi:10.1088/0031-9155/51/15/R01.

Rando, R. 2017. "The All-Sky Medium Energy Gamma-Ray Observatory." Journal of Instrumentation 12 (11): C11024–C11024. IOP Publishing. doi:10.1088/1748-0221/12/11/C11024.

Rando, R., S. Canevarolo, H. Xiao, and D. Bastieri. 2019. "Sensitivity to Gamma-Ray Bursts of a Nanosatellite MeV Telescope with a Silicon Tracker." The Astronomical Journal 158 (1): 42. American Astronomical Society. doi:10.3847/1538-3881/ab2454.

Richard, M.-H., M. Chevallier, D. Dauvergne, N. Freud, P. Henriquet, F. Le Foulher, J. M. Létang, et al. 2009. "Design Study of a Compton Camera for Prompt γ Imaging during Ion Beam Therapy." In 2009 IEEE Nuclear Science Symposium Conference Record (NSS/MIC), 4172–5. doi:10.1109/NSSMIC.2009.5402293.

———. 2011. "Design Guidelines for a Double Scattering Compton Camera for Prompt-γ Imaging during Ion Beam Therapy: A Monte Carlo Simulation Study." IEEE Transactions on Nuclear Science 58 (1): 87–94. doi:10.1109/TNS.2010.2076303.

Rohling, H., M. Priegnitz, S. Schoene, A. Schumann, W. Enghardt, F. Hueso-González, G. Pausch, and F. Fiedler. 2017. "Requirements for a Compton Camera For in Vivo Range Verification of Proton Therapy." Physics in Medicine and Biology 62 (7): 2795–811. IOP Publishing. doi:10.1088/1361-6560/aa6068.

Roser, J., E. Muñoz, L. Barrientos, J. Barrio, J. Bernabéu, M. Borja-Lloret, A. Etxebeste, et al. 2020. "Image Reconstruction for a Multi-Layer Compton Telescope: An Analytical Model for Three Interaction Events." Physics in Medicine and Biology 65 (14): 145005. IOP Publishing. doi:10.1088/1361-6560/ab8cd4.

Rudin, L. I., S. Osher, and E. Fatemi. 1992. "Nonlinear Total Variation Based Noise Removal Algorithms." Physica D: Nonlinear Phenomena 60 (1–4): 259–68. North-Holland.

Sakai, M., M. Yamaguchi, Y. Nagao, N. Kawachi, M. Kikuchi, and K. Torikai. 2018. "In Vivo Simultaneous Imaging with 99m Tc and 18 F Using a Compton Camera." Physics in Medicine and Biology 63 (20): 5006. IOP Publishing. doi: 10.1088/1361–6560/aae1d1

Sakai, M., R. K. Parajuli, Y. Kubota, N. Kubo, M. Kikuchi, K. Arakawa, and T. Nakano. 2020. "Improved Iterative Reconstruction Method for Compton Imaging Using Median Filter." PLOS ONE 15 (3): e0229366. Public Library of Science. doi:10.1371/journal.pone.0229366.

Sakai, M., Y. Kubota, R. K. Parajuli, M. Kikuchi, K. Arakawa, and T. Nakano. 2019. "Compton Imaging with 99mTc for Human Imaging." Scientific Reports 9 (1): 12906. Nature Publishing Group. doi:10.1038/s41598-019-49130-z.

Sato, Y., S. Ozawa, Y. Terasaka, K. Minemoto, S. Tamura, K. Shingu, M. Nemoto, and T. Torii. 2020. "Remote Detection of Radioactive Hotspot Using a Compton Camera Mounted on a Moving Multi-Copter Drone above a Contaminated Area in Fukushima." Journal of Nuclear Science and Technology 57 (6): 734–44. Taylor & Francis. doi:10.1080/00223131.2020.1720845.

Schoene, S., W. Enghardt, F. Fiedler, C. Golnik, G. Pausch, H. Rohling, and T. Kormoll. 2017. "An Image Reconstruction Framework and Camera Prototype Aimed for Compton Imaging for In-Vivo Dosimetry of Therapeutic Ion Beams." IEEE Transactions on Radiation and Plasma Medical Sciences 1 (1): 96–107. doi:10.1109/TNS.2016.2623220.

Schönfelder, V. 1991. "The Imaging Gamma-Ray Telescope COMPTEL Aboard GRO." Advances in Space Research 11 (8): 313–22. doi:10.1016/0273-1177(91)90183-K.

———. 2004. "Lessons Learnt from COMPTEL for Future Telescopes." New Astronomy Reviews, Astronomy with Radioactivities IV and Filling the Sensitivity Gap in MeV Astronomy 48 (1): 193–8. doi:10.1016/j.newar.2003.11.027.

Schönfelder, V., A. Hirner, and K. Schneider. 1973. "A Telescope for Soft Gamma Ray Astronomy." Nuclear Instruments and Methods 107 (2): 385–94. doi:10.1016/0029-554X(73)90257-7.

Schönfelder, V., H. Aarts, K. Bennett, H. Deboer, J. Clear, W. Collmar, A. Connors, A. Deerenberg, R. Diehl, and A. Von Dordrecht. 1993. "Instrument Description and Performance of the Imaging Gamma-Ray Telescope COMPTEL Aboard the Compton Gamma-Ray Observatory."Astrophysical Journal Supplement Series. American Astronomical Society.

Schönfelder, V., K. Bennett, H. Bloemen, R. Diehl, W. Hermsen, G. Lichti, M. McConnell, J. Ryan, A. Strong, and C. Winkler. 1996. "COMPTEL Overview: Achievements and Expectations." Astronomy and Astrophysics Supplement Series 120 (November): 13–21.

Schönfelder, V., K. Bennett, J. J. Blom, H. Bloemen, W. Collmar, A. Connors, R. Diehl, et al. 2000. "The First COMPTEL Source Catalogue." Astronomy and Astrophysics Supplement Series 143 (2): 145–79. EDP Sciences. doi:10.1051/aas:2000101.

Seo, H., J. H. Park, A. Ushakov, C. H. Kim, J. K. Kim, J. H. Lee, C. S. Lee, and J. S. Lee. 2011. "Experimental Performance of Double-Scattering Compton Camera with Anthropomorphic Phantom." Journal of Instrumentation 6 (01): C01024–C01024. IOP Publishing. doi:10.1088/1748-0221/6/01/C01024.

Seo, Y. 2019. "Quantitative Imaging of Alpha-Emitting Therapeutic Radio-pharmaceuticals." Nuclear Medicine and Molecular Imaging 53 (3): 182–8. doi:10.1007/s13139-019-00589-8.

Sharma, T., P. F. Bloser, J. M. Ryan, J. S. Legere, and M. L. McConnell. 2020. "Results from the Advanced Scintillator Compton Telescope (ASCOT) Balloon Payload." In Space Telescopes and Instrumentation 2020: Ultraviolet to Gamma Ray, 11444: 1144435. International Society for Optics and Photonics. doi:10.1117/12.2576108.

Shepp, L. A., and Y. Vardi. 1982. "Maximum Likelihood Reconstruction for Emission Tomography." IEEE Transactions on Medical Imaging 1 (2): 113–22. doi:10.1109/TMI.1982.4307558.

Sinclair, L. E., A. McCann, P. R. B. Saull, R. L. Mantifel, C. V. O. Ouellet, P. Drouin, A. M. L. Macleod, et al. 2020. "End-User Experience with the SCoTSS Compton Imager and Directional Survey Spectrometer." Nuclear Instruments and Methods in Physics Research Section A, Symposium on Radiation Measurements and Applications XVII, 954 (February): 161683. doi:10.1016/j.nima.2018.11.142.

Singh, M., and D. Doria. 1983. "An Electronically Collimated Gamma Camera for Single Photon Emission Computed Tomography. Part II: Image Reconstruction and Preliminary Experimental Measurements." Medical Physics 10 (4): 428–35. doi:10.1118/1.595314.

Singh, M., and R. R. Brechner. 1990. "Experimental Test-Object Study of Electronically Collimated SPECT." Journal of Nuclear Medicine 31 (2): 178–86. Citeseer.

Sitek, A.2008. "Representation of Photon Limited Data in Emission Tomography Using Origin Ensembles." Physics in Medicine and Biology 53 (12): 3201–16. IOP Publishing. doi:10.1088/0031-9155/53/12/009.

———. 2011. "Reconstruction of Emission Tomography Data Using Origin Ensembles." IEEE Transactions on Medical Imaging 30 (4): 946–56. doi:10.1109/TMI.2010.2098036.

———. 2012. "Data Analysis in Emission Tomography Using Emission-Count Posteriors." Physics in Medicine and Biology 57 (21): 6779–95. IOP Publishing. doi:10.1088/0031-9155/57/21/6779.

————. 2014. Statistical Computing in Nuclear Imaging. Boca Ratón, FL: CRC Press.

Sitek, A., and A. M. Celler. 2015. "Limitations of Poisson Statistics in Describing Radioactive Decay." Physica Medica 31 (8): 1105–7. doi:10.1016/j.ejmp.2015.08.015.

Smith, B. 2005. "Reconstruction Methods and Completeness Conditions for Two Compton Data Models." JOSA A 22 (3): 445–59. Optical Society of America. doi:10.1364/JOSAA.22.000445.

Studen, A., D. Burdette, E. Chesi, V. Cindro, N. H. Clinthorne, W. Dulinski, J. Fuster, et al. 2004. "First Coincidences in Pre-Clinical Compton Camera Prototype for Medical Imaging." Nuclear Instruments and Methods in Physics Research Section A, Proceedings of the 5th International Workshop on Radiation Imaging Detectors, 531 (1): 258–64. doi:10.1016/j.nima.2004.06.014.

Suzuki, Y., M. Yamaguchi, H. Odaka, H. Shimada, Y. Yoshida, K. Torikai, T. Satoh, et al. 2013. "Three-Dimensional and Multienergy Gamma-Ray Simultaneous Imaging by Using a Si/CdTe Compton Camera." Radiology 267 (3): 941–7. Radiological Society of North America. doi:10.1148/radiol.13121194.

Tajima, H., S. Watanabe, Y. Fukazawa, R. D. Blandford, T. Enoto, A. Goldwurm, K. Hagino, et al. 2018. "Design and Performance of Soft Gamma-Ray Detector Onboard the Hitomi (ASTRO-H) Satellite." Journal of Astronomical Telescopes, Instruments, and Systems 4 (2): 021411. International Society for Optics and Photonics. doi:10.1117/1.JATIS.4.2.021411.

Takada, A., H. Kubo, H. Nishimura, K. Ueno, K. Hattori, S. Kabuki, S. Kurosawa, et al. 2011. "Observation of Diffuse Cosmic and Atmospheric Gamma Rays at Balloon Altitudes with an Electron-Tracking Compton Camera." The Astrophysical Journal 733 (1): 13. American Astronomical Society. doi:10.1088/0004-637X/733/1/13.

Takada, A., T. Tanimori, Y. Mizumura, T. Takemura, K. Yoshikawa, Y. Nakamura, M. Abe, et al. 2020. "SMILE-3: Sky Survey in MeV Gamma-Ray Using the Electron-Tracking Compton Telescope Loaded on Balloons." In Space Telescopes and Instrumentation 2020: Ultraviolet to Gamma Ray, 11444: 1144467. International Society for Optics and Photonics. doi:10.1117/12.2560886.

Takeda, S., A. Harayama, Y. Ichinohe, H. Odaka, S. Watanabe, T. Takahashi, H. Tajima, et al. 2015. "A Portable Si/CdTe Compton Camera and Its Applications to the Visualization of Radioactive Substances." Nuclear Instruments and Methods in Physics Research Section A 787 (July): 207–11. doi:10.1016/j.nima.2014.11.119.

Takeda, S., H. Odaka, S. Ishikawa, S. Watanabe, H. Aono, T. Takahashi, Y. Kanayama, M. Hiromura, and S. Enomoto. 2012. "Demonstration of In-Vivo Multi-Probe Tracker Based on a Si/CdTe Semiconductor Compton Camera." IEEE Transactions on Nuclear Science 59 (1): 70–6. doi:10.1109/TNS.2011.2178432.

Tanaka, T., T. Mitani, S. Watanabe, K. Nakazawa, K. Oonuki, G. Sato, T. Takahashi, et al. 2004. "Development of an Si/CdTe Semiconductor Compton Telescope." In High-Energy Detectors in Astronomy, 5501: 229–40. International Society for Optics and Photonics. doi:10.1117/12.552600.

Tanimori, T., on behalf of SMILE Project. 2020. "MeV Gamma-Ray Imaging Spectroscopic Observation for Galactic Centre and Cosmic Background MeV Gammas by SMILE-2+ Balloon Experiment." Journal of Physics: Conference Series 1468 (1): 012046. IOP Publishing. doi:10.1088/1742-6596/1468/1/012046.

Tanimori, T., Y. Mizumura, A. Takada, S. Miyamoto, T. Takemura, T. Kishimoto, S. Komura, et al. 2017. "Establishment of Imaging Spectroscopy of Nuclear Gamma-Rays Based on Geometrical Optics." Scientific Reports 7 (1): 41511. Nature Publishing Group. doi:10.1038/srep41511.

Taya, T., J. Kataoka, A. Kishimoto, Y. Iwamoto, A. Koide, T. Nishio, S. Kabuki, and T. Inaniwa. 2016. "First Demonstration of Real-Time Gamma Imaging by Using a Handheld Compton Camera for Particle Therapy." Nuclear Instruments and Methods in Physics Research Section A, Proceedings of the 10th International "Hiroshima" Symposium on the Development and Application of Semiconductor Tracking Detectors, 831 (September): 355–61. doi:10.1016/j.nima.2016.04.028.

Terzioglu, F. 2015. "Some Inversion Formulas for the Cone Transform." Inverse Problems 31 (11): 115010. IOP Publishing. doi:10.1088/0266-5611/31/11/115010.

———. 2020. "Exact Inversion of an Integral Transform Arising in Compton Camera Imaging." Journal of Medical Imaging 7 (3): 032504. International Society for Optics and Photonics. doi:10.1117/1.JMI.7.3.032504.

Terzioglu, F., P. Kuchment, and L. Kunyansky. 2018. "Compton Camera Imaging and the Cone Transform: A Brief Overview." Inverse Problems 34 (5): 054002. IOP Publishing. doi:10.1088/1361-6420/aab0ab.

Todd, R. W., J. M. Nightingale, and D. B. Everett. 1974. "A Proposed γ Camera." Nature 251 (5471): 132–4. Nature Publishing Group. doi:10.1038/251132a0.

Tomitani, T., and M. Hirasawa. 2002. "Image Reconstruction from Limited Angle Compton Camera Data." Physics in Medicine and Biology 47 (12): 2129–45. IOP Publishing. doi:10.1088/0031-9155/47/12/309.

Tomono, D., T. Mizumoto, A. Takada, S. Komura, Y. Matsuoka, Y. Mizumura, M. Oda, and T. Tanimori. 2017. "First On-Site True Gamma-Ray Imaging-Spectroscopy of Contamination Near Fukushima Plant." Scientific Reports 7 (1): 41972. Nature Publishing Group. doi:10.1038/srep41972.

Tumer, O. T., A. Akyuz, D. Bhattacharya, S. C. Blair, G. L. Case, D. D. Dixon, C.- Liu, et al. 1995. "The TIGRE Instrument for 0.3-100 MeV Gamma-Ray Astronomy." IEEE Transactions on Nuclear Science 42 (4): 907–16. doi:10.1109/23.467770.

Uche, C. Z., W. H. Round, and M. J. Cree. 2011. "A Monte Carlo Evaluation of Three Compton Camera Absorbers." Australasian Physical & Engineering Sciences in Medicine 34 (3): 351–60. Springer.

Vetter, K., R. Barnowksi, A. Haefner, T. H. Joshi, R. Pavlovsky, and B. J. Quiter. 2018. "Gamma-Ray Imaging for Nuclear Security and Safety: Towards 3-D Gamma-Ray Vision." Nuclear Instruments and Methods in Physics Research Section A 878 (January): 159–68. doi:10.1016/j.nima.2017.08.040.

Wahl, C. G., W. R. Kaye, W. Wang, F. Zhang, J. M. Jaworski, A. King, Y. A. Boucher, and Z. He. 2015. "The Polaris-H Imaging Spectrometer." Nuclear Instruments and Methods in Physics Research Section A, Symposium on Radiation Measurements and Applications 2014 (SORMA XV), 784 (June): 377–81. doi:10.1016/j.nima.2014.12.110.

Wang, W., C. G. Wahl, J. M. Jaworski, and Z. He. 2012. "Maximum-Likelihood Deconvolution in the Spatial and Spatial-Energy Domain for Events with Any Number of Interactions." IEEE Transactions on Nuclear Science 59 (2): 469–78. doi:10.1109/TNS.2012.2183384.

Watanabe, S., H. Tajima, Y. Fukazawa, R. Blandford, T. Enoto, A. Goldwurm, K. Hagino, et al. 2016. "The Soft Gamma-Ray Detector (SGD) Onboard ASTRO-H." In Space Telescopes and Instrumentation 2016: Ultraviolet to Gamma Ray, 9905: 990513. International Society for Optics and Photonics. doi:10.1117/12.2231962.

Weinzierl, P., and G. Tisljar-Lentulis. 1958. "Two-Crystal-Compton-Spectrometer with Improved Geometrical Efficiency." Nuclear Instrument 3 (September). doi:10.1016/0369-643X(58)90019-7.

White, R. S. 1968. "An Experiment to Measure Neutrons from the Sun." In Bulletin of the American Physical Society, 13: 714.

White, R. S., D. A. Zych, and T. O. Tumer. 1990. "A New Double Compton Gamma Ray Scatter Telescope with Fiber Scintillator Array at the First Scatter to Track the Scattered Electron Direction." Proceedings of the 21st International Cosmic Ray Conference, 4: 175.

Wilderman, S.J., W.L. Rogers, G.F. Knoll, and J. C. Engdahl. 1997. "Monte Carlo Calculation of Point-Spread Functions of Compton Scatter Cameras." IEEE Transactions on Nuclear Science 44 (2): 250–254. doi: 10.1109/23.568817.

Wilderman, S. J., N. H. Clinthorne, J. A. Fessler, and W. L. Rogers. 1998. "List-Mode Maximum Likelihood Reconstruction of Compton Scatter Camera Images in Nuclear Medicine." In 1998 IEEE Nuclear Science Symposium Conference Record. 1998 IEEE Nuclear Science Symposium and Medical Imaging Conference (Cat. No.98CH36255), 3: 1716–20. doi:10.1109/NSSMIC.1998.773871.

Wilson, R. R. 1946. "Radiological Use of Fast Protons." Radiology 47 (5): 487–91. The Radiological Society of North America.

Xu, D., and Z. He. 2007. "Gamma-Ray Energy-Imaging Integrated Spectral Deconvolution." Nuclear Instruments and Methods in Physics Research Section A 574 (1): 98–109. doi:10.1016/j.nima.2007.01.171.

Yao, Z., Y. Xiao, and Z. Chen. 2017. "3D Multi-Focus Origin Ensembles Reconstruction Method for Compton Camera Imaging." In 2017 IEEE Nuclear Science Symposium and Medical Imaging Conference (NSS/MIC), 1–4. doi:10.1109/NSSMIC.2017.8532960.

Yao, Z., Y. Xiao, Z. Chen, B. Wang, and Q. Hou. 2019. "Compton-Based Prompt Gamma Imaging Using Ordered Origin Ensemble Algorithm with Resolution Recovery in Proton Therapy." Scientific Reports 9 (1): 1133. Nature Publishing Group. doi:10.1038/s41598-018-37623-2.

Yingzeng, Z., X. Qingpei, H. Fanhua, G. Xiaofeng, X. Yongchun, C. Chengsheng, Z. Jun, L. Fei, and Z. Rende. 2018. "Evaluation of Nuclear Warhead Symmetry Detection by Compton Camera." Nuclear Technology 204 (1): 83–93. Taylor & Francis. doi:10.1080/00295450.2018.1464839.

Yoshida, E., H. Tashima, K. Nagatsu, A. B. Tsuji, K. Kamada, K. Parodi, and T. Yamaya. 2020. "Whole Gamma Imaging: A New Concept of PET Combined with Compton Imaging." Physics in Medicine and Biology 65 (12): 125013. IOP Publishing. doi:10.1088/1361-6560/ab8e89.

Zhang, L., W. L. Rogers, and N. H. Clinthorne. 2004. "Potential of a Compton Camera for High Performance Scintimammography." Physics in Medicine and Biology 49 (4): 617–38. IOP Publishing. doi:10.1088/0031-9155/49/4/011.

Zhang, Y. 2020. "Recovery of Singularities for the Weighted Cone Transform Appearing in Compton Camera Imaging." Inverse Problems 36 (2): 025014. IOP Publishing. doi:10.1088/1361-6420/ab3cc8.

Zhang, Z. 2012. "Microsoft Kinect Sensor and Its Effect." IEEE Multimedia 19 (2): 4–10. IEEE.

Zheng, A., Z. Yao, and Y. Xiao. 2020. "GPU Accelerated Stochastic Origin Ensemble Method with List-Mode Data for Compton Camera Imaging in Proton Therapy." IEEE Transactions on Radiation and Plasma Medical Sciences 4 (2): 243–52. doi:10.1109/TRPMS.2019.2929423.

Ziock, K. P. 2018. "Principles and Applications of Gamma-Ray Imaging for Arms Control." Nuclear Instruments and Methods in Physics Research Section A 878 (January): 191–9. doi:10.1016/j.nima.2017.04.001.

Zoglauer, A., R. Andritschke, S. E. Boggs, R. Diehl, J. Greiner, D. H. Hartmann, G. Kanbach, and C. B. Wunderer. 2008. "Nuclear Astrophysics Capabilities of the GRIPS Telescope." New Astronomy Reviews, Astronomy with Radioactivities. VI 52 (7): 431–5. doi:10.1016/j.newar.2008.05.009.

Zych, A. D., T. J. O'Neill, D. Bhattacharya, C. Trojanowski, S. Wijeratne, C. Teichegaeber, and M. Mathews. 2006. "TIGRE Prototype Gamma-Ray Balloon Instrument." In Hard X-Ray and Gamma-Ray Detector Physics and Penetrating Radiation Systems VIII, 6319: 631919. International Society for Optics and Photonics. doi:10.1117/12.683838.

8

MRI-SPECT Multimodal Imaging

Jiri Zajicek

CONTENTS

8.1 Introduction

As Vandenberghe mentioned in [1] the idea of combining anatomical and functional imaging (PET and MRI) in a single system was first suggested in the early-mid 1990s [2, 3]. PET detectors capable of measuring in strong magnetic fields [4] and prototype MRI-compatible PET scanners capable of imaging small animals simultaneously with MRI started to appear soon afterward [5, 6]. Systems capable of sequential [7] or simultaneous PET and MRI acquisitions of the whole human body only became available commercially [8] after about 15 years of developments. The main reason for this slow progress is that the integration of PET and MRI is much more complex [9] than the evolution from standalone CT and standalone PET systems toward multimodality PET-CT, where the CT image is not only used for anatomical information but is also easily transformed to linear attenuation coefficients [1].

Demand for the combination of MRI with PET is growing and the methodology is highly developed. Most human PET-MRI scanners have been developed around 3 T MRI magnets. There are three different approaches shown in Figure 8.1 how to build the MIS based on the combination with

DOI: 10.1201/9781003218364-8

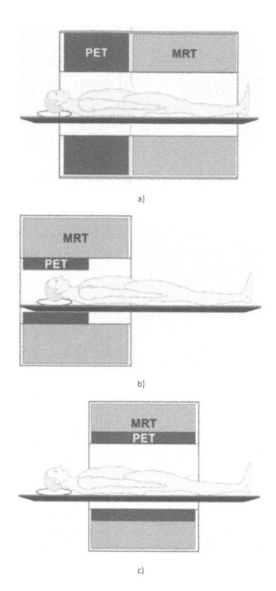

FIGURE 8.1

Potential approaches for combined MRI/PET in a clinical setting: (a) An in-line solution similar to current CT/PET scanners, which mechanically combines a standard MRI with a slightly modified PET scanner would be the simplest approach but also offers a limited variety of imaging protocols and would not allow simultaneous MR/PET imaging. In addition, the total imaging time for the sequential MRI and PET scans is prolonged. (b) Schematic diagram of the first commercial clinical MRI/PET scanner. The removable PET detector slip-fits into a standard clinical MRI system. This approach allows simultaneous data acquisition and flexibility in using the MRI scanner as a stand-alone device or as an MRI/PET device, but at the cost of a reduced MRI/PET field of view suitable only for brain imaging. (c) The most advanced but also the most technically challenging approach, providing the highest degree of flexibility in selecting imaging protocols, would be full integration of the PET detectors into the MRI gantry, adapted from Pichler (2010) [10].

MRI. However, for small animal systems there is a wide variety of MRI systems depending on the type of studies required, from low field 0.3 T systems up to 11 T, with 7 T being one of the most used field strengths. The lower field systems are more oriented toward anatomical imaging while the stronger fields are used for more challenging imaging tasks [1].

As mentioned in Ref. [11] in general, magnetic resonance imaging has a lower molar sensitivity than nuclear techniques, i.e., orders of magnitude higher concentrations must be present to generate a useful imaging signal. The complementary relationship between MRI and nuclear imaging is shown in Table 8.1.

These Most of the nuclear imaging methods commonly used today are based on the injection of a solution containing molecules labeled by some radioactive isotope. Compared to the radiotracers used in PET (^{18}F), solutions of ^{99m}Tc or some form of iodine are cheaper and also have a longer half-life. An overview of radioactive isotopes used for SPECT is in Table 8.2. SPECT imaging is mainly used in clinical practice for functional diagnostics of the brain, heart, and kidneys, or for the localization of tumors.

For patient examination, the activities of ^{99m}Tc are typically in the range from ~100 MBq to ~1 GBq. In experimental research on mice or phantoms, a much lower activity of ~10 MBq is typically used.

Preclinical MRI-SPECT combines the high-resolution molecular information from SPECT with the excellent soft tissue contrast of MRI, together with

TABLE 8.1

Complementary Nature of Magnetic Resonance Imaging (MRI) and Nuclear Imaging (NI). When the Two Modalities Are Combined, One Can Realize Complete Coverage of the Space between High Resolution, Anatomical Imaging, and Genetically Targeted Molecular Imaging, Adapted from Ref. [11]

MRI		High Resolution Low Contrast	ANATOMICAL
MRI		Low Resolution High Contrast	ANATOMICAL
MRI	NI	Ventilation	ANATOMICAL
MRI	NI	Vessel Flow	ANATOMICAL
MRI	NI	Perfusion	FUNCTIONAL
MRI	NI	Drainage	FUNCTIONAL
MRI	NI	Diffusion	FUNCTIONAL
MRI	NI	Oxygenation	FUNCTIONAL
MRI	NI	Metabolism	FUNCTIONAL
MRI	NI	Proliferation	FUNCTIONAL
MRI	NI	Cell Action (WBC)	FUNCTIONAL
MRI	NI	Cell Action (STEM)	FUNCTIONAL
MRI	NI	Nanoparticles	MOLECULAR
	NI	Antibodies	MOLECULAR
	NI	Receptor Density	MOLECULAR
	NI	Cytokines	MOLECULAR
	NI	Reporter genes	MOLECULAR

TABLE 8.2

Mostly Used Radiotracers in SPECT Imaging

Radiotracer	Half-life	Photon Energy (keV)
Tc-99m	6.01 hours	140.5
In-111	2.8 days	171.28
I-123	13.22 hours	159
I-131	8.02 days	364.49

localized chemical and physical information such as metabolite concentrations and water diffusion characteristics from MRI [12]. There is also the possibility of combining the chemical resolution of MRS with the sensitivity of the radiotracer. This would be a great tool in the study of drug pharmacokinetics and metabolism as well as therapeutic manipulations [13].

In contrast to SPECT, MRI provides exceptionally high spatial resolution of anatomical information (lesions). Moreover, thanks to functional MRI (fMRI), it is possible to localize chemical and physical information, such as metabolite concentrations and water diffusion characteristics. Multimodal imaging (based on a combination of MRI and SPECT) therefore brings new opportunities for more precise diagnostics of diseases like cancer, epilepsy, etc. [14].

One of the first work using semiconductor detectors (cadmium zinc telluride [CZT]) for combination of MRI and SPECT was presented by Wagenaar et al (2006) [15] followed by an MRI compatible SPECT system prototype based on similar CZT detectors (Meier et al 2009) [16] or development of an MR-compatible SPECT system for simultaneous data acquisition by Hamamura et al (2009) [17]. The MRC-SPECT-I detector ring consists of ten energy-resolving photon-counting (ERPC) CdTe detectors (Tan et al 2009) [18] and its ability to achieve a sub-500 μm resolution in simultaneous SPECT/MRI mode while working inside the Siemens Trio 3T clinical scanner was demonstrated by Lai et al (2015) [19]. Karel Deprez et al have constructed an MRI compatible aperture using rapid additive manufacturing with selective laser melting of tungsten powder [20]. The main advantage of tungsten is high density and subsequent easy mechanical processing when it is mixed with polymer resin.

There are several groups in the Europe developed both preclinical and clinical SPECT insert systems based on silicon photomultiplier (SiPM) detectors, e.g., Hutton et al (2016) [21] or Carminati et al [22] who using the same type of material for multi-pinhole collimator as I do. Lai et al. proposed an inverted compound eye (ICE) camera design and a second-generation MRI compatible SPECT system (MRC-SPECT-II) based on hybrid pixel-waveform (HPWF) detector design that has been developed by our group over the past several years [12].

Nowadays there are practically two groups leading the development both based on CdTe or CZT sensors and pinhole collimators. Johns Hopkins University has developed an MRI-compatible pixelated CZT detector (256 pixels) providing excellent energy resolution of 3.8% at 140.5 keV and spatial resolution of 4 mm [23]. Second team is The Company Gamma Medical Ideas, in collaboration with the University of California [24].

The first commercially available SPECT/MRI system is currently a pre-clinical in-line system by Mediso (nanoScan) combining 1 T MR with a traditional PMT-based SPECT [24]. Despite their excellent energy and spatial resolution, pixelated CdTe or CZT detectors are hardly scalable to static and MRI-compatible clinical systems because of cost and complexity (number of channels) issues. However, CdTe sensors with 55×55 μm of the pixel edge and 1 mm of thickness can reach detection efficiency of 35% for Technetium-99m energy (peak 140.5 keV). Nowadays CZT sensors with 110×110 μm of the pixel edge and 2 mm of the thickness could reach detection efficiency even of 55% for Technetium-99m energy (peak 140.5 keV).

8.2 Semiconductor Pixel Detectors

The structure of semiconductor detectors (Figure 8.2) based on PN junction structure is easy to create using modern manufacturing methods of semiconductor compartments (diffusion technique, transistors mesa, planar

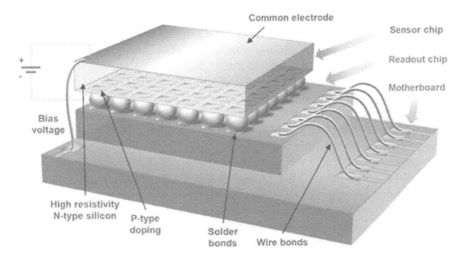

FIGURE 8.2
Principle of arrangement of hybrid semiconductor pixel detector: It is divided into an ASIC chip (readout chip) and the sensor which are connected by a bump-bonding technology, adapted from Ref. [26].

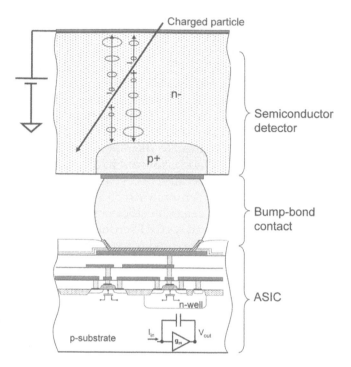

FIGURE 8.3
Principle of ionizing radiation detection and incision of detector, adapted from Ref. [27].

technology, etc.). The sensitive area is divided into individual PN junctions and each of these represents a separate detector. By this approach it is possible to create a sensitive area containing a large number of detectors (from hundreds to thousands). Such a sensor is possible through bump-bonding technology combined with an ASIC-integrated circuit (readout chip), which provides the basic processing of signals from the sensors [25].

There are many pixelated detectors based on the same principle (like Pilatus or CMS pixel detectors). In this section, the Medipix family of detectors (Medipix, Medipix2, Medipix2MXR, Timepix, and Medipix3) is described as an example of a hybrid detector.

The advantage of the hybrid structure is the possibility of separating the design and manufacture of the sensor and ASIC integrated circuit [25]. The sensor is generally made of silicon, but with hybrid technology, it is also possible to use other materials (GaAs or CdTe). The principle of ionizing radiation detection on the sensor material is shown in Figure 8.3.

8.2.1 Timepix Device

The application of hybrid pixel detector technology has become very popular outside the field of elementary particle physics. The Medipix chip in particular has proven very promising, and has been supported by large

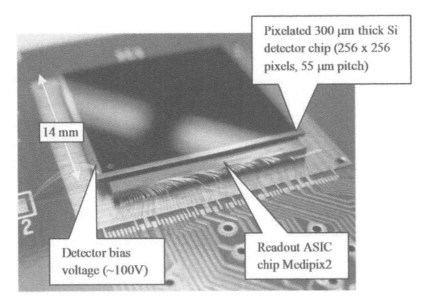

Pixelated 300 μm thick Si detector chip (256 x 256 pixels, 55 μm pitch)

14 mm

Detector bias voltage (~100V)

Readout ASIC chip Medipix2

FIGURE 8.4
Photograph of the detector generation Medipix2MXR with description, adapted from Ref. [27].

multinational collaborations Medipix (https://medipix.web.cern.ch/) [28]. This chip is very suitable for applications where single photon counting is required. It does not however provide information on electron arrival time. It allows measurements in arrival, time-over-threshold (TOT) and counting mode independently in each pixel. The external reference clock (*Ref_Clk*) generates the clock in each pixel that increments the counter depending on the selected mode [29].

The first detector was Medipix1 with 4096 square pixels, edge length of 170 μm and an active area of 1.2 cm^2. Each pixel contained a counter of 15 bits for registering events in the detector. The next detector was the Medipix2 detector, with 65536 pixels of 55 × 55 μm^2 surface area each. Each pixel contained a 13-bit "pseudo-random" counter. This detector led to the development of Medipix2MXR (Figure 8.4) with dimensions of 16.12 × 14.111 mm^2 and divided the area into 65536 pixels of 55 × 55 μm^2 (with active area 1.982 cm^2, which is 87.35% of total area). The detector is composed of individual cells for each PN junction and a data interface located in the lower part of the detector. The static power consumption of the detector is approximately 550 mW. The detector is made by 250 nm CMOS technology, which allows the use of about 500 transistors per pixel [25].

8.2.2 The Pixel Cell

The theory behind the Timepix device is described in detail by Llopart in [29]. The following paragraphs highlight the most important facts from this chapter, stated in the summary. Every pixel is divided into two large blocks

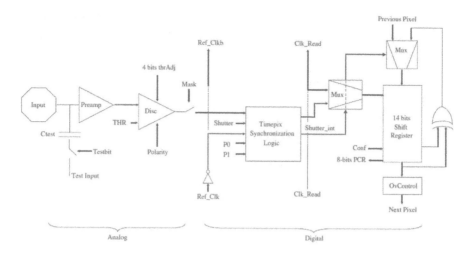

FIGURE 8.5
Timepix cell schematic, adapted from Ref. [29].

(the schematic of the Timepix cell is shown in Figure 8.5). The analog side consists of the preamplifier, the discriminator, and 4-bit threshold adjustment. The digital side consists of the Timepix Synchronization Logic (TSL), the 14-bit shift register, the overflow control logic, the *Ref_Clk* pixel buffer and an 8-bit Pixel Configuration Register (PCR). The PCR contains 4 bits for pixel threshold equalization: 1 bit for Masking (MaskBit), 1 bit for enabling the test pulse input (TestBit) and 2 bits for selecting the pixel operation mode (P_0 and P_1). Every pixel has two working states, depending on the shutter signal. This signal is applied to all the pixels of the matrix simultaneously with a precision of ~5 ns [29].

The pixel counter is incremented by the *Ref_Clk* depending on the settings of the pixel operation mode bits (P_0 and P_1):

1. *Event counting mode* ($P_0 = 0$ and $P_1 = 0$): Each event above threshold increments the counter by 1.
2. *TOT mode* ($P_0 = 1$ and $P_1 = 0$): The counter is incremented continuously while the input charge is over threshold.
3. *Arrival time mode* ($P_0 = 1$ and $P_1 = 1$): The counter is incremented from the moment the discriminator is activated until the global Shutter signal is set high [29].

8.2.3 Timepix Detector Modes

Counting mode (also called Medipix mode) is used for measurements of an effective threshold and electronic noise of a front-end chain (by s-curve method). Using a fixed threshold, an input charge is swept from a level where there are no counter counts (i.e., under threshold) to a level where 100% of the test pulses are counted, creating an s-shaped curve. In TOT mode, the counter is incremented continuously with *Ref_Clk* while the preamplifier output is above the threshold. Although the preamplifier output voltage pulse is only linear up to 50 ke$^-$ due to the limited power supply range (0–2.2 V) and the relative high gain (~16.5 mV/ke$^-$), the discriminator output pulse width is linear up to much larger input charges due to the constant current return to zero (controlled by the Ikrum DAC). When the chip is programmed in arrival time mode (also called Timepix mode), the pixel behaves as a time to digital converter (TDC). During acquisition, the counter is incremented from the first time the discriminator output goes high to the closing of the shutter. The quantization error can be up to two *Ref_Clk* periods since the start and end of the counting clock is synchronized to *Ref_Clk*. In systems where the charge deposition of a single event spreads over multiple pixels, e.g., gas electron multiplier GEMs), the matrix can be arranged in a chessboard-like pattern, whereby some pixels are configured in TOT mode and others in arrival time mode. This can be set up in order to compensate the off-line time-walk through knowledge of the input charge deposited in the neighboring pixels. The principle of ionizing radiation detection and summary of the working modes of the Timepix detector is shown in Figure 8.6.

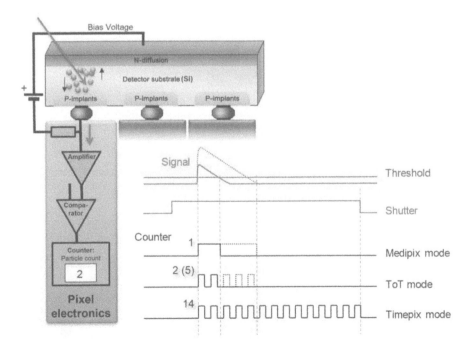

FIGURE 8.6
Principle of ionizing radiation detection and working modes of Timepix detector, adapted from Ref. [26].

Threshold equalization is used to compensate the pixel to pixel threshold variations due to local transistor threshold voltages and current mismatches or more global effects like on-chip power drops. This compensation is done by means of a 4-bit current DAC placed in the discriminator chain of each pixel [29].

For a further summary, operational modes are very well described as given by Jakubek [30]:

- **Medipix mode** – Counter counts incoming particles;
- **TimePix mode** – Counter works as a timer and measures time of the particle detection;
- **TOT mode** – Counter is used as Wilkinson type ADC allowing direct energy measurement in each pixel.

8.2.4 The Application of Timepix Detector in MRI System

CdTe Timepix detector functionality was first tested with neodymium magnets with magnetic field intensity of 1.2 T. The ^{57}Co spectrum was measured

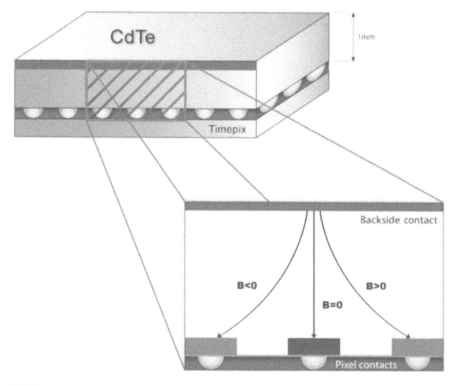

FIGURE 8.7
Illustration of the change of current direction for three values of magnetic field intensity and direction perpendicular to the plane of drawing, adapted from Ref. [31].

in the presence of different directions of magnetic field. Very small (~1%) but detectable shifts of spectral peaks positions were registered.

Secondly, we performed the measurement in an MRI system with high magnetic field of 4.7 T. The most important influence of the high magnetic field is a change of charge collection direction inside the sensor due to the Lorentz force acting on moving charge carriers (electrons). The charge is no longer collected along the electric field but is shifted sideways and collected by pixels on the side.

This effect is illustrated in Figure 8.7, where we observed a shift of pixels registering a signal caused by local leakage current flowing through several defective spots in the CdTe sensor. The leakage current is caused by local imperfections in the crystalline structure of the CdTe sensor. This current normally flows directly toward pixelated electrode where it is sensed by a group of pixels.

The magnetic field changes the direction of this current (Hall's effect), manifested as a shifting of the image (Figure 8.8).

We can expect the same image shift for detected gamma ray photons. The only difference is that gamma rays can create the charge in any depth of the sensor.

The deeper the interaction, the smaller the shift. The full image will therefore contain all possible shifts resulting in blurring in direction of the Lorentz force (perpendicular to both – magnetic and electric field intensity vectors). During experimental measurements with the pinhole set-up directional blurring was not observed due to smearing resulting from the large pinhole diameter (1.5 mm) [31].

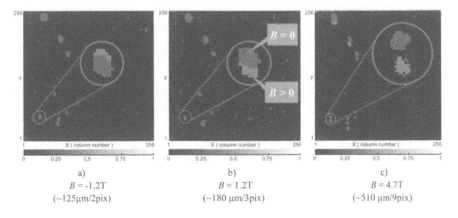

a)
$B = -1.2T$
(~125µm/2pix)

b)
$B = 1.2T$
(~180 µm/3pix)

c)
$B = 4.7T$
(~510 µm/9pix)

FIGURE 8.8
Influence of high magnetic field on charge collection in the CdTe prototype sensor is demonstrated by a shift of current leaking through several defective areas of the sensor (no radiation was involved). The direction of this current is affected by the Lorentz force and is known as Hall's effect. The images show leakage currents for different magnetic fields. The blue image corresponds to situation without magnetic field while red shows the situation with magnetic field: (a) $B = -1.2$ T (~125 µm shift), (b) $B = 1.2$ T (~180 µm shift), (c) $B = 4.7$ T (~510 µm shift), adapted from Ref. [31].

8.3 Collimators

Collimators are a well-understood problem. The basic design depends on three main parameters: (1) Stopping power means contrast at a given material thickness (material density). Materials with a high number of electrons ($Z = 72–82$) and high density (typically $9–19.25$ g·cm^{-3}) have high stopping power; (2) Angular resolution (geometry) and (3) Efficiency (ratio of transmitted vs. absorbed photons).

In the case of multimodal MRI-SPECT imaging, these collimators also have to be non-magnetic and non-conductive (MRI compatible) [31]. First approach is based on laser melting of wolfram with reaching density around 95% of the original material. This method is without discussion the best one; however, it is still very expensive. Next well known approach is the mixing of tungsten powder with polymer resin to reach density around 47% ($\rho = 9$ g·cm^{-3}) of the original material [32].

The choice of collimator materials depends on cost, required rigidity, and machining complexity. Different geometrical types of collimators include parallel-hole, fan beam, cone beam, pinhole, and multi-pinhole. The selection of materials further depends on the size and shape of the target object. In the case of pinhole/parallel-hole, the size and length of the holes are given by an optimization of the required sensitivity versus resolution. For most studies, parallel-hole collimators are the best choice. However, complex-shaped collimators for imaging of specific organs (brain, heart) or small objects (mice, rats) are also needed [33] in some applications. Fan-beam or cone-beam (converging collimators) collimators are suitable for use in cases when magnification of the measured object is required. One of the earliest designs was the pinhole collimator, which is very popular, especially for small animal imaging [34]. There are a few alternatives to conventional parallel-hole collimators, such as pinhole and slit-slat collimators. While it is straightforward to obtain a high-resolution image by pinhole collimators in clinical applications given a sufficiently small field of view (FOV), e.g., a mouse brain, the typically large FOV in combination with large detector renders the use of pinhole collimators inefficient [35].

Collimators in general are composed of materials with high atomic number (lead, tungsten, and even gold or uranium). The most commonly used parallel-hole, fan beam or cone beam collimators are made of lead foil sheets, folded in half-hexagonal holes or in a honeycomb structure. Collimators based on precise molding techniques are another option [33, 35].

According to Accorsia R. [34] the ideal SPECT detector would stop incoming photons with high efficiency, while measuring energy and position with high accuracy. For the detection of gamma rays in SPECT, inorganic scintillators such as NaI and CSI with photomultipliers (PMTs) or avalanche photodiodes (APDs) are mainly used. In commercial systems, sodium iodide doped with thallium (NaI[Tl]) is used, or in some cases, bismuth orthogermanate (BGO) which is more efficient but offers less spatial and energy resolution. The other option includes semiconductor detectors like Si, HPGe, or CdTe. However, Si detectors have very low detection efficiency for gamma rays of SPECT energies (about 7%) while HPGe requires cooling by liquid nitrogen. CZT detectors have also recently become commercially available. CdTe sensors have high detection efficiency for SPECT (ca. 60%) but suffer from noise and therefore air cooling is needed. Solid-state technology is expected to replace scintillators in most SPECT systems in the future. The main reason is that solid-state detectors effectively convert photons into current pulse and do not require a transformation to visible light. This has great potential for high spatial and energy resolution.

8.4 MRI-SPECT System

Combining two (like MRI with SPECT/PET) or more methods based on different imaging principles (anatomical and functional) has great potential in the diagnostics of diseases like cancer [36]. SPECT, like PET, is a powerful molecular imaging tool for applications in neurology [24].

In contrast to SPECT, MRI provides exceptionally high spatial resolution of anatomical information (lesions) as well as localized chemical and physical information such as metabolite concentrations and water diffusion characteristics [36]. Figure 8.9 shows the combination of MRI and SPECT for brain diagnostics in an epileptic patient.

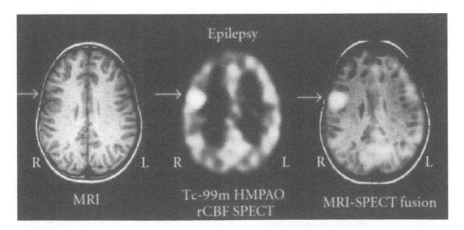

FIGURE 8.9
The example of MRI & SPECT applications, adapted from Ref. [38].

As mentioned earlier, the combination of MRI and SPECT is relatively new and presents many challenges. In the case of MRI-PET not only must detector compatibility be considered, but also the manufacturing of suitable MRI-compatible collimators. Currently, there are a few approaches for operating a SPECT system in an MR scanner. However, in most cases the detection process is affected by the high magnetic field. The incident gamma ray-induced charge carrier drifting path in particular is affected by the high magnetic field. This leads to projection shifting and position error or blurring in reconstruction [37].

There are several options for detecting gamma rays in a high-magnetic field. Use of a SiPM [39] brings many technical difficulties which will not be described in this work. It has been proved that CdTe detectors are MR-compatible and therefore are suitable for SPECT detection inside the high magnetic field. Similarly, CZT detectors used by Wagenaar et al [11] have proven appropriate. An MRI-compatible SPECT system for a 3T-MRI scanner based on pixel CdTe/CdZnTe detectors and multiple-pinhole aperture has been developed by the University of Illinois at Urbana-Champaign [37].

Due to the technological challenges of simultaneous MRI-SPECT imaging, this technology is still in expansion. Hamamura J. M. [24] states that due to the development of gamma ray detectors like APDs or semiconductor detectors (CZT, CdTe [24]) it is now possible to build fully magnetic-field-insensitive setups in magnetic induction fields up to 7 T.

The possibility of acquiring SPECT and MRI data simultaneously brings new opportunities in dynamic imaging using SPECT radionuclides and MRI contrast agents at the same time. Therefore, the development of bifunctional MRI-SPECT insert systems has great potential. The sensitivity of radiotracer techniques can also be combined with the chemical resolution of magnetic resonance spectroscopy (MRS) [36]. As in the case of PET, integrating SPECT with the MRI system demands minimal interference. Therefore, the components of the SPECT insert must be made of low magnetic susceptibility materials, including mechanical supporting, shielding, collimation materials and structure, to avoid the degradation of MR images. Another problem is the small space of the MR scanner bore which limits the placement of SPECT inside and therefore only compact detectors can be used [37].

Simultaneous acquisition of MRI and SPECT offers potential advantages in dynamic imaging of both structure and function at the same time with optimum spatial and temporal co-registration. There is also the possibility of combining the chemical resolution of MRS with the sensitivity of the radiotracer. This would be a great tool in the study of drug pharmacokinetics and metabolism as well as therapeutic manipulations. While there are many advantages in the simultaneous integration of MRI and SPECT, there are many technological challenges. The SPECT detector has to be fully functional during MRI sequence operation. The counting of incoming particles must not be influenced by the magnetic field. Also the resulting MR images should not be influenced by readout of the detector. In addition, simultaneous multimodal imaging reduces the overall scan time, avoids multiple anesthesia sessions, and prevents errors associated with co-registration [24, 40].

8.4.1 MRI-SPECT Issues

The main complication of using SPECT in combination with MRI is the presence of the high magnetic field. It is not possible to use any magnetic materials because of gradient fields which are switched rapidly at frequencies of the order 1 kHz. Lower frequencies have greater skin depth, but are more difficult to shield than the higher frequency RF (200 MHz at 4.7 T). These rapidly switching magnetic fields can induce eddy current loops (changing of magnetic flux creating eddy currents) in any conductive components introduced into the magnet bore, including SPECT circuitry. In addition to signal interference, this leads to heating and mechanical vibration. This would cause a loss of signal intensity in the MR image and increase of temperature. In the worst case scenario, it could lead to harm or destruction of the MRI system itself. Note also, that more intense magnetic fields provide not only stronger signals in MRI but also higher sensitivity to artifacts [31].

Any electronics situated within the magnet bore may be susceptible to RF interference generated by the MRI transmit coil. This effect is responsible for the drop in count rate that is observed in many MRI-compatible SPECT systems during MRI acquisition. RF shielding is, however, more effective at this higher frequency range and so in principle SPECT detectors and electronics within the magnet bore can be enclosed in a conducting shield to reduce RF interference.

The NMRI signals generated in the body in response to excitation by the MRI B_1 field are extremely weak, requiring the MRI receive coils to be of very high sensitivity and the complete MRI scanning room to be Faraday shielded. Any unwanted sources of RF within the MRI detection frequency range (e.g., 200 MHz at 4.7 T) occurring during the coil receive period, will degrade and distort the received signal and so degrade the final MRI images. Many frequencies present in modern digital electronics, e.g., clock pulses, lie in this frequency range so great care must be taken to minimize interference by shielding of the SPECT components. The same RF shielding works in both directions. The shielding material needs to be carefully chosen as the material itself may result in inhomogeneities and eddy currents. Unshielded power cables may also carry noise currents into the MRI coil circuitry.

Any small differences in magnetic susceptibility caused by SPECT system components within the magnet bore may result in inhomogeneities in the main magnetic B_0 field. It may also affect the linearity of the gradient fields beyond the level that can be accurately corrected by shimming. Examples include the scintillation crystals, any RF shielding materials, or dense gamma shielding materials such as lead or tungsten and in particular electronic components containing ferromagnetic materials. Gradients may induce eddy currents in for example shielding materials and electronics which in turn can distort the B_0 and gradient linearity [1].

Scintigraphic methods (PET, SPECT) often use scintillators with photomultipliers, which do not function in high magnetic fields. A possible solution of this issue is using scintillators with SiPMs [39] or hybrid semiconductor pixel detectors. In addition, SPECT systems mostly use heavy metal collimators (non-magnetic but conductive), which significantly disturbs the RF system of the MRI tomograph due to the occurrence of eddy currents causing significant loss of measurable signal, as mentioned earlier [31].

8.5 PET vs. SPECT Technology (Pros & Cons)

Both scintigraphic methods provide functional information of target organ with different radioisotopic sources. While PET systems require coincident detection of two gamma rays with energy of 511 keV, the SPECT method uses single gamma photons with lower energy. The electronics for PET is much more complex and energy to be detected is much higher than SPECT where gamma source ^{99m}Tc ($E = 140.5$ keV) is mostly used and also initial tests can be done with single detector only. The SPECT are appropriate for functional brain imaging, myocardial perfusion, and tumor imaging.

The comparison between PET and SPECT imaging is complicated due to the two methods having distinctly different detection physics. PET is more often the preferred clinical choice in nuclear imaging rather than SPECT. This is because of superior sensitivity, acquired resolution, and quantitative ability. However, it is unclear whether the same advantages apply for small animal imaging. The large difference in subject size between human patients and small animals requires a different physics for the optimization of cameras performance.

The distance from the object, number, diameter, and the location of each pinhole, and the number of projections play important roles in determining detection sensitivity in any tomographic imaging approach. Determination of the reconstructed resolution in SPECT depends on the combination of the multi-pinhole apertures specifications, the crystal intrinsic resolution, and the setup magnification. In the case of micro PET the reconstructed resolution depends on crystal pitch, annihilation photons non-collinearity caused by residual momentum of the positron, depth of interaction effects, and the reconstruction methodology [41].

PET provides a very good dynamism and sensitivity; scan times are short (approximately 30–40 minutes). The ability to perform quantitative measurements at the peak of the stress and speed is also advantageous, e.g., at cardiac examinations. However, selection of radioisotopes is limited by cost. ^{18}F is a widely used choice, with ^{82}Rb as a popular alternative, but they have short half-lives – in the case of ^{18}F, it is only 1.5 hours, posing a problem in long measurements. One of the biggest disadvantages of PET is a limitation in resolution of 1 mm; smaller objects cannot be imaged.

SPECT uses a wide selection of radioisotopes (e.g., ^{99m}Tc or some form of iodine, see Table 8.1) which are much cheaper and also have a longer half-life – in the case of ^{99m}Tc it is 6 hours. Furthermore, resolution limit is 275 µm, which is the biggest advantage of SPECT over PET. Unfortunately, in common old SPECT machines, the acquired resolution is usually worse than PET. This is partly due to the relatively slower technological progress of SPECT. Lastly, for SPECT it is usually necessary to use higher radiation dose with longer scanning time – some patients' examination could take as long as 2 hours.

However, these two points do not apply for Timepix detectors where scanning times are short even with low radiation dose. SPECT in general suffers from attenuation issues and a higher occurrence of artifacts is seen in resulting images.

Unlike in SPECT, PET determines line of response (LORs) by the coincident detection of two photons and does not need a collimator, thus making much more efficient use of the emitted photons. As a result, PET typically provides much higher sensitivity than SPECT. Other advantages include better resolution (with the exception of small animal imaging, especially as used in preclinical neuroscience) and a better potential for quantitative imaging. In some cases, SPECT and PET tracers are equivalent or in direct competition, but in many others, they serve complementary purposes. Despite its shortcomings, SPECT today remains the most common nuclear medicine technique because of its wider availability and reduced cost [42].

However, PET methods nowadays are generally better than SPECT, but only slightly. But radiolabeling of nanoparticles by chelation with ^{99m}Tc is still considerably easier and safer than the manual covalent attachment of ^{18}F. Regarding small animal cameras, this choice of chelation with ^{99m}Tc also results in trading lower sensitivity for higher resolution [41].

References

1. Vandenberghe S. et al, *PET-MRI: a review of challenges and solutions in the development of integrated multimodality imaging*, Phys. Med. Biol. **60** (2015) R115–R154, doi:10.1088/0031-9155/60/4/R115
2. Hammer B. E., *NMR-PET scanner apparatus*, Magn. Reson. Imag. **9** (1990) 4.
3. Hammer B. E., Christensen N. L. and Heil B. G., *Use of a magnetic field to increase the spatial resolution of positron emission tomography*, Med. Phys. **21** (1994) 1917–1920.
4. Shao Y., Cherry S. R., Siegel S., Silverman R. W. and Marsden P. K., *Feasibility study of high resolution PET detectors for imaging in high magnetic field environments*, J. Nucl. Med. **37** (1996) 330.

5. Christensen N. L., Hammer B. E., Heil B. G. and Fetterly K., *Positron emission tomography within a magnetic field using photomultiplier tubes and lightguides*, Phys. Med. Biol. **40** (1995) 691–697.
6. Shao Y., *Simultaneous PET and MR imaging*, Phys. Med. Biol. **42** (1997) 1965–1970.
7. Zaidi H., Ojha N., Morich M., Griesmer J., Hu Z., Maniawski P., Ratib O., Izquierdo-Garcia D., Fayad Z. A. and Shao L., *Design and performance evaluation of a whole-body ingenuity TF PET-MRI system*, Phys. Med. Biol. **56** (2011) 3091–3106.
8. Delso G., Furst S., Jakoby B., Ladebeck R., Ganter C., Nekolla S. G., Schwaiger M. and Ziegler S. I., *Performance measurements of the Siemens mMR integrated whole-body PET/MR scannerm*, J. Nucl. Med. **52** (2011) 1914–1922.
9. Cherry S. R., *Multimodality imaging: Beyond PET/CT and SPECT/CT*, Semin. Nucl. Med. **39** (2009) 348–353.
10. Pichler J. et al, *PET/MRI: The next generation of multi-modality imaging*, J. Nucl. Med. **51** (3) (2010), doi:10.2967/jnumed.109.061853
11. Wagenaar J. D. et al, Rationale for the Combination of Nuclear Medicine with Magnetic Resonance for Pre-Clinical Imaging, Technology in Cancer Research and Treatment, ISSN 1533-0346, **5** (4) August (2006).
12. Lai X. and Meng L.-J., *Simulation study of the second-generation MR-compatible SPECT system based on the inverted compound-eye gamma camera design*, Phys. Med. Biol. **63** (2018) 045008.
13. Bouziotis P. and Fiorini C., *SPECT/MRI: Dreams or reality?*, Clin. Transl. Imaging. **2** (2014) 571–573, doi:10.1007/s40336-014-0095-6.
14. Misri R. et al, *Development and evaluation of a dual-modality (MRI/SPECT) molecular imaging bioprobe*, Nanomedicine: NBM. **8** (2012) 1007–1016, doi:10.1016/j.nano.2011.10.013
15. Wagenaar D. et al, *Development of MRI-Compatible Nuclear Medicine Imaging Detectors*, in Proc. IEEE NSS/MIC Conf., San Diego, CA, USA (2006) 1825–1828.
16. Meier D. et al, *A SPECT Camera for Simultaneous SPECT/MRI*, in Proc. IEEE NSS/MIC Conf., Orlando, FL, USA (2009) 2313–2318.
17. Hamamura M. et al, *Development of an MR-compatible SPECT system (MRSPECT) for simultaneous data acquisition*, Phys. Med. Biol. **55** (6) (2010) 1563–1575.
18. Tan J. et al, *A prototype of the MRI-compatible ultra-high resolution SPECT for in vivo mice brain imaging*, in Proc. IEEE NSS/MIC Conf., Orlando, FL (2009) 2800–2805.
19. Lai X. et al, First sub-500 μm-resolution simultaneous SPECT/MRI imaging with the MRC- SPECT-I: An ultrahigh resolution MR-compatible SPECT system using highly pixelated Semiconductor detectors, IEEE Nuclear Science Symp. and Medical Imaging Conf. (IEEE) pp 1-4.
20. Deprez K. et al, *Rapid additive manufacturing of MR compatible multipinhole collimators with selective laser melting of tungsten powder*, Med. Phys. **40** (2013) 012501.
21. Hutton B. F. et al, *Development of clinical simultaneous SPECT/MRI*, Br. J. Radiol. **91** (1081) (2018).
22. Carminati M. et al, *SPECT/MRI INSERT compatibility: Assessment, solutions, and design guidelines*, IEEE Trans. Nucl. Sci. **2** (4) (2018) 369–379.
23. Meier D. et al, *A SPECT camera for combined MRI and SPECT for small animals*, Nucl. Instrum. Methods Phys. Res. A **652** (1) (2011) 731–734.
24. Hamamura J. M. et al, *Development of an MR-compatible SPECT system (MRSPECT) for simultaneous data acquisition*, Phys. Med. Biol., **55** (6) (2010) 1563–1575.
25. Kraus V., *Datové rozhraní pro řízení experimentu a čtení dat z pixelových detektorů*, Dissertation Thesis, Plzeň (2015).

26. Platkevič M., *Signal Processing and Data Read-Out from Position Sensitive Pixel Detectors*, Dissertation Thesis, Praha (2014).
27. Lopart X., *Design and Characterization of 64K Pixels Chips Working in Single Photon Processing Mode*, Mid Sweeden University, Doctoral Thesis 27, ISBN 978-91-85317-56-1 (2007).
28. Llopart X. et al, *First test measurements of a 64k pixel readout chip working in single photon counting mode*, IEEE Trans. Nucl. Sci. NS. **49** (5) (2002) 2279.
29. Llopart X. et al, *Timepix, a 65k programmable pixel readout chip for arrival time, energy and/or photon counting measurements*, Nucl. Instrum. Meth. A **581** (2007) 485–494, doi:10.1016/j.nima.2007.08.079
30. Jakubek J., *Pixel detectors for imaging with heavy charged particles*, Comput. Phys. Commun. **591** (1) (2008) 155–158, doi:10.1016/j.nima.2008.03.091
31. Zajicek J. et al, *Multimodal imaging with hybrid semiconductor detectors Timepix for an experimental MRI-SPECT system*, JINST. **8** (1) January (2013), doi:10.1088/1748-0221/8/01/C01022
32. Zajicek J. et al, *Experimental MRI-SPECT insert system with hybrid semiconductor detectors Timepix for animal scanner Bruker 47/20*, JINST. **12** (1) January (2017), PO1015, doi:10.1088/1748-0221/12/01/P01015
33. Deprez K. et al, *Rapid additive manufacturing of MR compatible multipinhole collimators with selective laser melting of tungsten powder*, Med. Phys. **40** (1) (2013) 012501.
34. Accorsia R., *Brain Single-Photon Emission CT Physics Principles*, Department of Radiology, The Children's Hospital of Philadelphia, University of Pennsylvania, Philadelphia, PA 19104.
35. Mahmood S. T. et al, *The potential for mixed multiplexed and non-multiplexed data to improve the reconstruction quality of a multi-slit–slat collimator SPECT system*, Phys. Med. Biol. **55** (8) (2010) 2247–2268.
36. Misri R. et al, *Development and evaluation of a dual-modality (MRI/SPECT) molecular imaging bioprobe*. Nanomedicine: NBM. **8** (2012) 1007–1016, doi:10.1016/j.nano.2011.10.013
37. Jiawei Tan, *The Development of MRI-Compatible SPECT System*, Dissertation Thesis, University of Illinois at Urbana-Champaign (2012).
38. http://www.hindawi.com/journals/ijmi/2011/813028/fig3/18/11/2020
39. Pichler J. et al, *State-of-the-Art Small Animal PET/MRI*, Laboratory for Preclinical and Imaging Technology of the Werner Siemens-Foundation. University of Tubingen, Germany.
40. Bouziotis P. and Fiorini C., *SPECT/MRI: Dreams or reality?*, Clin. Transl. Imaging. **2** (2014) 571–573, doi:10.1007/s40336-014-0095-6
41. Cheng D. et al, *A comparison of ^{18}F PET and ^{99m}Tc SPECT imaging phantoms and in tumored mice*, Bioconjug. Chem. **21** (8) August 18 (2010) 1565–1570, doi:10.1021/bc1001467
42. Accorsia R., Brain Single-Photon Emission CT Physics Principles, Department of Radiology, The Children's Hospital of Philadelphia, University of Pennsylvania, Philadelphia, PA 19104.

9

Possibilities of Uses of a CZT Detector in Security Applications

Anna Selivanova, Lubomír Gryc, and Jan Helebrant

CONTENTS

DOI: 10.1201/9781003218364-9

9.1 Introduction

Owing to new security and safety issues, new radiation detection systems with an undemanding operation are being widely developed. Among fields of uses of these systems belong, e.g., a radiation survey with unmanned aerial vehicles (Martin et al. 2016), radiation mapping with robots (Selivanova et al. 2020), radiation monitoring of persons/luggage using portal gates (Martin et al. 2020) or radiation measurements provided by citizens with portable radiation detectors (Brown et al. 2016) or even smartphone cameras (Drukier et al. 2011).

The chapter is focused on radiation monitoring using portal gates equipped with radiation detectors. Due to increasing risks of terrorism, research activities in this field are addressed to a design and development of new detection systems able to reveal hidden radioactive sources, radiological dispersal nuclear or explosive devices ("dirty bombs"). These detection systems should ensure safety of mass events (e.g., concerts or exhibitions) and public places, such as airports or railroad stations. In order to scan persons/luggage and to obtain 2D/3D images and, solid-state array detectors based on CdTe/CZT compounds could be used in such systems (Airport Suppliers 2020; Kromek Group PLC 2020a). These solutions have been already developing also within medical applications (Iniewski 2014).

Results presented in this chapter are related to a research project dedicated to the development of new portal monitors in the Czech Republic (RIV VaVaI 2020). The project was mainly engaged in uses of plastic scintillator detectors for fast scanning of passing persons (SÚRO 2020a). Moreover, scanning during driving with detectors placed in a car was considered (SÚRO 2020b). Nevertheless, different configurations of portal monitors with miscellaneous radiation sensors (including CZT/CdTe detectors) were designed and tested.

Semiconductor radiation detectors based on CdTe/CdZnTe (CZT) compounds seem to be promising devices for security/emergency applications. CdTe/CZT compounds have a wide band gap of 1.44–2.2 eV depending on Zn concentration (Toney, Schlesinger, and James 1999). Hence, compared to high-purity Germanium (HPGe) detectors, CdTe/CZT detectors do not require cooling, being operated at room temperature (Gazizov et al. 2017).

An energy resolution of spectrometric CZT detectors expressed as a full width at half maximum (FWHM) does not exceed 2.5% at 662 keV (^{137}Cs) (Kromek Group PLC 2020b), contrary to the energy resolution of, e.g., 8.5% at the same photon energy for NaI(Tl) detectors (Del Sordo et al. 2009; Mouhti et al. 2018).

9.2 Properties of the CZT Detector

Within the development of new portal monitors, a small CZT detector was selected as a potential spectrometric device due to its simple operation and good spectrometric properties. The model GR1-A+ of a coplanar-grid CZT

FIGURE 9.1
The GRI-A+ CZT detector with an AA battery to compare spatial parameters.

detector was employed in performed tests (Kromek Group PLC 2020c). The tests of the CZT detector included both Monte Carlo simulations and measurements with standard point sources.

The detector contains a CZT crystal with dimension of 10 mm × 10 mm × 10 mm and a total volume of 1 cm³. The device spatial parameters are 25 mm × 25 mm × 63 mm, while its weight is equal to 60 g. The detector energy resolution FWHM is up to 2.5% at 662 keV. The energy range of the detector is from 30 keV to 3 MeV. A detailed description of the CZT detector could be found on an official website of the manufacturer (Kromek Group PLC 2020c). The detector is depicted in Figure 9.1.

9.3 Monte Carlo Simulations

In order to prepare full energy peak (FEP) efficiency calibrations of the detector, the MCNP6.1 radiation transport Monte Carlo code was used (Goorley et al. 2012). In all simulation tasks, transport of electrons and photons was included in calculations (*mode p e*). In case of photons, simulations considered detailed physics with coherent scattering, while for electrons, the default settings were employed (Selivanova et al. 2019).

The detector responses and its FEP efficiencies for selected geometries of point sources and human phantoms were obtained using a pulse height tally (F8). The F8 tally could be used to score the energy distribution of pulses in a region of interest ("cell") corresponding to a physical detector (Goorley et al. 2012). In the used mathematical model of the CZT detector, the F8 tally was located inside a cell of the CZT crystal (and its active volume), excluding areas of incomplete charge collection, or dead layers (Selivanova et al. 2019).

9.3.1 Mathematical Models

On the basis of the manufacturer's specification of the GR1-A+ device and its X-ray images, the mathematical 3D model of the detector was created. A detailed description, spatial and material parameters of the detector could be found in a paper by Selivanova et al. (2019). Contrary to the first version of the detector model, an improvement to the model was added: a rear part of the detector was newly filled with a material approximating electronics instead of air (Rault et al. 2011). The material density was set to 6.28E-1 g cm^{-3}. The improved detector model is shown in Figure 9.2.

Another model of the same detector was created by Vichi et al. (2016), using the particle physics Monte Carlo simulation package FLUKA (Ferrari et al. 2005; Böhlen et al. 2014), or by Tajudin et al. (2019) and the EGS5 Monte Carlo code (Hirayama et al. 2005).

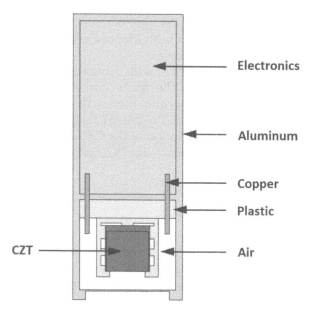

FIGURE 9.2
Scheme of improved mathematical model of CZT detector.

9.3.2 Regions of Incomplete Charge Collection

Due to significant discrepancies up to 41% between measurements and simulations (Selivanova et al. 2019), the model included dead layers, located on sides of the CZT crystal and on its top (in front of the coplanar anode). In simulations, a density of dead layers and their material composition was similar to an active volume of the CZT crystal. However, these regions and corresponding cells were excluded from scoring of the energy distribution of pulses.

In the MCNP model of the detector, regions of dead layers corresponded to expected areas of incomplete charge collection. According to Bolotnikov et al. (2012), presence of these areas is caused by defects in CZT crystals, leading to contributions to a Compton continuum instead of full-energy peaks and a subsequent decrease of FEP efficiencies. Moreover, the incomplete charge collection could contribute to a low-energy tail of full-energy peaks (Del Sordo et al. 2009; Bolotnikov et al. 2012).

In accordance with Luke et al. (2005), Monte Carlo simulations indicated a presence of a layer of 0.5 mm with the incomplete charge collection near the coplanar electrode surface. However, based on Monte Carlo simulations of the mathematical model, the side dead layer thickness was assessed to be 0.23 mm, while for the rear dead layer, the thickness was set to 1.25 mm. Therefore, the active volume of the CZT crystal was reduced from 1 cm^3 to 0.8 cm^3.

Using the model of the detector with dead layers, differences between measurements and simulations for point sources located at 25 cm in front of the detector entrance window in its axis did not exceed 5%. Thereafter, the mathematical model of the detector was validated using a ^{226}Ra needle a ^{60}Co standard point source. Within the validation, differences between simulated efficiencies and measured values were not greater than 4% (Selivanova et al. 2019).

Tajudin et al. (2019) also mentioned the required reduction of the detection volume of CZT crystal to eliminate large disagreements between simulations and measurements. In simulations, Tajudin et al. (2019) employed the CZT crystal active area of 0.8 cm × 0.8 cm instead of 1 cm × 1 cm, using side dead layers.

Hence, according to existing data and obtained results, dead layers, or regions with the incomplete charge collection, could be expected on surfaces of CZT crystals.

9.3.3 Variance Reduction Techniques

In order to improve statistics of Monte Carlo simulations alongside the calculation runtimes reduction, capabilities and specific tools of the MCNP6.1 radiation transport code were used. Therefore, several techniques of variance reduction presented below were implemented in the simulations.

9.3.3.1 Selective Electron Transport

All simulation tasks employed the selective electron transport, occurring in cells of the CZT detector only. The rest of the whole geometry assumed the turned off transport of electrons, using the importance card of cells (Goorley et al. 2012; Selivanova et al. 2019).

9.3.3.2 Adaptive Energy Cut-Off

In case of simulations with point sources and human phantoms, an adaptive energy cut-off (the ELPT card) was utilized. Using the ELPT card with thresholds for photons, energy cut-offs, histories of photons with energies below the threshold were terminated within calculations (Goorley et al. 2012).

The energy cut-offs were set slightly below energies of photons emitted from sources, being added to all cells, except cells of the CZT detector (Goorley et al. 2012). The described set-up could not be applied in simulation tasks where a total spectrum is simulated. However, the technique seemed to be very proper for simulations of FEP efficiencies only (Vrba 2016; Selivanova et al. 2019, 2020).

9.3.3.3 Nested DXTRAN Technique

Another implemented variance reduction method was the DXTRAN technique, or nested DXTRAN spheres. DXTRAN spheres placed in a region of interest, e.g., containing a cell with a tally, allow to improve sampling in the region (Goorley et al. 2012).

The method was mentioned as a very convenient technique of the variance reduction in case of simulations of large-scale NORM sources (Wallace 2013; Selivanova et al. 2019) or shielded sources (Selivanova et al. 2020). The technique could be also used if whole spectra are required (Selivanova et al. 2019). The detailed description of the method could be found in a publication by Booth et al. (2009).

Using the technique, two DXTRAN spheres, an inner sphere and an outer sphere, were employed in simulation tasks. Spheres radii were equal to 1 cm, resp. 2 cm. The center of both spheres was identical with the center of the CZT crystal. Hence, all cells of the CZT crystal, its active volume and dead layers, were surrounded by the DXTRAN spheres. In order to shorten calculation times, additional DXTRAN cut-offs were included in the set-up. For instance, in the vast majority of simulations presented in this contribution, an upper cut-off was set 100, while a lower cut-off was equal to 1E-11. A description of the cut-off settings process could be found, e.g., in a paper by Selivanova et al. (2019).

9.3.4 Portal Gates

Owing to preparations of FEP efficiency calibrations, a geometry of portal gates was proposed, using possible spatial parameters of existing solutions. Outer dimensions of the gates were equal to 2225 mm × 860 mm × 560 mm (height × width × depth), while inner dimensions were set to 2055 mm × 700 mm × 560 mm. A material of walls of the portal gates was polymethyl methacrylate (PMMA), material #182 from the Compendium of Material Composition Data for Radiation Transport Modeling (McConn Jr et al. 2011).

A thickness of walls of gates was set to 3 mm. A space inside was provisionally filled with air, material #4 (McConn Jr et al. 2011). The CZT detector was placed in close proximity to a center of an inner wall, in 280 mm to the gates entrance and exit (half of the depth). The detector was at 1 m above the ground. The detector entrance window was oriented to a passageway of gates. The proposed geometry could be seen in Figure 9.3, depicted with a point source in a center of the gates passageway (35 cm to both gates, half of the inner width) in the detector axis. The source was located at 1 m above the ground, as well as the CZT detector.

9.3.5 Sources

Geometries of radioactive sources at different distances to the gates were expected. Sources moved through the gates center, at the distance of 35 cm to both gates. Two types of sources were assumed, point sources (e.g., small source hidden in a luggage) and sources inside a human body (e.g., residual activity after nuclear treatments). For both types of geometries, FEP efficiencies were simulated.

9.3.5.1 Point Sources

In case of point sources, 11 positions of sources were anticipated, from −100 cm to +100 cm to the detector (or from −72 cm to +72 cm to the gates entrance and its exit), with an interval between two neighboring positions of 20 cm. Sources were placed at 1 m above the ground, as well as the CZT detector. The height of 1 m was selected due to possible position of a handbag/backpack/pockets with the small source inside.

Within Monte Carlo simulations, relevant radionuclides with photon energies from 46.54 keV (^{210}Pb) to 3 MeV (maximum of the detector energy range) were chosen. Altogether, 21 energies were simulated in corresponding separated tasks. Hence, 9 existing radionuclides (e.g., ^{137}Cs, ^{131}I or ^{60}Co) and, for completeness, 8 additional artificial energies (from 150 keV to 3 MeV) were considered. For this geometry, ideal conditions were assumed; therefore, possible source shielding (e.g., metallic boxes) was not included in simulations.

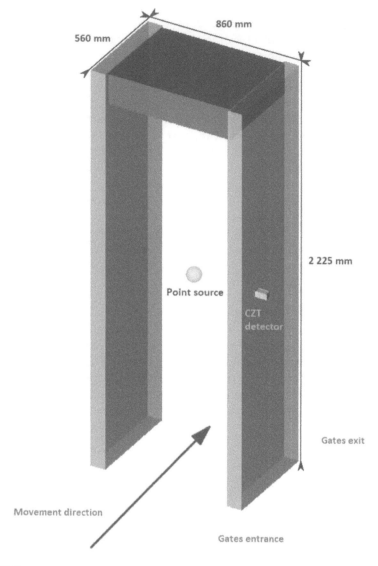

FIGURE 9.3
Proposed portal gates design with CZT detector inside and point source in passageway.

Graphical representation of a dependence of FEP efficiencies obtained in the Monte Carlo simulations on photon energies and positions of point sources is shown in Figure 9.4. Corresponding FEP efficiencies, selected values for the most expected radionuclides/energies, (point sources at considered

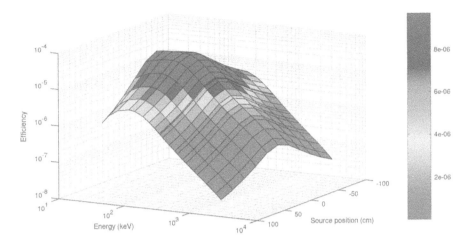

FIGURE 9.4
Dependence of FEP efficiencies of unshielded point sources on photon energies and source positions (distances to CZT detector).

distances to the CZT detector) are listed in Table 9.1. Statistical uncertainties of simulations were not greater than ±0.3%.

According to Figure 9.4, the detector responses (FEP efficiency curves) to unshielded point sources were symmetrical, both for the gates entrance side (from −100 cm to 0 cm) and for the exit side (from 0 cm to +100 cm), being similar to a bell shape. Hence, the response demonstrated maximum values for point sources in front of the detector (0 cm to the detector), decreasing with a growth of the source distance to the detector.

No other significant changes in the shape of FEP efficiency calibration curves were observed. Comparing FEP efficiency values for both directions of the source moving (Table 9.1), FEP efficiencies were almost similar. Therefore, considering ideal conditions of real measurements, the detector response, expressed, e.g., as counts per seconds, was also expected to be symmetrical.

9.3.5.2 Phantoms

Considering persons after nuclear treatments with residual activities in bodies, simulations with human phantoms passing through the gates were carried out. Two ORNL mathematical phantoms of an adult woman and a man were employed (Krstić and Nikezić 2007). Phantoms had a volume source containing ^{131}I homogeneously distributed in a thyroid (Figure 9.5). The source-emitted photons with energy of 364.5 keV only.

As well as in the case of simulations with point sources, phantoms were located at different distances to the gates (and the CZT detector), mimicking their movement through the gates. Contrary to previous simulations, larger distances (phantom-gates) were assumed, from −100 cm (the entrance

TABLE 9.1

FEP Efficiencies ε for Point Sources at Selected Distances to CZT Detector

Distance (cm)		−100	−80	−60	−40	−20	0	+20	+40	+60	+80	+100
Nuclide	Energy (keV)	E	ε	ε	ε	ε	E	ε	ε	ε	ε	ε
241Am	59.54	2.67E-06	4.39E-06	7.96E-06	1.61E-05	3.26E-05	4.71E-05	3.27E-05	1.61E-05	7.94E-06	4.34E-06	2.69E-06
133Ba	81.00	3.87E-06	6.13E-06	1.06E-05	2.01E-05	3.71E-05	4.86E-05	3.71E-05	2.00E-05	1.06E-05	6.12E-06	3.89E-06
152Eu	121.78	4.51E-06	6.97E-06	1.17E-05	2.12E-05	3.84E-05	4.84E-05	3.84E-05	2.12E-05	1.17E-05	6.97E-06	4.51E-06
-	150.00	4.25E-06	6.49E-06	1.07E-05	1.92E-05	3.49E-05	4.47E-05	3.49E-05	1.92E-05	1.07E-05	6.51E-06	4.24E-06
131I	364.49	1.17E-06	1.74E-06	2.79E-06	4.86E-06	8.76E-06	1.16E-05	8.75E-06	4.86E-06	2.79E-06	1.74E-06	1.17E-06
-	500.00	6.42E-07	9.58E-07	1.52E-06	2.65E-06	4.72E-06	6.24E-06	4.73E-06	2.64E-06	1.53E-06	9.56E-07	6.42E-07
134Cs	604.72	4.53E-07	6.75E-07	1.07E-06	1.85E-06	3.30E-06	4.36E-06	3.30E-06	1.86E-06	1.07E-06	6.75E-07	4.53E-07
137Cs	661.66	3.87E-07	5.76E-07	9.14E-07	1.58E-06	2.80E-06	3.70E-06	2.79E-06	1.58E-06	9.13E-07	5.74E-07	3.86E-07
134Cs	795.86	2.81E-07	4.18E-07	6.63E-07	1.14E-06	2.02E-06	2.66E-06	2.02E-06	1.14E-06	6.64E-07	4.17E-07	2.81E-07
60Co	1173.24	1.49E-07	2.22E-07	3.50E-07	5.99E-07	1.05E-06	1.39E-06	1.05E-06	6.00E-07	3.50E-07	2.21E-07	1.49E-07
60Co	1332.50	1.21E-07	1.80E-07	2.85E-07	4.89E-07	8.57E-07	1.13E-06	8.56E-07	4.89E-07	2.85E-07	1.80E-07	1.21E-07
-	3000.00	2.91E-08	4.31E-08	6.80E-08	1.15E-07	2.00E-07	2.62E-07	2.00E-07	1.14E-07	6.77E-08	4.29E-08	2.91E-08

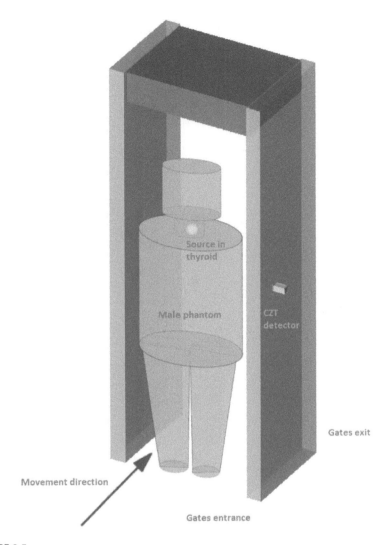

Source in thyroid

Male phantom

CZT detector

Gates exit

Movement direction

Gates entrance

FIGURE 9.5
Portal gates with CZT detector inside and male phantom with radioactive source in thyroid containing [131]I (not to scale).

side) to +100 cm (the exit side), while actual distances source-detector were greater. For instance, distances from phantoms to the CZT detector located the gates center were roughly from –128 cm to +128 cm, while distances from the source to the detector were approximately from –136 cm to +142 cm. An interval between two neighbor positions was equal to 20 cm.

FEP efficiencies were simulated for 11 positions of both phantoms. Results were summarized in Table 9.2. Statistical uncertainties of Monte Carlo

TABLE 9.2

FEP Efficiencies ε for ^{131}I Sources with Energy of 364.5 keV Homogeneously Distributed in Thyroids of Male and Female Phantom at Selected Distances to Gates (Not Detector)

Distance (cm)	−100	−80	−60	−40	−20	0	+20	+40	+60	+80	+100
Phantoms	*ε*	*ε*	*ε*	*ε*	*ε*	*ε*	*ε*	*ε*	*ε*	*ε*	*ε*
Male	9.37E-07	1.21E-06	1.56E-06	1.98E-06	2.35E-06	5.39E-07	4.75E-07	4.08E-07	3.30E-07	2.68E-07	2.21E-07
Female	9.78E-07	1.27E-06	1.68E-06	2.21E-06	2.77E-06	8.26E-07	5.86E-07	4.67E-07	3.60E-07	2.83E-07	2.28E-07

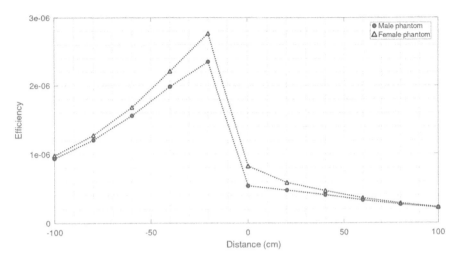

FIGURE 9.6
Dependence of FEP efficiencies of ^{131}I sources homogenously distributed in thyroids of phantoms (adult female and adult male) on phantom positions (distances to gates).

simulations did not exceed ±0.4%. For the simpler results representation, all assumed distances were presented as distances phantom-gates.

Corresponding FEP efficiency dependencies on both phantoms positions could be found in Figure 9.6. Negative values of distances corresponded to the location of phantoms in front of the gates entrance. In such position, both phantoms with sources of ^{131}I in their thyroids were oriented toward the gates. "Zero" position was a position, when phantoms were in the gates passageway. Positive values of distances (or the gates exit) corresponded to phantoms oriented outward the gates.

According to data in Table 9.2 and the corresponding graphical representation (Figure 9.6), FEP efficiencies were higher for the female phantom, compared to the male phantom. Differences between both sets of simulations were in a range from 4% (the most distant positions) up to 53% ("zero position"). These differences were probably caused by different spatial parameters of phantoms: their heights and the thyroid location inside bodies. Hence, the female phantom height was 163 cm, while the male height was 176 cm (Menzel, Clement, and DeLuca 2009). In case of the female phantom, the thyroid with ^{131}I inside was closer to the detector; therefore, the corresponding FEP efficiencies for all assumed geometries were greater.

On the basis of the graph in Figure 9.6, responses of the CZT detector on ^{131}I in both human phantoms were not symmetrical and differed for positions in front of the gates and behind them, contrary to ideal point sources (Figure 9.4). Moreover, maximum values of FEP efficiencies corresponded to the phantoms positions in front of the gates, but not for the phantoms in the passageway near the CZT detector (compared to ideal unshielded point sources, Section 9.3.5.1).

Hence, FEP efficiencies rapidly increased with the phantoms approaching the gates, while the subsequent FEP decrease started roughly inside the gates center and was much slower (Figure 9.6). For positions of phantoms inside the gates, their thyroids containing ^{131}I were shielded with bodies of the phantoms. Maximum depths of the phantoms were roughly 22 cm (the male phantom) and 24 cm (the female phantom). Anticipating phantoms outward the gates, their thyroids with sources of ^{131}I were shielded with tissues and organs of roughly 15 cm.

9.3.5.3 Test of Shielding Effects

In order to check simulations results and observed shielding effects in the previous paragraph, an additional test with analytical calculations was carried out. In the test, shielding of the ^{131}I point source (364.5 keV) with a layer of water with a thickness of 15 cm was assumed.

According to the XCOM database (Berger et al. 2010), a linear attenuation coefficient for photons with the energy of 364.5 keV in water was 1.10E-01 cm^{-1}. Therefore, considering water shielding of 15 cm, an intensity of photons of 364.5 keV would fall roughly to one fifth of the prior value.

Comparing FEP efficiencies for the phantoms on both sides from the gates, e.g., at distances of ±20 cm (Table 9.2), the same behavior could be observed. For the male phantom, the FEP efficiencies were 2.35E-6 (–20 cm) and 4.52E-7 (+20 cm). Comparing both FEP efficiency values, for the phantom oriented outward the gates, the FEP efficiency decreased approximately to one fifth, as well as in the analytical calculations discussed earlier. In case of the female phantom, the similar changes in the FEP efficiencies were occurred.

Therefore, in case of radioactive sources located in human bodies or hidden, e.g., inside clothing/accessories, shielding effects could be expected, as well as the non-symmetrical response of the CZT detector. However, assuming extra detection devices, these issues could be reduced.

9.3.6 Background

Owing to the subsequent calculation of minimum detectable activities (MDA), a background spectrum was required. In order to assess MDA values conservatively, the simulated spectrum of NORM radionuclides with higher specific activities was adopted from the preceding research (Selivanova et al. 2019).

The background spectrum was a sum of individual spectra of [40]K, [226]Ra/[238]U series and [232]Th series. Each spectrum was simulated as a separate task. In case of [226]Ra/[238]U and [232]Th series, a secular equilibrium was anticipated.

In the simulations, the CZT detector was at 1 m above the ground. NORM sources were set as soil cylinders with homogeneously distributed activities inside. The sources thickness was 1 m, while the radius was 30 m. A complete description of the source parameters could be found in the paper by Selivanova et al. (2019), as well as a process of their selection and results of corresponding tests.

Due to a simpler technical feasibility of simulations, additional simplifications were employed. For example, photons with energies with emission yields above chosen thresholds were simulated instead of full spectra. The same approach was used, e.g., by Wallace (2013), when appropriately selected thresholds of the emission yields should not lead to significant losses of accuracy.

Obtained spectra were rescaled in accordance with possible higher concentrations of NORM radionuclides in the Czech Republic (Matolín 2017). Hence, the ^{40}K specific activity was set 1409 Bq kg^{-1}. In case of ^{232}Th and ^{238}U/^{226}Ra series, specific activities were equal to 190 Bq kg^{-1}, resp. 211 Bq kg^{-1} (Selivanova et al. 2019). Cosmic radiation, radon contributions and its progeny were not included in simulations, as well as ^{137}Cs after Chernobyl.

The simulated background spectrum of NORM radionuclides for the CZT detector at 1 m above the ground is depicted in Figure 9.7. Significant full-energy peaks were marked using arrows and labels with information about radionuclides/series and corresponding energies. For example, such energies like 295.22 keV (^{226}Ra), 583.19 keV (^{232}Th), 609.31 keV (^{226}Ra) or 1460.82 keV (^{40}K) could be clearly identified. Hence, the simulated background spectrum was used to estimate MDA of selected artificial radionuclides and geometries of radioactive sources.

9.4 MDA Calculation

In order to estimate MDA values using spectrometric methods, the ISO 11929 standard was used (2010). Contrary to the Currie's formulation of MDA calculations (1968), the ISO 11929 standard allows to take into account systematic uncertainties. However, in cases of low systematic uncertainties, differences between both methods are almost negligible (Kirkpatrick, Venkataraman, and Young 2013; Selivanova et al. 2020). Nevertheless, statistical uncertainties of Monte Carlo simulations of FEP efficiencies and relative standard uncertainties of photon emissions were added to calculations (ISO 2010; Selivanova et al. 2020).

Within the MDA assessments, two source geometries were supposed: the point source geometry and ^{131}I sources in thyroids. Hence, simulated FEP efficiency values for the CZT detector (Sections 9.3.5.1 and 9.3.5.2) were used in calculations, considering different distances source-detector, resp. phantom-gates. The simulated background spectrum for the detector height of 1 m above the ground (Section 9.3.6) was employed in the MDA estimations. For

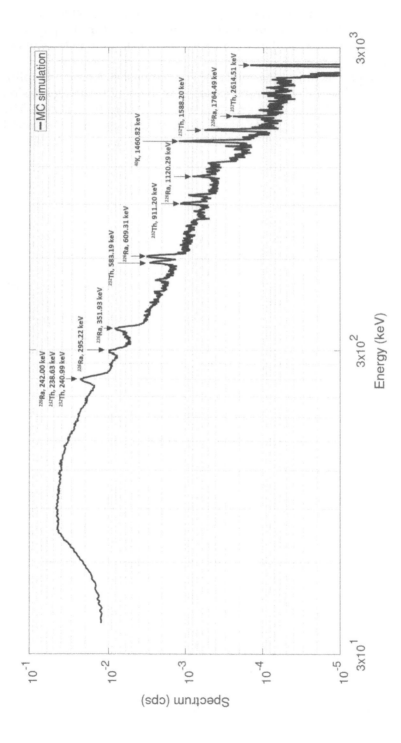

FIGURE 9.7
Simulated background spectrum of NORM radionuclides for CZT detector at 1 m above soil.

a completeness, in case of ^{60}Co and two energies (1173.24 keV and 1332.50 keV), both MDA values were calculated.

The gates sampling time (the acquisition time) of one measurement by the CZT detector was set to 1 second. The anticipated background spectrum acquisition time was 20 minutes.

9.4.1 Point Sources

MDA values of unshielded point sources of ^{131}I, ^{137}Cs and ^{60}Co at distances from −100 cm to +100 cm to the CZT detector are summarized in Table 9.3. The corresponding graphical representation is shown in Figure 9.8. For a better readability of the graph, the y axis was presented in a logarithmic scale.

In accordance with Table 9.3, MDA values were in a range from several tenths of MBq to tens of MBq. For instance, MDAs for ^{131}I and the energy of 364.5 keV were in an interval from roughly 0.7 MBq (for the center position inside the gates) up to 6.7 MBq (the most distant position). In case of ^{137}Cs, MDA values were from approximately 1.4 MBq to 13.8 MBq. For ^{60}Co source with two energies (1173.24 keV and 1332.50 keV), MDAs for the same source position were mutually resembling (Table 9.3), being in a range from roughly 3 MBq to 30 MBq. The both ^{60}Co MDA values did not differ by more than approximately 10%. Hence, for rough estimation, the lower values could be used.

Based on the calculated intervals of the detectable activities, the proposed portal gates with the CZT detector inside could reveal even small calibration sources or sources used in brachytherapy (Domenech 2016).

On the grounds of MDAs in Table 9.3 and graphs in Figure 9.8, maximum MDA values corresponded to the most distant position source-detector, while the MDA minima agreed with sources placed in close proximity to the detector, or in the gates center. Therefore, MDAs decreased with shortening of the distance source-detector, being inversely related to the growth of the FEP efficiencies (Section 9.3.5).

On the basis of the graphical representation in Figure 9.8, MDA values were symmetrical relatively "zero" position (the gates center), being very similar for both sides of the gates: the gates entrance (negative values of distances) and the gates exit (positive values of distances). Moreover, a shape of the dependence of point source MDAs on their distances to the detector was in a good agreement with the symmetrical and bell-shaped dependence of FEP efficiencies of point sources (Figure 9.4).

Obtained MDA values of point could be used as orientational levels of detectable activities for cases of unshielded small sources at different distances to the CZT detector. In order to assess MDAs, the spectrometric method was employed (ISO 2010), using information about background contributions in regions of interest (ROI) corresponded to areas of full-energy peaks. However, in case of non-spectrometric evaluations, following changes in whole spectra, MDA values could be lower.

TABLE 9.3

MDA Values of Unshielded Point Sources at Different Distances to CZT Detector

Distance (cm) Nuclide	Energy (keV)	−100 MDA (MBq)	−80 MDA (MBq)	−60 MDA (MBq)	−40 MDA (MBq)	−20 MDA (MBq)	0 MDA (MBq)	+20 MDA (MBq)	+40 MDA (MBq)	+60 MDA (MBq)	+80 MDA (MBq)	+100 MDA (MBq)
^{131}I	364.49	6.72	4.50	2.81	1.62	0.90	0.68	0.90	1.61	2.81	4.50	6.73
^{137}Cs	661.66	13.82	9.28	5.84	3.39	1.91	1.44	1.91	3.39	5.85	9.30	13.84
^{60}Co	1173.24	27.33	18.36	11.64	6.79	3.86	2.93	3.86	6.78	11.64	18.41	27.29
^{60}Co	1332.50	30.18	20.33	12.84	7.49	4.27	3.25	4.27	7.49	12.83	20.37	30.15

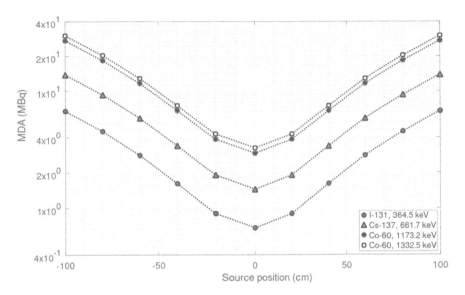

FIGURE 9.8
Dependence of MDA values of unshielded point sources on different distances to CZT detector.

9.4.2 Phantoms

Considering adult male and female phantoms at distances from −100 cm to +100 cm to the gates (not source-detector), MDA values for all expected positions of both phantoms are listed in Table 9.4. The phantoms MDA dependence on their location is depicted in Figure 9.9.

According to data in Table 9.4, MDA values for phantoms could vary roughly from 3 MBq (phantoms position in front of the gates entrance) up to 36 MBq (the most distant position, 100 cm from the gates exit), depending on their positions. Hence, persons with ^{131}I in thyroids after nuclear treatments could be clearly revealed, when residual activities of ^{131}I in thyroids could be tens or hundreds of MBq (Mathieu et al. 1999).

Comparing MDAs for both phantoms, MDA values for the female phantom were lower than values for the male phantom. Differences between both sets of MDAs were in a range of roughly from 4% up to 53%, corresponding to differences in the FEP efficiencies (Table 9.2). Lower MDAs for the female phantom were caused by its higher FEP efficiency values and smaller spatial parameters of the phantom, being discussed in Section 9.3.5.2.

Contrary to MDA values of ideal unshielded point sources (Figure 9.8), MDAs for phantoms were not symmetrical relatively the gates center (Figure 9.9), as well as for FEP efficiencies for both phantoms (Figure 9.6).

TABLE 9.4

MDA Values of Adult Male and Female Phantoms with ^{131}I Distributed in Thyroid on Different Distances to Gates

Distance (cm)	−100	−80	−60	−40	−20	0	+20	+40	+60	+80	+100
Phantom	MDA (MBq)	MDA (MBq)	MDA (MBq)	MDA (MBq)	MDA (MBq)	MDA (MBq)	MDA (MBq)	MDA (MBq)	MDA (MBq)	MDA (MBq)	MDA (MBq)
Male	8.38	6.51	5.03	3.96	3.34	14.58	16.52	19.27	23.77	29.29	35.60
Female	8.03	6.17	4.67	3.55	2.84	9.51	13.40	16.81	21.81	27.72	34.50

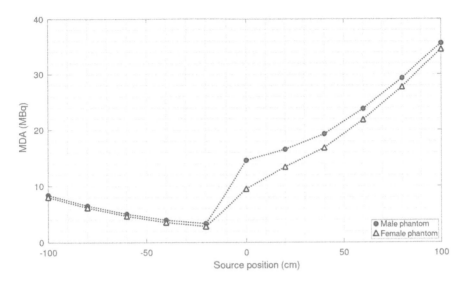

FIGURE 9.9
Dependence of MDA values of adult male and female phantoms with ^{131}I distributed in thyroid on different distances to gates.

9.5 Measurements

9.5.1 Geometry

In order to test the CZT detector as a possible complementary radiation detector, several measurement series were carried out. Within the tests, the CZT detector was placed at 1 m above the ground and hidden behind a PMMA slab with spatial parameters of 1000 mm × 1000 mm × 25 mm (height × width × thickness), approaching the proposed gates geometry. Owing to a better stability of a whole construction mimicked the gates, the slab was mounted on a plastic pallet (Figure 9.10).

Standard point sources, ^{137}Cs, ^{60}Co and ^{131}I, were used in the measurements. Hence, three corresponding independent measurement series were performed. Owing to the main goal of measurements, imitation of real conditions, sources were hidden inside a small paper box. The paper box was fixed at 1 m above the ground on a robot for indoor/outdoor uses, Morpheus (Zalud 2016). In case of ^{131}I and its easier manipulation, the source was additionally placed inside a plastic case (Figure 9.10). Moreover, in the same conditions, measurements of background were carried out.

An acquisition time (live time) of one measurement was set to 1 second. Spectra were acquired using the MultiSpect Analysis gamma spectroscopy software (Kromek Group PLC 2020d). Dead times were almost negligible (roughly tenths of a percent) in all measurements.

FIGURE 9.10
Geometry of measurements: CZT detector behind PMMA slab mimicking portal gates and radioactive source (^{131}I) hidden in paper box mounted on robot.

9.5.2 Robot

The robot, Morpheus, was created under Orpheus-X/AC series of robotic platforms (Zalud 2016). These robotic platforms could be used both for indoor and outdoor measurements, such as radiation monitoring or exploration of disaster areas (Burian et al. 2014; Lazna et al. 2019).

The robot was employed in the measurements due to its smooth running with an almost constant speed, allowing to approximate a continuous movement of measured persons through the gates. The robot was operated using a console by a trained operator (SÚRO 2020c).

The robot's speed was 0.25 m s^{-1}. A length of its total path was 2 m (200 cm), corresponding to anticipated distances from –100 cm to +100 cm to the detector, as well as in the assumed geometries of the Monte Carlo simulations with point sources (Section 9.3.5.1). Hence, in each measurement series, eight values were obtained, corresponding to eight measurement points, roughly 29 cm distant from each other.

9.5.3 Activities

According to calculated MDA values of unshielded point sources (Table 9.3), point standard sources of ^{131}I, ^{137}Cs and ^{60}Co with activities above MDAs were selected. For ^{131}I, the activity on the day of measurements was 7.9 MBq, being above MDAs for all considered distances, from –100 cm to +100 cm to the CZT detector (Table 9.3). The activity of ^{137}Cs was 7.7 MBq on the day of measurements. This value was above MDA values for source positions roughly from –70 cm to +70 cm (Figure 9.8) to the CZT detector. In case of ^{60}Co, the activity was 3.7 MBq on the day of measurements. According to Table 9.3, the point source of ^{60}Co could be detected spectrometrically at distances approximately from –20 cm to +20 cm to the detector.

9.5.4 Count Rates

Within the evaluation of obtained spectra, both total count rates in whole spectra and count rates in ROIs, or corresponding full-energy peaks, were investigated. All count rates were corrected to dead times (although dead times were practically insignificant). For a completeness, count rates from measurements with radioactive sources were compared with background count rates in the lab.

9.5.4.1 Total Count Rate

For non-spectrometric assessments, changes in total count rates (whole spectra) depending on the source position for measurement series with point sources of ^{131}I, ^{137}Cs and ^{60}Co could be found in Figure 9.11. The dependence of the CZT detector response, resembling a bell, was similar to the shape of simulated FEP efficiencies described in Section 9.3.5.1 (Figure 9.4). Total count rates increased with the distance shortening to the CZT detector and then decreased, when sources moved away from the detector.

As mentioned in Section 9.5.1, background spectra were additionally acquired in the lab, where all measurements were carried out. The total background count rate was in a range of 3–16 cps, while an average value was equal to 9 cps.

According to Figure 9.11, the lowest total count rate was roughly 30 cps and corresponded to the point source of ^{60}Co at ±100 cm to the CZT detector. Hence, the total count rate for ^{60}Co was approximately three times as high as the average background level (9 cps) even for the most distant positions of the source, where the source activity on the day of the measurements was below the MDA (Table 9.3).

Therefore, in case of the rough and quick non-spectrometric evaluation, comparing measurements of radioactive sources and the background, total count rates for the sources were several times above the average background count rate. Afterward, the sources with lower activities (below MDA)

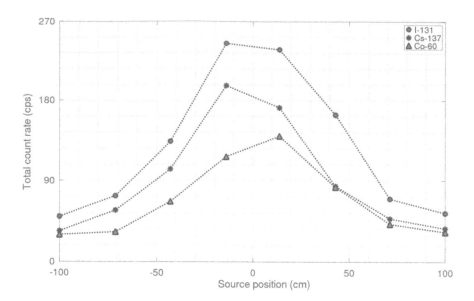

FIGURE 9.11
Total count rates of moving point sources of ^{131}I, ^{137}Cs and ^{60}Co at distances from −100 cm to +100 cm to CZT detector in experiment with geometry mimicking portal gates.

could be revealed even at distances from −100 cm to +100 cm using non-spectrometric methods.

9.5.4.2 ROI Count Rate

Count rates in full-energy peak areas, or regions of interest (ROI), for all three radionuclides, ^{131}I, ^{137}Cs and ^{60}Co and four corresponding energies, 364.49 keV, 661.66 keV, 1173.24 keV, resp. 1332.50 keV, could be found in Figure 9.12.

Contrary to total count rates (Figure 9.11), the shape of ROI count rates was less symmetrical due to worse statistics caused by low activities of employed sources and the short sampling time (1 second). However, in case of sources close to the detector, clear increases in ROI count rates could be observed.

Minimum values of ROI count rates in measurements with radioactive sources were in an interval of 0–2 cps. Background count rates in the similar ROIs were approximately several tenths of cps. Nevertheless, significant contributions to assumed ROIs agreed with the forecasted positions, where the used activities were above MDAs (discussed in Section 9.5.3).

Therefore, MDAs estimated spectrometrically on the basis of Monte Carlo simulations agreed with results of the experiment with standard point sources and low activities. Nevertheless, although spectrometric methods provided information about MDA for different radionuclides, in case of the non-spectrometric assessments, changes in the detector responses were more significant and easier for the evaluation.

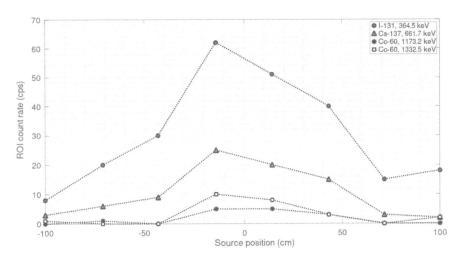

FIGURE 9.12
Count rates of moving point sources of ^{131}I, ^{137}Cs and ^{60}Co in regions of interest (ROI) at distances from –100 cm to +100 cm to CZT detector in experiment with geometry mimicking portal gates.

9.6 Conclusion

The CZT detector was tested within the development of portal gates. Field tests of the detector were based on Monte Carlo simulations. Obtained results of measurements demonstrated good agreement with forecasted estimations. Moreover, FEP efficiency calibrations for cases of unshielded point sources or radioactive sources inside human bodies were performed. According to subsequent MDA calculations, persons with residual activities in bodies after nuclear treatments or small sources with lower activities could be revealed. Hence, the tested CZT detector seemed to be appropriate for concealed radiation monitoring due to good spectrometric and detection properties, simple operation and very small pocket-size.

However, regardless the CZT detector's advantages, portal gates/monitors arisen within the project were based on large-volume plastic detectors (Grisa, Sas, and Gryc 2020) with one additional neutron detector only, a PMMA slab covered with ZnS:Ag and [6]LiF. At last, simple manufacturing of plastic detectors and their inexpensive costs belonged among determinative factors for the final development. The new detection systems were designed as line dividers, mounting portal monitors around doors, turnstiles or standard portal gates (SÚRO 2020a). Additionally, another version of the detection system for cars, e.g., Skoda Octavia Combi or Ford Transit, was created within the project (RIV VaVaI 2019).

Nevertheless, results of tests with the selected CZT detector will be used in the subsequent research in safety/security applications.

Acknowledgments

Miloš Drahokoupil is acknowledged for the robot operation and the technical support during measurements in the laboratory in the National Radiation Protection Institute (SÚRO), Prague. The research was supported by the project of Ministry of the Interior of the Czech Republic no. VI20172020104 – The new generation of portal monitors to ensure the security of the population.

References

Airport Suppliers. 2020. "Airport Radiation Detection Solutions – Kromek Group Plc." https://www.airport-suppliers.com/supplier/kromek-group-plc/.

Berger, M. J., J. H. Hubbell, S. M. Seltzer, J. Chang, J. S. Coursey, R. Sukumar, D. S. Zucker, and K. Olsen. 2010. *XCOM: Photon Cross Section Database (Version 1.5)*. Gaithersburg, MD: National Institute of Standards and Technology (NIST). http://physics.nist.gov/xcom.

Böhlen, T. T., F. Cerutti, M. P. W. Chin, A. Fassò, A. Ferrari, P. G. Ortega, A. Mairani, P. R. Sala, G. Smirnov, and V. Vlachoudis. 2014. "The FLUKA Code: Developments and Challenges for High Energy and Medical Applications." *Nuclear Data Sheets* 120 (June): 211–214. doi:10.1016/j.nds.2014.07.049.

Bolotnikov, A. E., G. S. Camarda, Y. Cui, G. De Geronimo, J. Fried, R. Gul, A. Hossain, et al. 2012. "Rejecting Incomplete Charge-Collection Events in CdZnTe and Other Semiconductor Detectors." *Nuclear Instruments and Methods in Physics Research Section A* 664 (1): 317–323. doi:10.1016/j.nima.2011.10.066.

Booth, T. E., K. C. Kelley, and S. S. McCready. 2009. "Monte Carlo Variance Reduction Using Nested Dxtran Spheres." *Nuclear Technology* 168 (3): 765–767. doi:10.13182/NT09-A9303.

Brown, Azby, Pieter Franken, Sean Bonner, Nick Dolezal, and Joe Moross. 2016. "Safecast: Successful Citizen-Science for Radiation Measurement and Communication after Fukushima." *Journal of Radiological Protection* 36 (2): S82–S101. doi:10.1088/0952-4746/36/2/S82.

Burian, F., L. Zalud, P. Kocmanova, T. Jilek, and L. Kopecny. 2014. "Multi-Robot System for Disaster Area Exploration." *Flood Recovery, Innovation and Response IV* 184: 263–274. doi:10.2495/FRIAR140221.

Currie, Lloyd A. 1968. "Limits for Qualitative Detection and Quantitative Determination. Application to Radiochemistry." *Analytical Chemistry* 40 (3): 586–593. doi:10.1021/ac60259a007.

Del Sordo, Stefano, Leonardo Abbene, Ezio Caroli, Anna Maria Mancini, Andrea Zappettini, and Pietro Ubertini. 2009. "Progress in the Development of CdTe and CdZnTe Semiconductor Radiation Detectors for Astrophysical and Medical Applications." *Sensors* 9 (5): 3491–3526. doi:10.3390/s90503491.

Domenech, Haydee. 2016. *Radiation Safety: Management and Programs. Radiation Safety: Management and Programs*. 1st ed. Cham, Switzerland: Springer International Publishing. doi:10.1007/978-3-319-42671-6.

Drukier, Gordon A., Eric P. Rubenstein, Peter R. Solomon, Marek A. Wojtowicz, and Michael A. Serio. 2011. "Low Cost, Pervasive Detection of Radiation Threats." In *2011 IEEE International Conference on Technologies for Homeland Security (HST)*, 365–371. IEEE. doi:10.1109/THS.2011.6107897.

Ferrari, A., P. R. Sala, A. Fasso, and J. Ranft. 2005. *FLUKA: A Multi-Particle Transport Code*. Menlo Park, CA.: Stanford Linear Accelerator Center (SLAC). doi:10.2172/877507.

Gazizov, I. M., A. A. Smirnov, V. G. Fedorkov, Yu. P. Kharitonov, V. S. Khrunov, and V. M. Zaletin. 2017. "Uncooled CdTe and CdZnTe Based Detectors for γ-Radiation Spectrometry." *Atomic Energy* 121 (5): 365–370. doi:10.1007/s10512-017-0213-4.

Goorley, T., M. James, T. Booth, F. Brown, J. Bull, L. J. Cox, J. Durkee, et al. 2012. "Initial MCNP6 Release Overview." *Nuclear Technology* 180 (3): 298–315. doi:10.13182/NT11-135.

Grisa, Tomas, Sas, Daniel, and Lubomir Gryc. 2020. "On a multidimensional data processing method for radiation portal monitors." *Nuclear Technology and Radiation Protection* 35 (3): 235–243. doi:10.2298/NTRP2003235G.

Hirayama, Hideo, Yoshihito Namito, Alex F. Bielajew, Scott J. Wilderman, U. Michigan, and Walter R. Nelson. 2005. *The EGS5 Code System*. Menlo Park, CA.: Stanford Linear Accelerator Center (SLAC). doi:10.2172/877459.

Iniewski, K. 2014. "CZT Detector Technology for Medical Imaging." *Journal of Instrumentation* 9 (11): C11001–C11001. doi:10.1088/1748-0221/9/11/C11001.

ISO. 2010. *ISO11929 Standard. Determination of the Characteristic Limits (Decision Threshold, Detection Limit and Limits of the Confidence Interval) for Measurements of Ionizing Radiation – Fundamentals and Application*.

Kirkpatrick, J. M., R. Venkataraman, and B. M. Young. 2013. "Minimum Detectable Activity, Systematic Uncertainties, and the ISO 11929 Standard." *Journal of Radioanalytical and Nuclear Chemistry* 296 (2): 1005–1010. doi:10.1007/s10967-012-2083-5.

Kromek Group PLC. 2020a. "EV3500 Scalable X-Ray CZT Linear Array." https://www.kromek.com/product/ev3500-x-ray-scalable-linear-array/.

Kromek Group PLC. 2020b. "GR1 Family Range – CZT Based Gamma-Ray Spectrometers." https://www.kromek.com/nuclear/gr1-ctz-gamma-ray-detectors/.

Kromek Group PLC. 2020c. "GR1 CZT Gamma-Ray Detector Spectrometer." https://www.kromek.com/product/gamma-ray-detector-spectrometers-czt-based-gr-range/.

Kromek Group PLC. 2020d. "MultiSpect Analysis Gamma Spectroscopy Software." https://www.kromek.com/product/multispect-analysis-gamma-spectroscopy-software/.

Krstić, D., and D. Nikezić. 2007. "Input Files with ORNL-Mathematical Phantoms of the Human Body for MCNP-4B." *Computer Physics Communications* 176 (1): 33–37. doi:10.1016/j.cpc.2006.06.016.

Lazna, Tomas, Ota Fisera, Jaroslav Kares, and Ludek Zalud. 2019. "Localization of Ionizing Radiation Sources via an Autonomous Robotic System." *Radiation Protection Dosimetry* 186 (2–3): 249–256. doi:10.1093/rpd/ncz213.

Luke, P. N., M. Amman, J. S. Lee, and C. Q. Vu. 2005. "Pocket-Size CdZnTe Gamma-Ray Spectrometer." *IEEE Transactions on Nuclear Science* 52 (5): 2041–2044. doi:10.1109/TNS.2005.856732.

Martin, P. G., O. D. Payton, J. S. Fardoulis, D. A. Richards, Y. Yamashiki, and T. B. Scott. 2016. "Low Altitude Unmanned Aerial Vehicle for Characterising Remediation Effectiveness Following the FDNPP Accident." *Journal of Environmental Radioactivity* 151 (January): 58–63. doi:10.1016/j.jenvrad.2015.09.007.

Martin, P. G., Yannick Verbelen, Elia Sciama Bandel, Mark Andrews, and T. B. Scott. 2020. "Project Gatekeeper: An Entrance Control System Embedded Radiation Detection Capability for Security Applications." *Sensors* 20 (10): 2957. doi:10.3390/s20102957.

Mathieu, Isabelle, Jacques Caussin, Patrick Smeesters, André Wambersie, and Christian Beckers. 1999. "Recommended Restrictions after 131I Therapy." *Health Physics* 76 (2): 129–136. doi:10.1097/00004032-199902000-00004.

Matolín, Milan. 2017. "Verification of the Radiometric Map of the Czech Republic." *Journal of Environmental Radioactivity* 166: 289–295. doi:10.1016/j.jenvrad.2016.04.013.

McConn Jr, R. J., C. J. Gesh, R. T. Pagh, R. A. Rucker, and R. G. Williams III. 2011. "Compendium of Material Composition Data for Radiation Transport Modeling – Revision 1, PIET-43741-TM963, PNNL-15870 Rev. 1." http://www.pnnl.gov/main/publications/external/technical_reports/PNNL-15870Rev1.pdf.

Menzel, Hans-Georg, Christopher Clement, and Paul DeLuca. 2009. "ICRP Publication 110. Realistic Reference Phantoms: An ICRP/ICRU Joint Effort. A Report of Adult Reference Computational Phantoms." *Annals of the ICRP* 39 (2): 3–5. doi:10.1016/j.icrp.2009.09.001.

Mouhti, I., A. Elanique, M. Y. Messous, B. Belhorma, and A. Benahmed. 2018. "Validation of a NaI(Tl) and LaBr 3 (Ce) Detector's Models via Measurements and Monte Carlo Simulations." *Journal of Radiation Research and Applied Sciences* 11 (4): 335–339. doi:10.1016/j.jrras.2018.06.003.

Rault, E., S. Staelens, R. Van Holen, J. De Beenhouwer, and S. Vandenberghe. 2011. "Accurate Monte Carlo Modelling of the Back Compartments of SPECT Cameras." *Physics in Medicine and Biology* 56 (1): 87–104. doi:10.1088/0031-9155/56/1/006.

RIV VaVaI. 2019. "Prototyp Mobilní Portálový Detektor Do Vozu Octavia Combi [Prototype Mobile Portal Detector for Octavia Combi]." *Rejstřík Informací o Výsledcích – Informační Systém [Information Register of R&D Results – R&D Information System]*. https://www.isvavai.cz/riv?s=jednoduche-vyhledavani&ss=detail&n=0&h=RIV%2F25506331%3A_____%2F19%3AN0000007.

RIV VaVaI. 2020. "VI20172020104 – Nová Generace Portálových Monitorů pro Zajištění Bezpečnosti Obyvatelstva (PoMoZ) [The New Generation of Portal Monitors to Ensure the Security of the Population]." *Rejstřík Informací o Výsledcích – Informační Systém [Information Register of R&D Results – R&D Information System]*. https://www.rvvi.cz/cep?s=jednoduche-vyhledavani&ss=detail&n=0&h=VI20172020104.

Selivanova, Anna, Jiří Hůlka, Daniel Seifert, Václav Hlaváč, Pavel Krsek, Vladimír Smutný, Libor Wagner, et al. 2020. "The Use of a CZT Detector with Robotic Systems." *Applied Radiation and Isotopes* 166 (December): 109395. doi:10.1016/j.apradiso.2020.109395.

Selivanova, Anna, Jiří Hůlka, Tomáš Vrba, and Irena Češpírová. 2019. "Efficiency Calibration of a CZT Detector and MDA Determination for Post Accidental Unmanned Aerial Vehicle Dosimetry." *Applied Radiation and Isotopes* 154 (December): 108879. doi:10.1016/j.apradiso.2019.108879.

SÚRO. 2020a. "Signalizační Detektory pro Rychlý Scan Velkého Počtu Procházejících Osob [Signal Detectors for Fast Scanning of a Large Number of Passers-By]." Prague, Czech Republic. https://youtu.be/iKZ4oeV_U7k.

SÚRO. 2020b. "Mobilní Portálový Monitor ve Vozidlech s Variabilním Použitím [Mobile Portal Monitor in Vehicles with Variable Use]." Prague, Czech Republic. https://youtu.be/QRXxn_7DW70.

SÚRO. 2020c. "Dálkově Ovládaný Pozemní Mobilní Systém pro Měření Pole IZ a Hot Spots [Remotely Controlled Ground Mobile System for Measuring Fields of Ionizing Radiation and Hot Spots]." Prague, Czech Republic. https://youtu.be/pUehP9HtWu4.

Tajudin, M. S., Yoshihito Namito, Toshiya Sanami, and Hideo Hirayama. 2019. "Full-Energy Peak Efficiency and Response Function of 1 cm³ CdZnTe Detectors." *Malaysian Journal of Fundamental and Applied Sciences* 15 (4): 580–584. doi:10.11113/mjfas.v15n4.1254.

Toney, J. E, T. E Schlesinger, and R. B James. 1999. "Optimal Bandgap Variants of Cd1–xZnxTe for High-Resolution X-Ray and Gamma-Ray Spectroscopy." *Nuclear Instruments and Methods in Physics Research Section A* 428 (1): 14–24. doi:10.1016/S0168-9002(98)01575-7.

Vichi, Sara, Angelo Infantino, Gianfranco Cicoria, Davide Pancaldi, Domiziano Mostacci, Filippo Lodi, and Mario Marengo. 2016. "An Innovative Gamma-Ray Spectrometry System Using a Compact and Portable CZT Detector for Radionuclidic Purity Tests of PET Radiopharmaceuticals." *Radiation Effects and Defects in Solids* 171 (9–10): 726–735. doi:10.1080/10420150.2016.1253090.

Vrba, T. 2016. "Crucial Parameters for Proper Simulation of the Detector Used in in vivo Measurements." *Radiation Protection Dosimetry* 170 (1–4): 359–363. doi:10.1093/rpd/ncv448.

Wallace, J. D. 2013. "Monte Carlo Modelling of Large Scale NORM Sources Using MCNP." *Journal of Environmental Radioactivity* 126 (December): 55–60. doi:10.1016/j.jenvrad.2013.06.009.

Zalud, Ludek. 2016. "MORPHEUS." http://www.ludekzalud.cz/morpheus/.

10

Xenon-Based EL-TPCs for 4π Detection of X-Rays in the Energy Range 10–100 keV

Angela Saa Hernández, Diego González-Díaz, and Carlos D. R. Azevedo

CONTENTS

10.1 History of X-Ray Gaseous Detectors at Synchrotron Light Sources

Fundamental discoveries in the history of particle physics are inextricably connected to gaseous detectors, given their versatility, imaging capabilities, and large area coverage [1]. The ionization chamber, perhaps the simplest of all gas-filled radiation detectors, enabled the detection of subatomic particles (Rutherford and Geiger, 1908). Few years later, Geiger built a gas-discharge particle counter, that was used in experiments leading to the identification of the alpha particle with the helium nucleus and to the development of Rutherford's model of the atom. Further development of proportional counters provided a means to identify particles based on their ionization ability. The first position-sensitive detector (PSD) for particle tracking was the "cloud" chamber, built by Wilson in 1912, a sealed environment containing a supersaturated vapor of water or alcohol, in which the passage of charged particles was revealed through their ionization trails, optically imaged. Later, spark chambers, consisting of a stack of metal plates with gas interleaved, were developed to visualize particle tracks in gases at atmospheric pressure. Spark chambers heralded a new era in which the emerging silicon-based electronics would play a crucial role. Thus, in 1968, G. Charpak introduced the multiwire proportional chamber (MWPC), a large area proportional counter that provided high resolution position measurements by collecting the signals from the individual wires, recorded with state-of-the-art amplification electronics. With the need for an ever-increasing detection volume at collider experiments, Charpak's invention played a crucial role and enabled fundamental discoveries such as the J/ψ meson by Ting and Richter and the W and Z bosons by Rubbia. Large-area proportional gas detectors, reconceived with sophisticated micropattern designs (dots, strips, meshes, or holes), continue to be instrumental in high energy nuclear and particle physics, representing the contemporary golden standard of gaseous PSDs [2].

In the field of condensed matter, the development of gaseous PSDs grew rapidly during the 80s in spite of their greater complexity relative to the traditional silver bromide x-ray emulsion, also known as x-ray film. Apparently, the growth was motivated by the increasing availability of digital processing power and, specially, by the establishment of central material-science facilities such as the synchrotron light sources [3] that required detector compliance with x-rays rates up to 10^5–10^6 ph/s. At the beginning of the 80s, 1D gas proportional detectors with position readout [4], based on the MWPC principles, were installed at light sources for small angle scattering investigations and protein crystallography research using synchrotron radiation [5, 6]. R&D in gaseous detectors for applications in synchrotrons continued, resulting a decade later in the development of 2D detectors covering an active area of 10×10 cm^2 and providing a position resolution of 100 µm full width at half maximum (FWHM) [7]. Compared to the slow AgBr films, the use of these novel PSDs enabled new possibilities, such as time-resolved structure studies. In fact, during the decade of the 90s, gas-based PSDs dominated over those based on scincillation or semiconductors. Bateman argues in reference [3] that this was due to: (i) their manufacturing flexibility (they were developed in a great variety of shapes and sizes, adapting to the different applications, e.g., radial drift devices, parallel gap devices, multistep avalanche devices, etc.), and (ii) the availability of materials used for their construction (wires, meshes, and other conventional materials in the workshops of the facilities). On the other hand, scintillation and semiconductor detectors depended on sophisticated hardware, and were only available from specialized high technology industry. Indeed, the development of high position resolution semiconductor detectors took off boosted by the pressure of the silicon technology developments for electronics, and by the development of miniaturized readout chips (VLSI). In the early days, silicon photodiodes were used for applications in x-ray absorption spectroscopy [8, 9], but were limited by the radiation damage produced under the direct exposure to x-rays above 5 keV. A second line of development involved pixel devices, with the invention of the charged coupled device (CCD), the first solid state imaging device by Boyle and Smith in 1970, which began to be successfully implemented in synchrotron radiation experiments toward the end of the 90s [10]. Due to the limited imaging area available in a CCD, of the order of few cm^2, and their radiation-soft materials, they were often used optically coupled to a scintillation screen, with sizes of the order of hundreds of cm^2.

Detectors at contemporary synchrotron light sources have to deal most often with x-rays in the medium-to-high energy range (10–100 keV) and an extremely variable flux, depending on the application, from very low (~1 ph/s) to very high rates (~10^{12} ph/s). Compatibility with very high dose rates stemming from x-ray ionization, of the order of MGy/s, is thus a relatively frequent requirement. Nowadays, the most commonly used x-ray detectors at these facilities rely on the semiconductor industry. Their working principle is based on the generation by the incoming radiation of electron-hole pairs in the semiconductor, which can then be measured by readout electronics. While silicon has been the reference material for this kind of detectors, their limited energy range, up to 15–20 keV, and low radiation hardness leading to degradation in sensor performance, has driven the research and development of other semiconductors (Ge, GaAs, Cd(Zn)Te, etc.). Being semiconductor detectors compact, fast, and efficient, besides relatively affordable, gaseous detectors have become rare guests at synchrotron light sources throughout the 00s and 10s.

Recent developments enabling the advent of the 4th generation of synchrotron light sources (also known as diffraction-limited light sources, despite being diffraction limited only for x-rays below 10 keV), is currently opening up a new era in x-ray science. The leap in performance is based on: (i) a multi-bend achromat lattice design, that reduces the horizontal emittance by various orders of magnitude relative to today's synchrotrons, increasing the photon beam brightness correspondingly, (ii) new insertion devices optimized for brightness and flux, including in some cases superconducting undulators that produce significantly higher photon flux at higher energies, and (iii) beamline improvements (e.g., new optics) to take advantage of the vastly improved source performance. Among the upgraded facilities, those accelerating electron beams above 5 GeV, such as ESRF-EBS [11], the projected APS-U [12], Petra IV [13], and SPring-8-II [14], will provide dramatically enhanced brightness and coherent flux, at least two orders of magnitude beyond today's capability, at photon energies where the inelastic (Compton) interactions dominate, opening a unique opportunity to use this type of scattering in ways that were not conceived before.

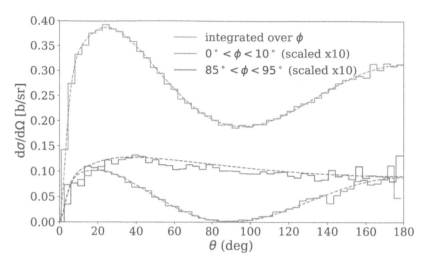

FIGURE 10.1
Differential cross section for Compton-scattered photons on DNA (in barn per stereoradian), for a linearly polarized x-ray beam of 64 keV as obtained with Monte Carlo simulations (using Geant4 [15]) and tabulated values [16] (dashed lines), for different azimuthal regions: $\phi = [0-10]°$ (green), $\phi = [85-95]°$ (blue), and integrated over ϕ (red). ϕ indicates the angle relative to the direction of the polarization vector. Reproduced from [47] with permission of the International Union of Crystallography.

To efficiently exploit Compton-scattered x-ray photons, a detector with a high-stopping power and near-4π coverage is required, as illustrated in Figure 10.1. This poses a formidable challenge for current detection technologies, which are costly and have detection areas below the required size. Clearly, the availability of a 4π/high energy x-ray detector would soon become an essential asset at any next generation facility, if it can be implemented in a practical way.

10.2 Fundamentals of X-Ray Interactions in Gaseous Media

A brief summary of the properties of gases of interest to the detection of high energy x-rays, namely: argon (Ar), krypton (Kr), and xenon (Xe), is presented from the point of view of their ability to absorb radiation and transform the absorbed energy into charge carriers and/or photon emission. The resulting position and energy resolution are discussed in this Section.

When x-rays of energies of the order of few keV impinge on a gaseous medium, interaction takes place primarily through photoelectric absorption, with just a small probability of Compton scattering. Indeed, photoelectric absorption dominates in the energy range from 10 to 100 keV in the case of Xe and Kr, while Compton scattering becomes the most probable interaction in Ar for energies above 75 keV, as shown in Figure 10.2 (top). The photon mean free path is represented in Figure 10.2 (bottom), as a function of their energy, indicating the positions corresponding to the characteristic K- and L-shell emission lines [17].

Upon interaction with a noble gas atom, the released photoelectron creates a cloud of secondary ionization with a size that depends on the

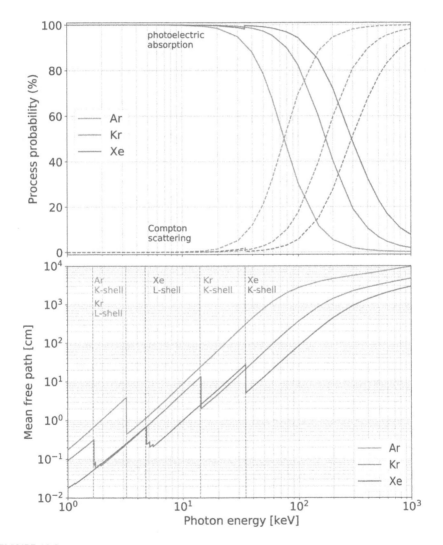

FIGURE 10.2
Top: probability of photoelectric absorption (solid line) and Compton scattering (dashed line) for incoming photons in the energy range from 1 keV to 1 MeV, in different gas media: argon (red), krypton (green), and xenon (blue). Bottom: characteristic K- and L-shell emission lines and photon mean free path in the range 1 keV to 1 MeV.

TABLE 10.1

Energies for Production of Electron-ion Pairs (Measured with Fast Electrons) [18], and for Production of Scintillation Photons (Measured with α-Particles for Pressures in the Range 1–2 bar) [19]

	W_I [eV]	W_{sc} [eV]
Ar	26.4	~50
Kr	24.2	~42
Xe	22.1	~30

photon energy and the gas itself. Ionization electrons and scintillation photons are liberated at an average energy expenditure (W-value) of few 10s of eV (Table 10.1).

The intrinsic position resolution of a gaseous detector stems from the ability to reconstruct the barycenter of the ionization trail, thus depending on the photoelectron range, multiple scattering, and the distance of thermalization of the ionization electrons in the gas. Additionally, if the photon energy is slightly above that of the K-shell, characteristic emission will ensue, and the position resolution will reach minimum values. This beneficial effect can be understood as a result of the x-ray interaction with the atom inner shells on the one hand, and the high x-ray fluorescence yield typical of heavy elements on the other. The energy carried away by characteristic emission results in less energetic photoelectrons and, therefore, in a smaller charge cloud. As studied by Azevedo et al. in reference [20] and shown in Figure 10.3, 1 cm-thick

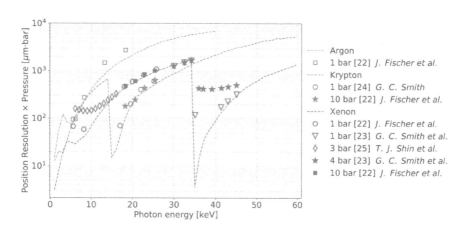

FIGURE 10.3

Position resolution in a $10 \times 10 \times 1$ cm^3 gaseous detector filled with xenon, argon, or krypton, for a temperature around 293 K, and photon energies in the range of 1–60 keV. Simulations in reference [20] are shown as dashed lines.

detectors can make an optimal use of this feature, by enhancing the escape probability of the secondary x-ray emission. This plot conveys the intuitive notion that the intrinsic position resolution of a gas is inversely proportional to its density (or to its pressure, at constant temperature) as shown in reference [21], hence the y-axis displays the product of both. A compilation of experimental data from references [22–25] is shown, together with simulation results obtained with the DEGRAD transport code [26].

An unavoidable limitation at reaching the intrinsic position resolution of the gas is the random motion by the thermalized ionization electrons during the charge collection process.[1] Although charges can be drifted swiftly into the amplification region upon application of an external field, such a random motion (stemming largely from elastic scattering with the noble gas), is governed by an asymmetric diffusion law. As a function of the drift distance, z, the original ionization cloud is convoluted with a Gaussian detector response, characterized through the widths:

$$\sigma_{z(x,y)} = D^*_{L(T)}\sqrt{z} \qquad (10.1)$$

where D^*_L and D^*_T are the field-dependent longitudinal (parallel to the field) and transverse (perpendicular to the field) diffusion coefficients, respectively. Simulations performed with the electron transport codes Magboltz/Pyboltz [27, 28], indicate that the presence in the gas of a small amount of a molecular additive, typically CO_2 or CH_4, cools down the electrons effectively, reducing the contribution from diffusion well below that in the pure noble gas (Figure 10.4). Minimizing the size of the ionization cloud is important not only for position reconstruction but for x-ray counting as well.

Ultimately, the accuracy at reconstructing the energy of the primary photon is limited by the Fano factor (F), and the fluctuations introduced by the signal amplification process (characterized through f), once accounting for the photon energy ε and the average energy to create an ionization electron, W_I (see Table 10.1). The expression for the energy resolution (FWHM) reads then:

$$\mathcal{R} = 2.355\sqrt{F+f}\sqrt{\frac{W_I}{\varepsilon}} \qquad (10.2)$$

Thanks to the electroluminescence (EL) readout, the second term can be brought close to zero and an energy resolution close to the intrinsic one (from F), stemming from fluctuations in the partition between excited and ionized states, obtained. A compilation of experimental data from various systems [29–39] is shown in Figure 10.5, together with the Fano-limited energy resolution derived in reference [40]. Results in EL mode are given in blue and avalanche mode results are given in red.

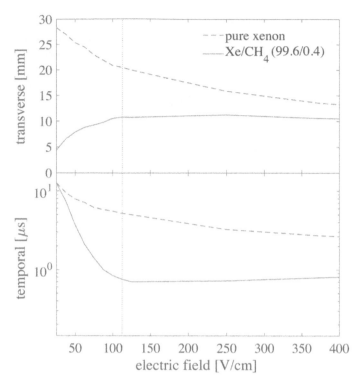

FIGURE 10.4
Top: transverse size of a point-like ionization cluster after drifting a distance $z = 50$ cm, obtained from Magboltz. Bottom: longitudinal size of a point-like ionization cluster (in time units), in the same conditions. Results for pure xenon and a mixture based on Xe/CH$_4$ are shown for comparison. Reproduced from [46] with permission of the International Union of Crystallography.

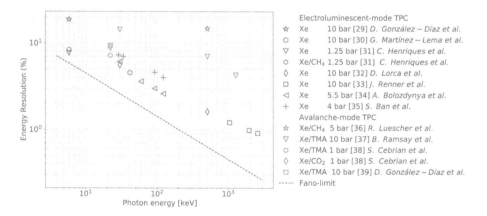

FIGURE 10.5
Compilation of experimental measurements of energy resolution as a function of the photon energy, for electroluminescent mode (blue) and avalanche mode (red) TPCs. The Fano-limited energy resolution (dashed line) is shown for comparison.

10.3 Achieving Photon-Counting in 4π for High Energy, High Intensity X-Ray Experiments

TPCs, introduced by D. Nygren in 1974 [41, 42], are nowadays ubiquitous in particle and nuclear physics, chiefly used for reconstructing interactions at high track multiplicities [43], and/or when very accurate event reconstruction is needed [44–46]. They have not been used so far for the detection of synchrotron light, despite a TPC shaped as a hollow cylinder would naturally provide a 4π-geometry, especially suited for the detection of Compton scattered photons from a sample placed at its center.

With this simple concept in mind, a new imaging technology based on an EL TPC for Compton x-ray scattering was introduced in reference [47]. The main characteristics of the particular TPC-flavor proposed can be summarized as: (i) efficient to high energy x-rays thanks to the use of xenon as the active medium, (ii) continuous readout mode with a time sampling around $\Delta T_s = 0.5$ µs, (iii) typical temporal extent of an x-ray signal (at mid-chamber): $\Delta T_{x-ray} = 1.35$ µs, (iv) about 2000 readout pixels/pads, and (v) single-photon counting capability, with a Fano-limited energy resolution potentially down to 2% FWHM for 60 keV x-rays, thanks to the EL mode. Importantly, however, the main advantage of using EL instead of conventional avalanche multiplication is the suppression of ion space charge, traditionally a shortcoming of TPCs operated under high rates.

The design is inspired by the proposal in reference [48] that has been successfully adopted by the NEXT collaboration in order to measure neutrinoless double-beta decay [49], but three main simplifications are proposed: (i) operation at atmospheric pressure, to facilitate the integration and operation at present x-ray sources, (ii) removal of the photomultiplier-based energy-plane, and (iii) introduction of a compact all-in-one EL structure, purposely designed for photon-counting experiments [29].

10.3.1 Detection Concept

The working principle, depicted in Figure 10.6, is the following:

1. X-rays that Compton-scatter at the sample interact with the xenon gas and give rise to ionization clusters with a typical (1σ) size of 0.25–1 mm, as shown in Figure 10.7 for two photon energies: 30 and 64 keV. These two energies will be used in the following to characterize the detector. The former, sitting just below the *K*-shell energy of xenon, is a priory the most convenient for counting due to the absence of characteristic x-ray emission inside the chamber. The latter represents the theoretical optimum in terms of dose at the sample, when applied to scanning Compton x-ray microscopy (SCXM)

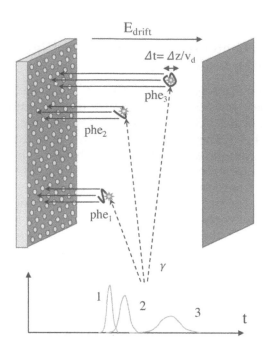

FIGURE 10.6
Schematic representation of the working principle of the EL-TPC. Photons scattered at the sample reach the xenon gas, creating ionization clusters that drift, while diffusing, toward the anode plane, where they induce electroluminescence. Reproduced from [47] with permission of the International Union of Crystallography.

[50], a technique under which the performance of this detector will be discussed in Section 10.4 of this chapter.

2. The ionization clusters drift, while diffusing, toward the EL/anode plane. For a drift field $E_{drift} = 110$ V/cm, the cluster's longitudinal size can be kept at the $\sigma_z = 4$ mm level even for a 50-cm long drift, corresponding to a temporal spread of $\sigma_t = 0.75$ μs, while the transverse size approaches $\sigma_{x,y} = 10$ mm, as shown in Figure 10.4. The electron drift velocity is $v_d = 5$ mm/μs.

3. Provided sufficient field focusing can be achieved at the EL structure, the ionization clusters will enter a handful of holes, creating a luminous signal. A silicon photomultiplier (SiPM) will be situated right underneath every hole, thus functioning, in effect, as a pixelated readout. A close-up of the pixelated readout region, that relies on the recent developments on large-hole acrylic multipliers [29] is displayed in Figure 10.8. Given the relatively large x-ray mean free path of around 20 cm in xenon at 1 bar, a sparse distribution of clusters is formed, which can be conveniently recorded with 10 mm-size pixels/pads, on a readout area of around 2000 cm² ($N_{pix} = 2000$).

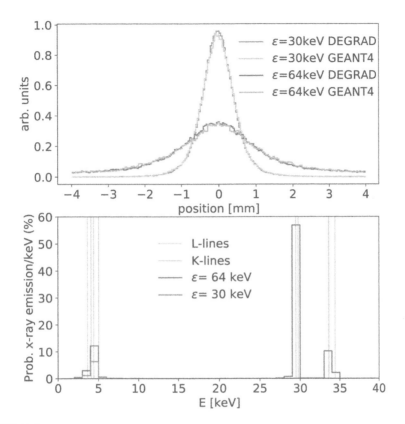

FIGURE 10.7
Top: ionization distributions in xenon gas, stemming from x-rays interacting in an infinite volume. They are obtained after aligning each x-ray ionization cloud by its barycenter, and projecting it over an arbitrary axis. Calculations from Geant4 are compared with the microscopic code DEGRAD [26]. Bottom: probability of characteristic x-ray emission in xenon for an incident photon energy of 30 keV (red) and 64 keV (blue), in Geant4. All K-shell (green) and L-shell (orange) lines, as tabulated in reference [17], are shown for comparison. Reproduced from [47] with permission of the International Union of Crystallography.

From the FWHM per x-ray cluster at about mid-chamber: $\Delta_{x,y}|_{x-ray} = 2.35 / \sqrt{2} \cdot \sigma_{x,y} = 16$ mm, an average multiplicity M of around 4 per cluster may be assumed if resorting to 10 mm×10 mm pixels/pads. The temporal spread, on the other hand, can be approximated by: $\Delta T_{x-ray} = 2.35 / \sqrt{2} \cdot \sigma_z / v_d = 1.35$ μs. Taking as a reference an interaction probability of $P_{int} = 2.9 \times 10^{-4}$ (5 μm water-equivalent cell, 10 mm of air), a 70% detection efficiency ε, and an $m = 20\%$ pixel occupancy, this configuration yields a plausible estimate of the achievable counting rate as:

$$r_{max} = \frac{1}{\varepsilon P_{int}} \frac{m \cdot N_{pix}}{M} \frac{1}{\Delta T_{x-ray}} = 3.6 \times 10^{11} \text{ (ph/s)} \qquad (10.3)$$

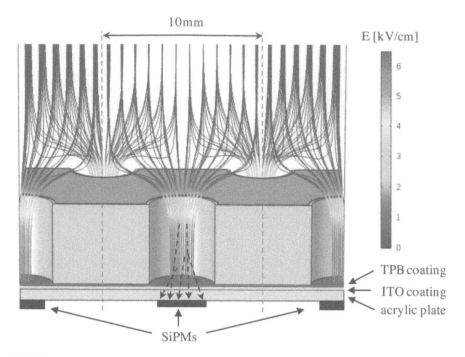

FIGURE 10.8
Close-up of the electroluminescence region, based on the recently introduced acrylic-based electroluminescence multipliers, developed in collaboration between IGFAE and the CERN-RD51 workshops. Reproduced from [47] with permission of the International Union of Crystallography.

compatible a priori with the beam rates for hard x-rays foreseen at 4th generation light sources.

10.3.2 Geometry

The suitability of the TPC technology depends primarily on the ability to detect x-ray photons within a realistic gas volume, in the absence of pressurization. The mean free path in xenon for x-rays in the range 10–100 keV varies between 1 and 84 cm at 1 bar, as shown in Figure 10.2 (bottom). Specifically, the mean free path is very similar for photon energies of 30 and 64 keV aiming at SCXM, ~20 cm; therefore, the most natural 4π-geometry adapting to this case is a hollow cylinder with a characteristic scale of around half a meter. On the other hand, the geometrical acceptance is a function of $\arctan(2R_i / L)$, with L being the length and R_i the inner radius of the cylinder. In order to place the necessary sample setup (namely the sample holder and associated mechanics), an inner bore of $R_i = 5$ cm is kept. Finally, the xenon thickness ($R_o - R_i$), that is the difference between the outer and inner TPC radii, becomes the main factor for the detector efficiency, as shown in Figure 10.9.

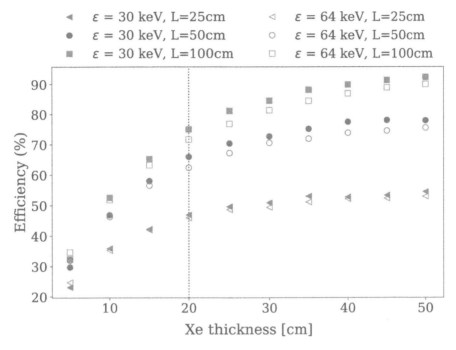

◄ $\varepsilon = 30$ keV, L=25cm   ◁ $\varepsilon = 64$ keV, L=25cm
● $\varepsilon = 30$ keV, L=50cm   ○ $\varepsilon = 64$ keV, L=50cm
■ $\varepsilon = 30$ keV, L=100cm   □ $\varepsilon = 64$ keV, L=100cm

FIGURE 10.9
Efficiency as a function of the thickness of the xenon cylinder $(R_o - R_i)$ for different lengths, at energies of 30 and 64 keV. The dotted line indicates the benchmark geometry considered in text, for a length $L = 50$ cm. Reproduced from [47] with permission of the International Union of Crystallography.

A realistic geometry for this detector would consist of an inner cylinder shell made out of 0.5 mm-thick aluminum walls, with 2 mm HDPE (high density polyethylene), 50 μm kapton and 15 μm copper, sufficient for making the field cage of the chamber, that is needed to minimize fringe fields (inset in Figure 10.10). The HDPE cylinder can be custom-made and the kapton-copper laminates are commercially available and can be adhered to it by thermal bonding or even epoxied, for instance. The external cylinder shell may well have a different design, but it has been kept symmetric for simplicity. A configuration that enables a good compromise in terms of size and flexibility is assumed in the following: $L = 50$ cm and $R_0 = 25$ cm. The geometrical acceptance nears in this case 80%. Additional 10 cm would be typically needed, axially, for instrumenting the readout plane and taking the signal cables out of the chamber, and another 10 cm on the cathode side, for providing sufficient isolation with respect to the vessel, given that the voltage difference will near 10 kV. Although those regions are not discussed here in detail, and have been replaced by simple covers, the reader is referred to reference [49] for possible arrangements. With these choices, the vessel geometry considered in simulations is shown in Figure 10.10, having a weight below 10 kg.

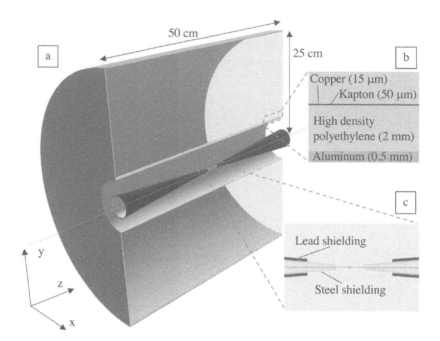

FIGURE 10.10

(a) TPC geometry in Geant4, aimed at providing nearly 4π-coverage for Compton-scattered x-rays from a sample placed in the center. (b) Detail of the region faced by x-rays when entering the detector, that includes the vessel and field cage. (c) Detail of the sample region and the shielding cones. Reproduced from [47] with permission of the International Union of Crystallography.

The beam enters the experimental setup from the vacuum pipes (not included in the figure) into two shielding cones (made of stainless steel and covered with lead shields) and from there into the sample region. Our case study is that of a 33-nm DNA feature inside a 5 μm cell, and 5 mm air to and from the shielding cones. The conical geometry is conceived not to crop the angular acceptance of the x-rays scattered on-sample, providing enough space to the focusing beam, and enabling sufficient absorption of stray x-rays from beam-air interactions along the pipes. In a 4π-geometry as the one proposed here, the sample holder and associated mechanics should ideally be placed along the polarization axis, where the photon flux is negligible. The necessary structural material of the walls and the presence of air in the sample region reduce the overall efficiency, presented in Figure 10.9, from 62.8% to 58.5% (64 keV) and from 64.5% to 40.0% (30 keV). Replacing the air by a helium atmosphere in the sample region could reduce the number of stray x-ray fields and increase the detector efficiency. However, its implementation has been considered technically complicated at this stage and therefore it is not studied further in this work.

10.3.3 Image Formation

The parameters used for computing the TPC response rely largely on the experience accumulated during the NEXT R&D program. A voltage of –8.5 kV at the cathode and 3 kV across the EL structure, with the anode sitting at ground, is considered, a situation that corresponds to fields around $E_{drift} = 110$ V/cm and $E_{el} = 6$ kV/cm in the drift and EL regions, respectively. The gas consists of Xe/CH$_4$ admixed at 0.4% in volume in order to achieve a 40-fold reduction in cluster size compared to operation in pure xenon (Figure 10.4). The EL plane will be optically coupled to a SiPM matrix, at the same pitch, forming a pixelated readout. The optical coupling may be typically done with the help of a layer of indium-tin oxide (ITO) and tetraphenyl butadiene (TPB) deposited on an acrylic plate, following reference [49]. This ensures wavelength shifting to the visible band, where SiPMs are usually more sensitive. The number of SiPM-photoelectrons per incoming ionization electron, n_{phe}, that is the single most important figure of merit for an EL-TPC, can be computed from the layout in Figure 10.8, after considering: an optical yield $Y = 250$ ph/e/cm at $E_{el} = 6$ kV/cm [29], a TPB wavelength-shifting efficiency $WLSE_{TPB} = 0.4$ [51], a solid angle coverage at the SiPM plane of $\Omega_{SiPM} = 0.3$ and a SiPM quantum efficiency $QE_{SiPM} = 0.4$. Finally, according to measurements in reference [52], the presence of 0.4% CH$_4$ reduces the scintillation probability by $P_{scin} = 0.5$, giving, for a $h = 5$ mm-thick structure:

$$n_{phe} = Y \cdot h \cdot WLSE_{TPB} \cdot \Omega_{SiPM} \cdot QE_{SiPM} \cdot P_{scin} = 3 \qquad (10.4)$$

Since the energy needed to create an electron-ion pair in xenon is $W_I = 22$ eV, each 30–64 keV x-ray interaction will give raise to a luminous signal worth 4000–9000 photoelectrons (phe), spanning over 4–8 pixels, hence well above the SiPM noise. The energy resolution (FWHM) is obtained from reference [52] as:

$$\mathcal{R}(\varepsilon = 64 \text{ keV}) \simeq 2.355 \sqrt{F + \frac{1}{n_{phe}}\left(1 + \frac{\sigma_G^2}{G^2}\right)} \sqrt{\frac{W_I}{\varepsilon}} = 3.1\% \qquad (10.5)$$

with σ_G / G being the width of the single-photon distribution (around 0.1 for a typical SiPM) and $F \simeq 0.17$ the Fano factor of xenon. For comparison, a value compatible with R($\varepsilon = 64$ keV) = 5.5% was measured for acrylic-hole multipliers in reference [29]. In the following, the contribution of the energy resolution is included as a gaussian smearing in the TPC response.

Finally, the time response function of the SiPM is included as a Gaussian with a 7-ns width, convoluted with the transit time of the electrons through the EL structure $\Delta T_{EL} = 0.36$ μs, being both much smaller in any case than the typical temporal spread of the clusters (dominated by diffusion). The sampling time is taken to be $\Delta T_s = 0.5$ μs as in reference [49], and a matrix of

1800 10 mm-pitch SiPMs is assumed for the readout. Images are formed after applying a 10-phe threshold to all SiPMs.

A fully processed TPC image for one time slice ($\Delta T_s = 0.5$ µs), obtained at a beam rate of $r = 3.7 \times 10^{10}$ ph/s for a photon energy $\varepsilon = 64$ keV, is shown in Figure 10.11. The main clusters have been marked with crosses, by resorting to "Monte Carlo truth," i.e., they represent the barycenter of each primary ionization cluster in Geant4. The beam has been assumed to be continuous, polarized along the x-axis, impinging on a 5-µm water cube surrounded by air, with a 33-nm DNA cubic feature placed at its center. The Geant4 simulations are performed at fixed time, and to simulate a scanning technique the x-ray interaction times are subsequently distributed uniformly within the dwell time corresponding to different scan positions along the sample. It must be noted that interactions taking place at about the same time may be recorded at different times depending on the z-position of each interaction (and viceversa, clusters originating at different interaction times, may eventually be reconstructed in the same time slice). This scrambling (unusual under typical TPC operation) renders every time slice equivalent for the purpose of counting. In principle, the absolute time and z-position can be disambiguated from the size of the cluster, using the diffusion relation in Equation (10.1), thus allowing photon-by-photon reconstruction in time, space, and energy. A demonstration of the strong correlation between z-position and cluster width, for 30 keV x-ray interactions, can be found, for instance, in reference [39].

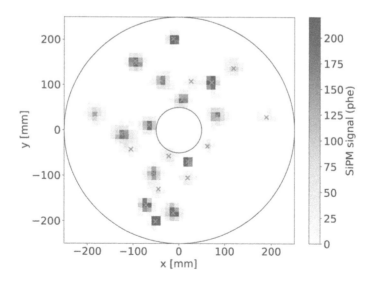

FIGURE 10.11

A typical TPC image reconstructed from the SiPM signals (in phe), as recorded in one time-slice ($\Delta T_s = 0.5$ µs), for a beam rate of $r = 3.7 \times 10^{10}$ s^{-1}. The crosses show the clusters' centroids, obtained from "Monte Carlo-truth" information. Reproduced from [47] with permission of the International Union of Crystallography.

The design parameters used in this subsection are compiled in the Annexure.

10.3.4 Photon Counting

10.3.4.1 Ideal Counting Limit

The photon counting capabilities of a realistic detector implementation are limited by the attenuation in the structural materials, re-scatters, characteristic emission, as well as the detector inefficiency. These intrinsic limitations can be conveniently evaluated from the signal-to-noise ratio, defined from the relative spread in the number of ionization clusters per scan step, as obtained in Monte Carlo (n_{MC}):

$$S/N = n_{MC}/\sigma_{n_{MC}} \tag{10.6}$$

Figure 10.12 shows the deterioration of the S/N for 64 keV photons, as the realism of the detector increases. It has been normalized to the relative spread in the number of photons scattered on-sample per scan step, $\sqrt{N_0}$, so that it equals 1 for a perfect detector:

$$S/N^* \equiv \frac{1}{\sqrt{N_0}} \cdot S/N \tag{10.7}$$

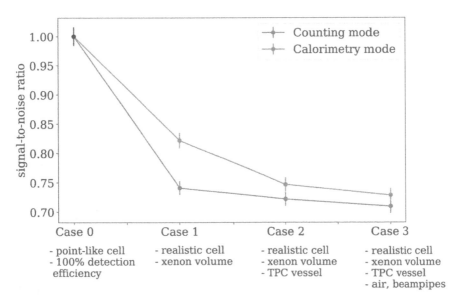

FIGURE 10.12
Intrinsic counting performance (using Monte Carlo-truth information) for 64 keV x-ray photons, characterized by the signal-to-noise ratio (relative to case 0). Photon counting (green) and calorimetric mode (red) are displayed as a function of the realism of the simulations. Reproduced from [47] with permission of the International Union of Crystallography.

The figure also shows the S/N^* in "calorimetric mode," with the counting performed by simply integrating the total collected light per scan step (ε_{tot}), instead of photon-by-photon. S/N^* is defined in that case, equivalently, as: $S/N^* = (\varepsilon_{tot}/\sigma_{\varepsilon_{tot}})/\sqrt{N_0}$. The values obtained are just slightly below the ones expected considering detector inefficiency alone:

$$S/N^* \simeq \sqrt{\varepsilon} \qquad (10.8)$$

therefore suggesting a small contribution from re-scatters in the materials or other secondary processes.

10.3.4.2 Real Counting

Photon-by-photon counting, contrary to the calorimetric mode, enables $x, y, t + t_{drift}$ and ε determination, and arguably the interaction time t and z-position can be obtained from the study of the cluster size, as it has been demonstrated earlier in reference [39] for 30 keV x-rays at near-atmospheric pressure. Given the nature of the detector data (Figure 10.11), consisting of voxelized ionization clouds grouped forming ellipsoidal shapes, generally separable, and of similar size, the K-means clustering method [53] is selected to perform cluster counting. The K-means method evaluates the partition of N observations (the ionization clouds) in n clusters, so as to minimize the inertia I, defined as the sum of the squared distances of the observations to their closest cluster center. The cluster counting algorithm has been implemented as follows: (i) the "countable" clusters are first identified time-slice by time-slice using Monte Carlo-truth information, as those producing a signal above a certain energy threshold (ε_{th}) in that slice. The energy threshold is chosen to be much lower than the typical cluster energies. In this manner, only small clusters are left out of the counting process when most of their energy is collected in adjacent time-slices from which charge has spread out due to diffusion, and where they will be properly counted once the algorithm is applied there; (ii) a weighted inertia distribution is formed, as conventionally done in K-means, and a threshold (δI_{th}) is set to the variation of the inertia with the number of clusters counted by the algorithm (n). The threshold is optimized for each beam rate condition. Two beam rates for which the average efficiency and purity of the cluster identification in 2D slides is larger than 80% are illustratively depicted in Figure 10.13. The counting efficiency and purity can been defined, as customary, as:

$$\varepsilon_{counting} = \frac{n_{matched}}{n_{MC}} \qquad (10.9)$$

$$p_{counting} = \frac{n_{matched}}{n} \qquad (10.10)$$

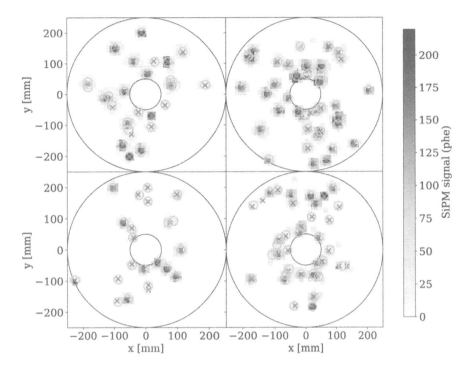

FIGURE 10.13
Cluster counting performance for typical $\Delta T_s = 0.5\ \mu s$ time-slices, for different energies (ε) and beam rates (r). Crosses indicate the cluster centroids from MC and circles are the clusters found by K-means. The average counting-efficiency and purity along the detector are given below in brackets. Top left: $\varepsilon = 64$ keV and $r = 3.7 \times 10^{10}$ ph/s ($\varepsilon_{counting} = 88.2\%$, $p_{counting} = 86.9\%$). Top right: $\varepsilon = 64$ keV and $r = 7.5 \times 10^{10}$ ph/s ($\varepsilon_{counting} = 84.2\%$, $p_{counting} = 83.2\%$). Bottom left: $\varepsilon = 30$ keV and $r = 6.5 \times 10^{10}$ ph/s ($\varepsilon_{counting} = 87.9\%$, $p_{counting} = 87.5\%$). Bottom right: $\varepsilon = 30$ keV and $r = 1.3 \times 10^{11}$ ph/s ($\varepsilon_{counting} = 83.9\%$, $p_{counting} = 83.1\%$). For $\varepsilon = 30$ keV only about half of the clusters are produced, which enables measuring at higher beam rates than $\varepsilon = 64$ keV, at comparable efficiency and purity. Reproduced from [47] with permission of the International Union of Crystallography.

where $n_{matched}$ is the number of counted clusters correctly assigned to MC clusters, n_{MC}. The K-means optimization parameters have been chosen to simultaneously maximize the counting efficiency while achieving $n \simeq n_{MC}$, therefore $\varepsilon_{counting} \simeq p_{counting}$.

Figure 10.14 (top) shows the performance of the counting algorithm, presenting the average number of clusters counted per 2D slice as a function of beam rate, with ε_{th} and δI_{th} optimized for each case as described earlier (green line). Red lines indicate the predictions outside the optimized case that illustrate the consistent loss of linearity as the beam rate increases. Figure 10.14 (bottom) shows the relative spread in the number of counted clusters σ_n / n, and comparison with Monte Carlo truth. These results can be qualitatively understood if recalling that, by construction, the threshold inertia is strongly correlated with the average number of clusters and its size. Therefore, a simple K-means algorithm will inevitably bias the number of counted clusters to match its expectation on I, if no further considerations are made. Thus, once δI_{th} has been adjusted to a certain beam rate, there will be systematic overcounting for lower beam rates, and undercounting for higher ones, as reflected by Figure 10.14 (top). In present conditions, a 2nd order polynomial is sufficient to capture this departure from proportionality introduced by the algorithm. A similar (although subtler) effect takes place for the cluster distributions obtained slice-by-slice, where this systematic overcounting-undercounting effect makes the cluster distribution marginally (although systematically) narrower, as seen in Figure 10.14 (bottom). As a consequence, the directly related magnitude S/N^* (Equations (10.6) and (10.7)), is not deteriorated by the counting algorithm. On the other hand, proportionality is lost, and its impact needs to be addressed, depending on the application.

Finally, the photon-counting efficiency (Equation (10.9)) can be assessed through Figure 10.16-top, where it is displayed as a function of the beam rate on target. It can be seen how, for the case of 30 and 64 keV photons, its value exceeds 85% for rates up to 10^{11} ph/s and 0.5×10^{11} ph/s, respectively. At these high beam rates, counting capability suffers from event pile-up while, at low beam rates, it is limited by the presence of low-energy

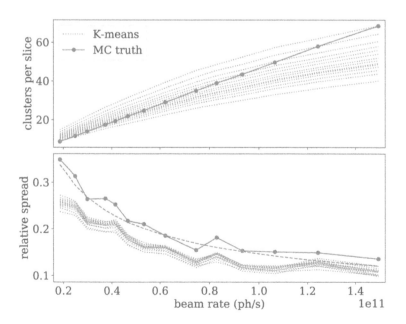

FIGURE 10.14

Top: counting performance characterized through the average number of clusters counted per 2D time-slice as a function of the beam rate for $\varepsilon = 64$ keV. Bottom: relative spread of the number of clusters per 2D time-slice from Monte Carlo truth and counted with K-means. The $\sim 1/\sqrt{r}$ expectation (dashed) is shown for comparison. Reproduced from [47] with permission of the International Union of Crystallography.

deposits (corresponding to x-ray interactions for which most of the energy is collected in adjacent slices). It must be recalled, at this point, that a complete reconstruction requires combining 2D time-slices as the ones studied here, in order to unambiguously identify clusters in 3D. Given that clusters extend over 4–6 slices due to diffusion, and are highly uncorrelated, a 3D-counting efficiency well above 90% can be anticipated in the earlier conditions.

10.4 Application to Scanning Compton X-Ray Microscopy

The proposed EL-TPC is characterized here in light of its performance as a cellular microscope, through the study of the smallest resolvable DNA-feature in a typical cell, as a function of the scan time (ΔT_{scan}). In a scanning, dark-field, configuration, the ability to resolve a feature of a given size (d) embedded in a medium can be studied through the schematic representation shown in Figure 10.15, that corresponds to an arbitrary step within a 2D-scan.

Three main assumptions lead to this simplified picture: (i) the dose fractionation theorem [54], based on which one can expect 3D reconstruction capabilities at the same resolution (and for the same dose) than in a single 2D-scan, (ii) the ability to obtain a focal spot, d', down to a size comparable to (or below) that of the feature to be resolved, d, and (iii) a depth of focus exceeding the dimensions of the sample under study, l. The study case in Figure 10.15 is adopted as the benchmark case, and the Rose criterion [55] is

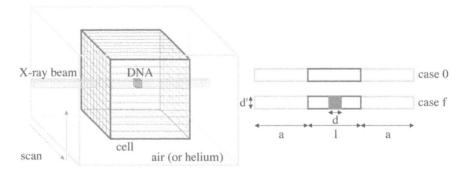

FIGURE 10.15
Study case for SCXM: a cubic DNA feature (size d) is embedded in a cubic water cell ($l = 5$ μm), surrounded by air ($a = 5$ mm). The photon beam scans regions containing only water (case 0), or water and DNA (case f). These two cases are used to evaluate the resolving power of SCXM at a given dose. Reproduced from [47] with permission of the International Union of Crystallography.

used as the condition needed to discern case f (feature embedded within the scanned volume) from case 0 (no feature), that reads in the Poisson limit as:

$$\frac{|N_f - N_0|}{\sqrt{\sigma_{N_f}^2 + \sigma_{N_0}^2}} = \frac{|N_f - N_0|}{\sqrt{N_f + N_0}} \geq 5 \tag{10.11}$$

with N being the number of scattered photons. Substitution of physical variables in Equation (10.11) leads directly to a required fluence of:

$$\phi \geq \phi_{min} = 25 \frac{(2l-d) \cdot \lambda_w^{-1} + d \cdot \lambda_f^{-1} + 4 \cdot a \cdot \lambda_a^{-1}}{d'^2 \cdot d^2 \cdot (\lambda_f^{-1} - \lambda_w^{-1})^2} \tag{10.12}$$

Here λ_w, λ_f, and λ_a are the Compton-scattering mean free paths of x-rays in water, DNA, and air (or helium), respectively (Table 10.2), dimensions are defined in Figure 10.15, and $d' \simeq d$ can be assumed. The dose imparted at the feature in these conditions is approximated by:

$$D = \phi_{min} \cdot \varepsilon \cdot \frac{N_A}{M_f} \cdot \left| \sigma_{ph} + \int \frac{d\sigma_c}{d\Omega} \cdot \left(1 - \frac{1}{1 + \frac{\varepsilon}{m_e c^2}(1 - \cos\theta)} \right) d\Omega \right| \tag{10.13}$$

where σ_{ph} is the photoelectric cross section and $d\sigma_c / d\Omega$ is the differential cross section for Compton scattering, both evaluated at the feature. M_f is the feature molar mass, N_A the Avogadro number, ε the photon energy and θ its scattering angle. The dose inherits the approximate l/d^4 behavior displayed in Equation (10.12).

On the other hand, the smallest resolvable DNA-feature as a function of the scan time (ΔT_{scan}), according to reference [47], is given by:

$$d = \left(R^2 2l^2 \frac{(l\lambda_w^{-1} + 2a\lambda_a^{-1})}{(\lambda_f^{-1} - \lambda_w^{-1})^2} \frac{1}{C_l(r)^2 \cdot S / N^{*2} \cdot r \cdot \Delta T_{scan}} \right)^{1/4} \tag{10.14}$$

TABLE 10.2

Mean Free Path for Different Materials at the Studied Energies 30 and 64 keV, According to NIST

Mean Free Path	30 keV	64 keV	Material
λ_w [cm]	5.47	5.69	Water
λ_f [cm]	3.48	3.54	DNA
λ_a [cm]	4950.49	4945.60	Air

where R equals 5 under the Rose criterion and the rate-dependent coefficient $C_l < 1$ depends on the deviation of the counting algorithm from the proportional response. Since the smallest resolvable feature size ($d^\dagger$) is ultimately determined by the dose imparted at it when structural damage arises, the necessary scan time to achieve such performance ($\Delta T^\dagger_{scan}$) can be readily obtained:

$$\Delta T^\dagger_{scan} = R^2 2 l^2 \frac{(l\lambda_w^{-1} + 2a\lambda_a^{-1})}{(\lambda_f^{-1} - \lambda_w^{-1})^2} \frac{1}{C_l(r)^2 \cdot S / N^{*,2} \cdot r \cdot (d^\dagger)^4} \tag{10.15}$$

The dose-limited resolutions for the present case have been calculated in reference [47]: $d^\dagger = 36$ nm for $\varepsilon = 64$ keV, and $d^\dagger = 44$ nm for $\varepsilon = 30$ keV.

The limiting scan time (i.e., above which structural damage will appear) can be hence assessed from the behavior of Equation (10.15) with beam rate, as shown in Figure 10.16-bottom. For 64 keV, the loss of linearity of the counting algorithm at high rates results in a turning point at 9.3×10^{10} ph/s, above

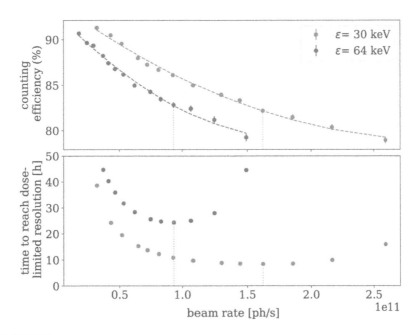

FIGURE 10.16
Top: efficiency of the cluster counting process as a function of the beam rate for x-rays of 30 and 64 keV. Bottom: time to reach the dose-limited resolution as a function of the beam rate. A minimum is reached when the product of $C_l^2 \cdot r$ reaches a maximum, i.e., the time decreases with beam rate until the effect of the non-proportional counting becomes dominant. The optimum beam rate and corresponding counting efficiency are marked with a dotted line for both energies. Reproduced from [47] with permission of the International Union of Crystallography.

which an increase in rate stops improving the ability to resolve an image. For 30 keV, due to the absence of characteristic emission, only about half of the clusters are produced and the optimum rate is found at a higher value, $r = 1.6 \times 10^{11}$. The counting efficiency and purity in these conditions is in the range 82–84%.

It is now possible to evaluate Equation (10.14) under different scenarios: (i) a relatively simple calorimetric mode (total energy is integrated), for which a hard x-ray beam rate typical of the new generation of synchrotron light sources as $r = 10^{12}$ ph/s is assumed, and (ii) a rate-limited photon-by-photon counting scenario, for the optimum rates $r = 9.3 \times 10^{10}$ ph/s (64 keV) and $r = 1.6 \times 10^{11}$ ph/s (30 keV), obtained earlier. Values for $C_l(r)$ are extracted from 2nd-order fits as described in reference [47]. The remaining parameters are common to both modes: $S/N^* = 0.71$, efficiency $\varepsilon = 58.5\%$ (64 keV), $S/N^* = 0.63$, $\varepsilon = 40.0\%$ (30 keV), $l = 5\ \mu m$, $a = 5\ mm$, $R = 5$, with the mean free paths (λ) taken from Table 10.2. Results are summarized in Figure 10.17. At 64 keV, the dose-limited resolution $d^{\dagger} = 36$ nm can be achieved in approximately 24 hours while, at 30 keV, $d^{\dagger} = 44$ nm is reached in just 8 hours. In the absence of systematic effects, operation in calorimetric mode would bring the scan time down to ≤ 1 hour in both cases, although abandoning any photon-by-photon counting capabilities.

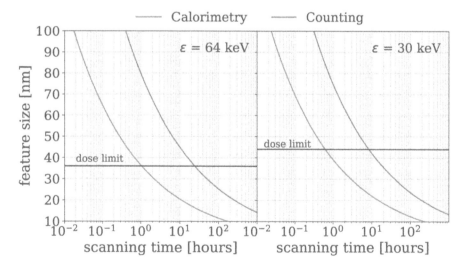

FIGURE 10.17
Resolution achievable with a 64 keV photon beam (left) and a 30 keV photon beam (right) as a function of the scan time for a cell of 5 μm (green line). The red line shows the limit in which a calorimetric measurement is performed and photon-by-photon counting is abandoned. The horizontal line shows the dose-limited resolution in each case, prior to inducing structural damage. Reproduced from [47] with permission of the International Union of Crystallography.

10.5 Summary: Limits and Scope of the Technology

The results presented illustrate the potential of the proposed technology for high coverage/high energy x-ray detection at high-brightness synchrotron light sources, exemplified through a new technique for cellular microscopy (SCXM). To derive them, some simplifications, that should be superseded in future work, were adopted and are analyzed here:

1. *Availability of photon-by-photon information*: cluster reconstruction with high efficiency and purity can help at removing backgrounds not accounted for, as well as any undesired systematic effect (beam or detector related). Since this technique provides a parallax-free measurement, the concept may be extended to other applications, e.g., x-ray crystallography. The presence of characteristic emission from xenon will unavoidably create confusion, so if unambiguous correspondence between the ionization cluster and the parent x-ray is needed, one must consider operation at ≤ 30 keV.

2. *Data processing and realism*: photon-by-photon counting at a rate nearing 5×10^7 ph/s over the detector ($\equiv 10^{11}$ ph/s over the sample), as proposed here, is a computer intensive task. Achieving this with sufficient speed and accuracy will require the optimization of the counting algorithm, something that will need to be accomplished, ultimately, with real data. To this aim, both the availability of parallel processing as well as the possibility of simultaneous operation in calorimetric mode are desirable features.

3. *Simplicity and compactness*: the detector geometry proposed here has been conceived as a multipurpose permanent station. A portable device focused purely on SCXM, on the other hand, could simply consist of a cubic $25\,\text{cm} \times 25\,\text{cm} \times 25\,\text{cm}$ vessel that may be positioned, e.g., on top of the sample (at a distance of about ~5 cm). The geometry would thus have an overall efficiency around 30% for 64 keV photons. For SCXM, and given that $S/N^* \simeq \sqrt{\varepsilon}$ as shown in this work, a loss of efficiency can be almost fully compensated by means of the corresponding increase in beam rate, at the price of a deteriorated value for the dose limited resolution $d^{\dagger}$. In this case, a value corresponding to $d^{\dagger} = 41$ nm could be achieved in 12 hours, for our test study.

4. *Feasibility*: the technology proposed comes from the realm of high energy physics, with an inherent operational complexity that might not be affordable at light source facilities. A further possibility could be considered, by resorting to ultra-fast (1.6 ns resolution) hit-based TimePix cameras (e.g., [56, 57]) with suitable VUV-optics, allowing 256×256 pixel readout at 80 MHit/s, and thus abandoning completely

the SiPM readout. The vessel would just house, in such a case, the acrylic hole multiplier and cathode mesh, together with the power leads; it would be filled with the xenon mixture at atmospheric pressure and interfaced to the outside with a VUV-grade viewport. This would compromise partly the ability to disentangle clusters by using time information, as well as energy information, since only the time over threshold would be stored and not the temporal shape of each cluster, or its energy. On the other hand, it would enhance the spatial information by a factor of 30 relative to the SiPM matrix proposed here (the hole pitch of the acrylic hole-multiplier should be reduced accordingly). Indeed, TimePix cameras are regularly used nowadays for photon and ion counting applications [58, 59], but have not been applied to x-ray counting yet, to the best of our knowledge. The counting and signal processing algorithms could be in this way directly ported, given the similarity with the images taken in those applications. The readiness of such an approach, aiming at immediate implementation, represents an attractive and compelling avenue.

Annexure: EL-TPC Parameters

Compilation of the main parameters used for the simulation of the TPC response, together with additional references when needed.

TABLE 10.3

Parameters of the TPC Vessel

R_i	5	cm	inner radius
R_o	25	cm	outer radius
L	50	cm	length

TABLE 10.4

Main Gas Parameters (Xenon + 0.4% CH_4)

In the drift/collection region			
E_c	110	V/cm	collection field
V_{cat}	−8.5	kV	cathode voltage
F	0.15		Fano factor [48]
W_I	22	eV	energy to create an e^--ion pair [48]
D_T^*	0.548	mm/$\sqrt{cm}$	transverse diffusion coefficient [28]
D_L^*	1.52	mm/$\sqrt{cm}$	longitudinal diffusion coefficient [28]
v_d	5.12	mm/µs	drift velocity [28]
In the electroluminescence (EL) region			
E_{EL}	6	kV/cm	EL field
V_{gate}	−3	kV	voltage at FAT-GEM entrance ("gate")
$v_{d,EL}$	13.7	mm/µs	drift velocity [28]

TABLE 10.5

Parameters of the Electroluminescent Structure

r_h	3	mm	hole radius
t	5	mm	thickness
p_h	10	mm	hole-to-hole pitch
m_{opt}	250	ph/e/cm	optical gain [29]
P_{scin}	0.5		scintillation probability [52]

TABLE 10.6

Parameters of the Readout

P_{si}	10	mm	pitch of SiPM matrix
ΔT_s	0.5	μs	time sampling/time per slice
σ_t	7	ns	temporal width of SiPM signal [60]
σ_G / G	0.1		relative spread of single phe charge in SiPM [60]
Ω_{TPB}	0.3		geometrical acceptance of SiPM after wavelength shifter
QE_{wls}	0.4		quantum efficiency of wavelength shifter [51]
QE_{si}	0.4		quantum efficiency of SiPM [60]

Note

1. We use the word "thermalization," for simplicity, to refer to a steady-state electron swarm drifting in statistical equilibrium. For ions both concepts (thermal motion and statistical equilibrium) are interchangeable for most practical purposes, due to the higher interaction rate compared to electrons. The assumption of Gaussian diffusion used in the following (characterized through expression 1) is exact under the assumption of statistical equilibrium, even if the energy distribution would not be exactly thermal.

References

1. E. Aprile, A. E. Bolotnikov, A. I. Bolozdynya, and T. Doke, 'Noble gas detectors', Wiley–VCH, 2006.
2. F. Sauli, 'Micro-pattern gaseous detectors: Principles of operation and applications', World Scientific Publishing Co Pte Ltd, 2020.
3. J. E. Bateman, 'Detectors for condensed matter studies', Nucl. Instr. Meth. A 273 (1988) 721–730.
4. R. A. Boie, J. Fischer, Y. Inagaki, F. C. Merritt, V. Radeka, L. C. Rogers, and D. M. Xi, 'High resolution X-ray gas proportional detectors with delay line position sensing for high counting rates', Nucl. Instr. Meth. A 201 (1982) 93–115.
5. Ng. H. Xuong, S. T. Freer, R. Hamlin, C. Nielson, and W. Vernon, 'The electron stationary picture method for high-speed measurement of reflection intensities from crystals with large unit cells', Acta Cryst. A 34 (1978) 289.

6. A. R. Faruqui and H. Andrews, 'A high-resolution multiwire area detector for X-ray scattering', Nucl. Instr. Meth. A 283 (1989) 445–447.
7. G. C. Smith, B. Yu, J. Fischer, V. Radeka, and J. A. Harder, 'High rate, high resolution, two dimensional gas proportional detectors for X-ray synchrotron radiation experiments', Nucl. Instr. Meth. A 323 (1992) 78–85.
8. A. Jucha, D. Bonin, E. Dartyge, A. M. Flank, A. Fontaine, and D. Raoux, 'Photodiode array for position-sensitive detection using high X-ray flux provided by synchrotron radiation', Nucl. Instr. Meth. A 226 (1984) 40–44.
9. H. Oyanagi, T. Matsushita, U. Kaminaga, and H. Hashimoto, 'Linear detector for time-resolved EXAFS in deispersive mode', Journal de Physique Colloques 47 (C8) (1986) 139–142.
10. B. Rodricks and C. Brizard, 'Programmable CCD imaging system for synchrotron radiation studies', Nucl. Instr. Meth. 311 (1992) 613–619.
11. G. Admans, P. Berkvens, A. Kaprolat, and J.-L.Revol (Eds.), 'ESRF upgrade programme phase II (2015–2022)', Technical Design Study, http://www.esrf.eu/Apache_files/Upgrade/ESRF-orange-book.pdf
12. 'Advanced Photon Source Upgrade Project', Final Design Report APSU-2.01-RPT-003 (2019).
13. C. G. Schroer, R. Röhlsberger, E. Weckert, R. Wanzenberg, I. Agapov, R. Brinkmann, and W. Leemans (Eds.), 'PETRA IV: Upgrade of PETRA III to the ultimate 3D X-ray microscope', Conceptual Design Report (2019), https://bib-pubdb1.desy.de/record/426140/files/DESY-PETRAIV-Conceptual-Design-Report.pdf
14. Y. Asano et al., 'SPring-8-II', Conceptual Design Report (2014), http://rsc.riken.jp/eng/pdf/SPring-8-II.pdf
15. S. Agostinelli et al., 'Geant4—a simulation toolkit', Nucl. Instr. Meth. 506 (2003) 250–303.
16. J. H. Hubbell et al., 'Atomic form factors, incoherent scattering functions, and photon scattering cross sections', J. Phys. Chem. Ref. Data 4 (1975) 471.
17. A. C. Thompson, D. Vaughan, et al. 'X-ray data booklet', Table 1-3. https://xdb.lbl.gov/xdb.pdf
18. S. P. Ahlen, 'Theoretical and experimental aspects of the energy loss of relativistic heavily ionizing particles', Rev. Mod. Phys. 52 (1980) 121–173.
19. K. Saito, H. Tawara, T. Sanami, E. Shibamura, and S. Sasaki, 'Absolute number of scintillation photons emitted by alpha particles in rare gases', IEEE Transactions on Nuclear Science 49 (4) (2002) 1674–1680.
20. C. D. R. Azevedo, S. Biagi, R. Veenhof, P. M. Correia, A. L. M. Silva, L. F. N. D. Carramate, and J. F. C. A. Veloso, 'Position resolution limits in pure noble gaseous detectors for X-ray energies from 1 to 60keV', Phys. Lett. B 741 (2015) 272–275.
21. C. D. R. Azevedo, D. González-Díaz, P. M. M. Correia, S. Biagi, A. L. M. Silva, L. F. N. D. Carramate, and J. F. C. A. Veloso, 'Pressure effects on the X-ray intrinsic position resolution in noble gases and mixtures', JINST 11 (12) (2016) P12008.
22. J. Fischer et al., 'X-ray position detection in the region of 6 µm RMS with wire proportional chambers, Nucl. Instrum. Meth. A 252 (1986) 239.
23. G. C. Smith et al., 'X-ray position resolution in proportional chambers in the region of 100 µm (FWHM) above the xenon K-edge', Nucl. Instrum. Meth. A 350 (1994) 621.
24. G. C. Smith, Lecture on gaseous detectors for science (and discussion), https://indico.cern.ch/event/179611/session/4/contribution/0/material/slides/0.pdf

25. T. J. Shin et al., 'Two-dimensional multiwire gas proportional detector for X-ray photon correlation spectroscopy of condensed matter', Nucl. Instrum. Meth. A 587 (2008) 434.
26. S. Biagi, 'DEGRAD: An accurate auger cascade model for interaction of photons and particles with gas mixtures in electric and magnetic fields', http://indico. cern.ch/event/245535/session/5/contribution/14/material/slides/, 2013. http:// degrad.web.cern.ch/degrad/
27. S. Biagi, 'Monte Carlo simulation of electron drift and diffusion in counting gases under the influence of electric and magnetic fields', Nucl. Instr. Meth. A 421 (1999) 234–240. http://magboltz.web.cern.ch/magboltz/
28. B. Al Atoum, S. F. Biagi, D. Gonzalez-Diaz, B. J. P Jones, and A. D. McDonald, 'Electron transport in gaseous detectors with a python-based Monte Carlo simulation code', Comput. Phys. Commun. 254 (2020) 107357.
29. D. González-Díaz et al., 'A new amplification structure for time projection chambers based on electroluminescence', Journal of Physics: Conference Series, Volume 1498, Micro-Pattern Gaseous Detectors Conference (2019) 5–10, arXiv:1907.03292.
30. G. Martínez-Lema et al., 'Calibration of the NEXT-white detector using ^{83m}Kr decays', JINST 13 (2018) 10, P10014.
31. C. Henriques, C. Monteiro, D. González-Díaz, et al., 'Electroluminescence TPCs at the thermal diffusion limit', J. High Energy Phys. 27 (2019).
32. D. Lorca et al., 'Characterisation of NEXT-DEMO using xenon K_α X-rays', JINST 9 (2014) P10007.
33. J. Renner et al., 'Energy calibration of the NEXT-white detector with 1% resolution near $Q_{\beta\beta}$ of ^{136}Xe', J. High Energy Phys. 230 (2019).
34. A. Bolozdynya et al., 'A high pressure xenon self-triggered scintillation drift chamber with 3D sensitivity in the range of 20-140 keV deposited energy', Nucl. Instr. Meth. A 385 (1997) 225–238.
35. S. Ban et al., 'Electroluminescence collection cell as a readout for a high energy resolution Xenon gas TPC', Nucl. Instr. Meth. A 875 (2017) 185–192.
36. R. Luescher et al, 'Search for $\beta\beta$ decay in ^{136}Xe: New results from the Gotthard experiment', Phys. Lett. B 434 (1998) 407.
37. B. Ramsay and P. C. Agrawal, 'Xenon-based penning mixtures for proportional counters', Nucl. Instr. Meth. A 278 (1989) 576–582.
38. S. Cebrian et al., 'Micromegas-TPC operation at high pressure in xenon-trimethylamine mixtures', JINST 8 (2013) P01012.
39. D. González-Díaz et al., 'Accurate γ and MeV-electron track reconstruction with an ultra-low diffusion Xenon/TMA TPC at 10 atm', Nucl. Instr. Meth. A 804 (2015) 8–24.
40. D. Nygren, 'High-pressure xenon gas electroluminescent TPC for 0-$\nu\beta\beta$-decay search', Nucl. Instr. Meth. A 603 (2009) 337–348.
41. D. Nygren, 'Proposal to investigate the feasibility of a novel concept in particle detection', Tech. report 2-22-74, Lawrence Berkeley Lab (1974).
42. D. Nygren, 'Origin and development of the TPC idea', Nucl. Instr. Meth. A 907 (2018) 22–30.
43. J. Alme et al., 'The ALICE TPC, a large 3-dimensional tracking device with fast readout for ultra-high multiplicity events', Nucl. Instr. Meth., A 622 (2010) 316–367.
44. N. S. Phan, R. J. Lauer, E. R. Lee, D. Loomba, J. A. J. Matthews, and E. H. Miller, 'GEM-based TPC with CCD imaging for directional dark matter detection', Astrop. Phys. 84 (2016) 82–96.

45. R. Acciarri et al., 'Long-Baseline Neutrino Facility (LBNF) and Deep Underground Neutrino Experiment (DUNE) Conceptual Design Report, Volume 4 The DUNE Detectors at LBNF', arXiv:1601.02984.

46. D. González-Dí-az, F. Monrabal, and S. Murphy, 'Gaseous and dual-phase TPCs for imaging rare processes', Nucl. Instr. Meth. A 878 (2018) 200–255.

47. A. Saa Hernandez, D. González-Díaz, P. Villanueva, C. Azevedo, and M. Seoane, 'A new imaging technology based on Compton X-ray scattering', J. Synchrotron Rad. 28, doi:10.1107/S1600577521005919, (2021).

48. D. Nygren, 'Optimal detectors for WIMP and $0-\nu\beta\beta$ searches: Identical high-pressure xenon gas TPCs?', Nucl. Instr. Meth. A 581 (2007) 632–642.

49. F. Monrabal et al. (the NEXT collaboration), 'The Next White (NEW)', JINST 13 (2018) P12010.

50. P. Villanueva-Perez, S. Bajt, and H. N. Chapman, 'Dose efficient Compton X-ray microscopy', Optica 5 (2018) 450–457.

51. C. Benson, G. Orebi Gann, and V. Gehman, 'Measurements of the intrinsic quantum efficiency and visible reemission spectrum of tetraphenyl butadiene thin films for incident vacuum ultraviolet light', Eur. Phys. J. C. 78 (2018) 329.

52. C. Henriques, 'Studies of xenon mixtures with molecular additives for the NEXT electroluminescence TPC', PhD Thesis, Departamento de Física da Faculdade de Ciências e Tecnologia da Universidade de Coimbra (2019).

53. J. MacQueen, 'Some methods for classification and analysis of multivariate observations', Proc. of the 5th Berkeley Symposium on Mathematical Statistics and Probability, Volume 1: Statistics, University of California Press (1967) 281–297.

54. R. Hegerl and W. Hoppe, 'Influence of electron noise on three-dimensional image reconstruction', Z. Naturforsch. 31 (1976) 1717–1721.

55. A. Rose, 'A unified approach to the performance of photographic film, television pickup tubes, and the human eye', J. Soc. Motion Pict. Eng. 47 (1946) 273–294.

56. https://www.amscins.com/tpx3cam/

57. A. Nomerotski, 'Imaging and time stamping of photons with nanosecond resolution in Timepix based optical cameras', Nucl. Instr. Meth. A 937 (2019) 26–30.

58. L. M. Hirvonen, M. Fisher-Leving, K. Suhling, and A. Nomerotski, 'Photon counting phosphorescence lifetime imaging with TimepixCam', Rev. Sci. Instrum. 88 (2017) 013104.

59. M. Fisher-Levine, R. Boll, F. Ziaee, C. Bomme, B. Erk, D. Rompotis, T. Marchenko, A. Nomerotski, and D. Rolles, 'Time-resolved ion imaging at free-electron lasers using TimepixCam', J. of Synchrotron Radiat. 25 (2018) 336–345.

60. https://www.hamamatsu.com/eu/en/product/optical-sensors/mppc/mppc_mppc-array/index.html

Index

Note: Locators in *italics* represent figures and **bold** indicate tables in the text.